Diabetes and Viruses

Keith W. Taylor • Heikki Hyöty
Antonio Toniolo • Arie J. Zuckerman
Editors

Diabetes and Viruses

Springer

Editors
Keith W. Taylor
† (Deceased)

Antonio Toniolo
Laboratory of Medical Microbiology
University of Insubria
and Ospedale di Circolo
Varese, Italy

Heikki Hyöty
Department of Virology
University of Tampere
and Tampere University Hospital
Tampere, Finland

Arie J. Zuckerman
University College London Medical School
London, UK

ISBN 978-1-4614-4050-5 ISBN 978-1-4614-4051-2 (eBook)
DOI 10.1007/978-1-4614-4051-2
Springer New York Heidelberg Dordrecht London

Library of Congress Control Number: 2012942861

Printed on acid-free paper

Springer is part of Springer Science+Business Media (www.springer.com)

Foreword

Every house has many builders and is never finished—Paavo Haavikko (Finnish poet, 1931–2008).

To the memory of Keith W. Taylor (Shropshire, 1930—Rye, 2012) who set a wheel in motion in diabetic research with original investigations on the possible role of viral infections. His perspective and judgement contributed greatly to this work, and his spirit pervades this volume. We also wish to recall the wonderful and continuous support given to him by his wife Margareth, his daughter Ann and especially Nick, his son.

This book is also dedicated in everlasting loving memory to Alice Zuckerman (neé Adamson; 28 January 1932 to 16 January 2011), who devoted her life, love and energy to her husband Arie, and children Mark and Jane; and who encouraged, supported and inspired them to excel in the science and art of Medicine.

Finally, we acknowledge the generous contribution of Gianni Valcavi, Attorney, and Cariplo Foundation (Milan) without which diabetes research in Varese (Italy) would have been not possible. It is also a pleasure to acknowledge the skilful help and pleasant cooperation of our secretaries Ms. Tanya Shennan, Mrs. Irene Smith and Mrs. Stefania Triballi. Lastly, we gratefully recognize the distinguished skill and patience of Arthur Smilios, Ms. Fabian Shalini and the entire Springer's staff during the preparation of this book.

The Editors

Preface

While the term "the global epidemic of diabetes" is used frequently both by the popular media and in the medical literature, it is not used in the context of infection. The late Keith Taylor reflects on the historical background of the relationship between viruses and diabetes noting that the association between mumps and diabetes was described in the middle of the nineteenth century, but it was not until 1927 that the Norwegian Army physician Edvard Gundersen published a paper in the *Journal of Infectious Diseases* entitled "Is diabetes of infectious origin?". The subsequent history of virus infection and diabetes in humans and animals is described eloquently in the Chap. 1 of this book, which contains precisely what is stated in the title; that is, information on diabetes and viruses.

The Editors recruited a galaxy of leading researchers and physicians from many countries including, in alphabetical order, Australia, Cuba, Finland, France, Italy, Japan, Sweden, the UK and the USA, who accepted the challenge to produce rapidly an authoritative account of the current knowledge and research in progress on this important topic, for which the Editors are most grateful.

Many topics are reviewed expertly including the role of autoimmunity, molecular mimicry, genetic factors, immune mechanisms, environmental factors (an ever popular topic on virtually every aspect of human activity), and with a particular emphasis on a number of viruses affecting the pancreas in animals and humans. The text is written in a way that we hope will be understood by general physicians, clinical specialists in diabetes, researchers—especially those involved in immunology and virology—senior nurses, public health workers and medical students. We also hope that the pharmaceutical industry is listening. Throughout we attempted to avoid the description of excessively complex techniques and molecular porn, and simplify technical jargon.

Finally, there is an old military maxim "never attack a revolution", and—in the context of this book—we should not ignore the direct or indirect role of viruses in the aetiology of diabetes mellitus, but rather continue to explore this intriguing association.

London, UK — Arie J. Zuckerman

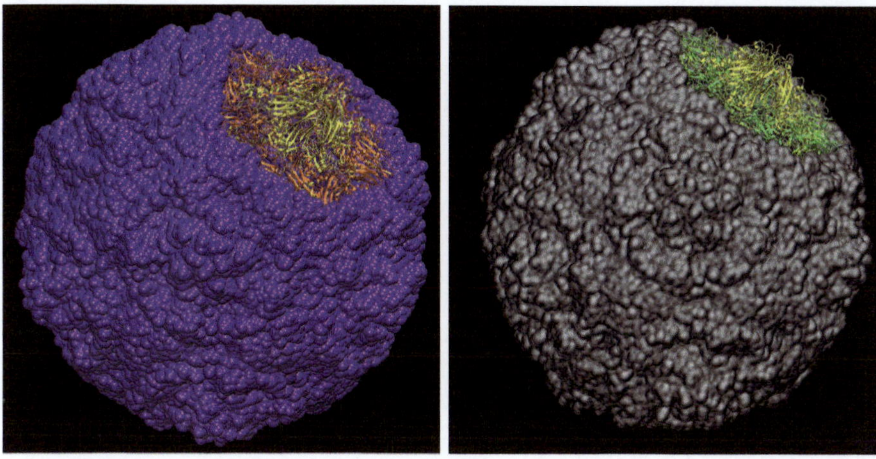

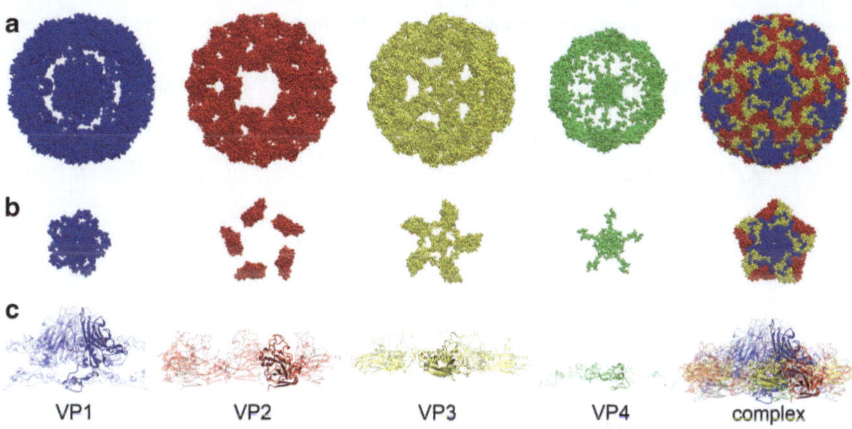
a
b
c
VP1
VP2
VP3
VP4
complex

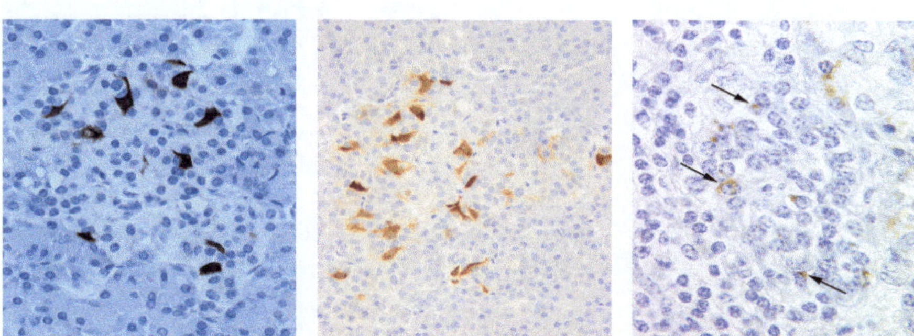

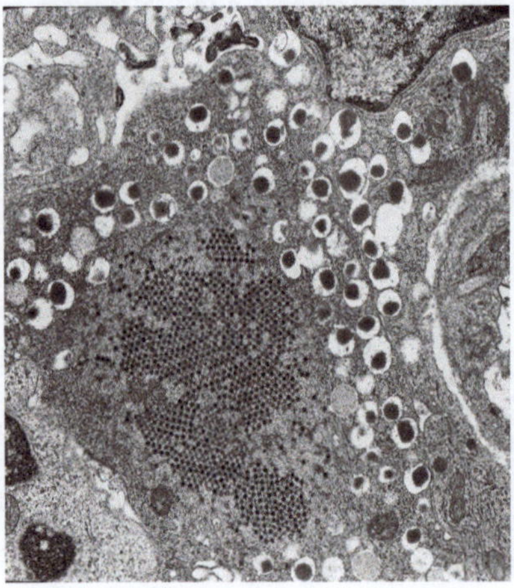
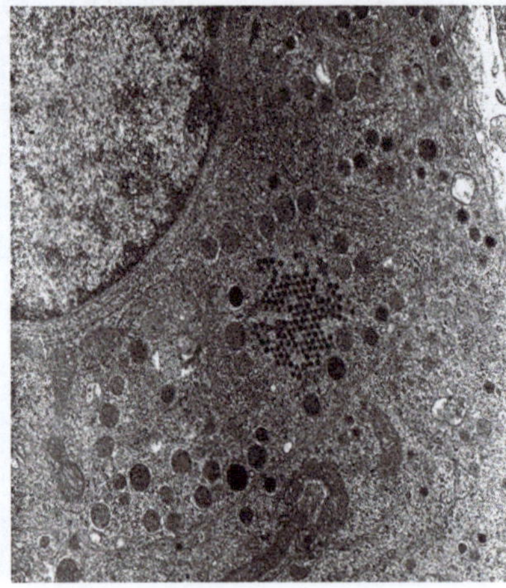
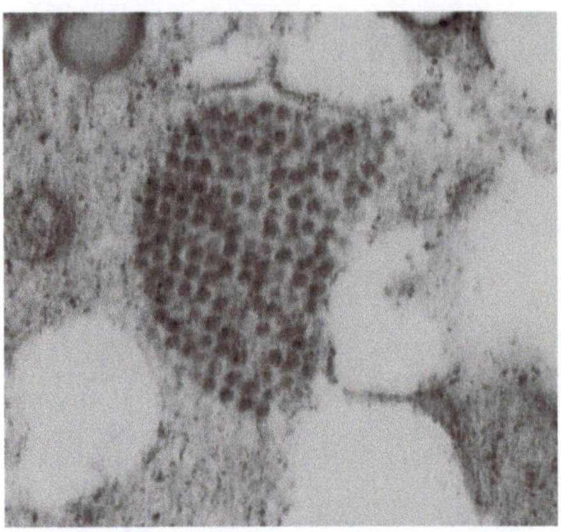
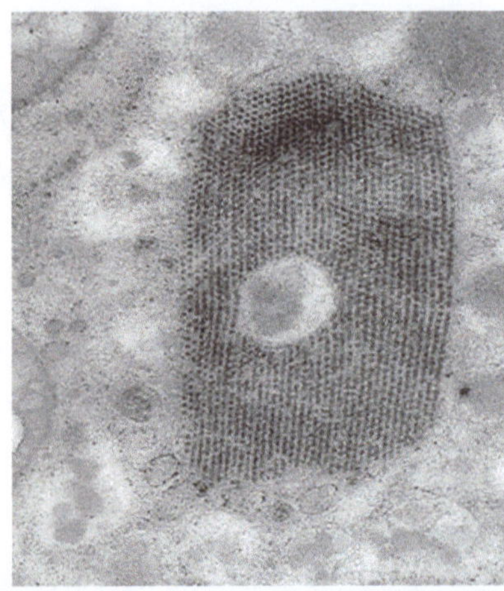
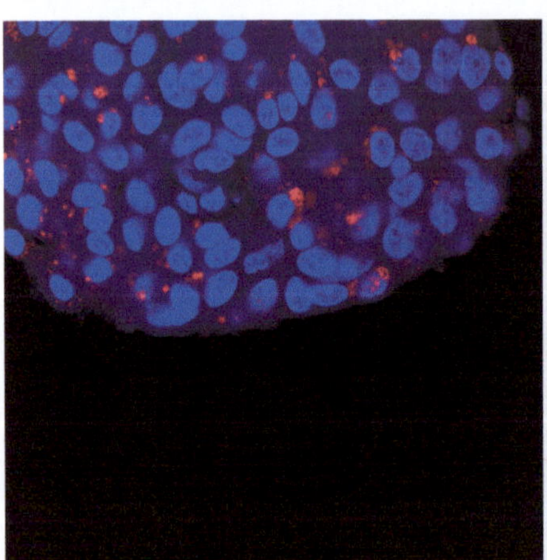
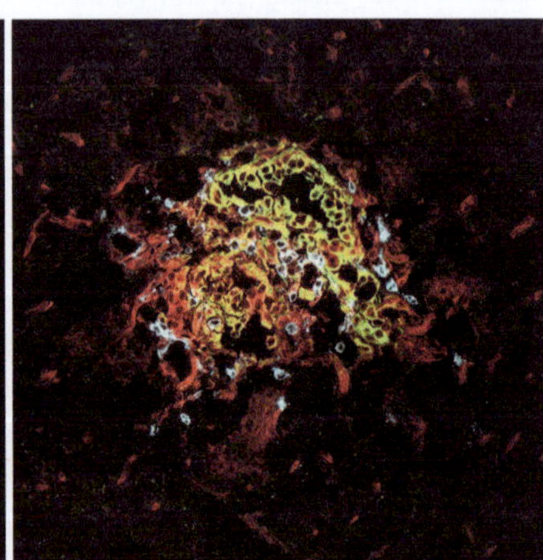

Figure Legends

Figs. 1 and 2 Three-dimensional model of an enterovirus. One pentamer of capsid proteins is shown in detail (two different orientations of the virus particle). The remaining part of the capsid surface is shown as Van der Waals spheres. Reconstruction based on the X-ray analysis of coxsackievirus A9 at 1.2 Å resolution has been performed using the VMD 1.8.7 program (Protein Data Bank access code 1D4M). Courtesy of Vesa Hytönen, University of Tampere, Finland.

Fig. 3 Assembly of the enterovirus capsid. A) The four capsid proteins represented as Van der Waals spheres (VP1, blue; VP2, red; VP3, yellow; VP4, green) are shown as assembled in the capsid. B) The four capsid proteins are shown as assembled in a pentamer. C) Side view of capsid proteins assembled in a single pentamer. As in Figures 1 and 2, the models are based on the X-ray structure of coxsackievirus A9 (courtesy of Vesa Hytönen, University of Tampere, Finland).

Fig. 4 Immunohistochemical detection of a capsid protein of an undefined type of enterovirus in the cytoplasm of islet cells of a recent case of type 1 diabetes (courtesy of S. Richardson and A. Foulis).

Fig. 5 Immunohistochemical detection of a capsid protein of an undefined type of enterovirus in the cytoplasm of islet cells of a case of type 1 diabetes of short duration (courtesy of H. Hyoty and S. Oikarinen).

Fig. 6 Expression of bovine diarrhea virus (BDV) antigen in the cytoplasm of pancreatic islet cells of a cattle with type 1 diabetes (courtesy of K. Matsuda and H. Taniyama).

Fig. 7 Crystalline array of 70 nm virus particles in a beta cell (note insulin-containing granules) of a mouse neonatally infected with reovirus type 1 (courtesy of T. Onodera, A. Toniolo, A.L. Notkins).

Fig. 8 Crystalline array of 70 nm virus particles in a alpha cell (note glucagon-containing granules) of a mouse neonatally infected with reovirus type 1 (courtesy of T. Onodera, A. Toniolo, A.L. Notkins).

Fig. 9 Crystalline array of enterovirus particles (approximately 30 nm) in a cultured human pancreatic islet that had been infected in vitro with coxsackievirus B5 (courtesy of G. Frisk).

Fig. 10 Crystalline array of enterovirus particles (approximately 30 nm) in the cytoplasm of a beta cell of cultured human pancreatic islets that had been infected in vitro with coxsackievirus B5 (courtesy of G. Frisk).

Fig. 11 Enterovirus capsid antigen (red) in the cytoplasm of cultured human pancreatic islets infected in vitro with coxsackievirus B3 (nuclei in blue; courtesy of M. Craig and M. Poon).

Fig. 12 Pancreatic islet from a patient with type 1 diabetes of 3 years duration. Frozen pancreas section stained for insulin (green), HLA class I (red), and CD8 (cyan blue). Note the infiltration of CD8(+) T cells and the hyperexpression of HLA class I molecules (confocal microscopy; courtesy of K. Coppieters, T. Kay, M. von Herrath).

Contributors

Mark A. Atkinson Departments of Pathology and Pediatrics, The University of Florida College of Medicine, Gainesville, FL, USA

Andreina Baj Laboratory of Microbiology, Department of Experimental Medicine, University of Insubria Medical School, Varese, Italy

Huriya Beyan Blizzard Institute, Queen Mary College, University of London, London, UK

Elizabeth P. Blankenhorn Department of Immunology and Microbiology, Drexel University College of Medicine, Philadelphia, PA, USA

Martin Blixt Department of Medical Cell Biology, Uppsala University, Uppsala, Sweden

A.J. Bone University of Brighton, Sussex, UK

Eduardo Cabrera-Rode Department of Immunology of Diabetes, National Institute of Endocrinology, Havana, Cuba

Martha Campbell-Thompson Department of Pathology, Immunology, and Laboratory Medicine, Gainesville, FL, USA

Urs Christen Pharmazentrum Frankfurt/ZAFES, Goethe University Hospital, Frankfurt am Main, Germany

Nora M. Chapman Enterovirus Research Laboratory, Department of Pathology and Microbiology, University of Nebraska Medical Center, 986495 Nebraska Medical Center, Omaha, NE, USA

Serena Wai-Yan Chiu Blizzard Institute of Cell and Molecular Science, Queen Mary College, University of London, London, UK

Maria E. Craig Institute of Endocrinology and Diabetes, The Children's Hospital at Westmead, Sydney, NSW, Australia

Discipline of Paediatrics and Child Health, The University of Sydney, Sydney, NSW, Australia

School of Women's and Children's Health, University of New South Wales, Sydney, NSW, Australia

Rachel Desailloud UPJV CHU, Service d'Endocrinologie-Diabétologie-Nutrition, Amiens, France

Oscar Díaz-Horta Department of Immunology of Diabetes, National Institute of Endocrinology, Havana, Cuba

Francesco Dotta University of Siena Medical School, Policlinico "Le Scoette", Siena, Italy

Martin Eichmann Department of Immunobiology, King's College London School of Medicine, London, UK

Giovanni Federico Unit of Pediatric Endocrinology and Diabetes, Department of Reproductive Medicine and Child Development, University of Pisa Medical School, Pisa, Italy

A.K. Foulis Glasgow University, Glasgow, UK

Gun Frisk Department of Women's and Children's Health, Uppsala University, Akademiska Hospital, Uppsala, Sweden

Letizia Galleri Diabetes Unit, Department of Internal Medicine, Endocrine and Metabolic Sciences and Biochemistry, University of Siena, Siena, Italy

Umberto Di Mario ONLUS Research Foundation, Siena, Italy

Anne Goffard Université Lille 2, CHRU, Laboratoire de Virologie/EA3610, Loos-lez-Lille, France

Fabio Arturo Grieco Diabetes Unit, Department of Internal Medicine, Endocrine and Metabolic Sciences and Biochemistry, University of Siena, Siena, Italy

Umberto Di Mario ONLUS Research Foundation, Siena, Italy

Toshiaki Hanafusa Department of Internal Medicine (I), Osaka Medical College, Takatsuki, Japan

Len Harrison Diabetes Research Center, The Walter & Eliza Hall Institute of Medical Research, Parkville, VIC, Australia

Robert Hermann Immunogenetics Laboratory, University of Turku, Turku, Finland

Toshiharu Hayashi Department of Veterinary Pathology, Yamaguchi University, Yamaguchi, Japan

Lily Ho-Le Blizzard Institute, Queen Mary College, University of London, London, UK

Didier Hober Laboratory of Virology/EA3610, University Lille 2, CHRU Lille Institut Hippocrate, Loos-Lez-Lille, France

Margo C. Honeyman Diabetes Research Center, The Walter & Eliza Hall Institute of Medical Research, Parkville, VIC, Australia

Heikki Hyoty Department of Virology, University of Tampere and Tampere University Hospital, Tampere, Finland

Jorma Ilonen Immunogenetics Laboratory, University of Turku, Turku, Finland

Department of Clinical Microbiology and Immunology, University of Eastern Finland, Kuopio, Finland

Akihisa Imagawa Department of Metabolic Medicine, Graduate School of Medicine, Osaka University, Suita, Japan

Richard Insel Center for Human Genetics and Molecular Pediatric Disease, University of Rochester Medical Center, Rochester, NY, USA

Juvenile Diabetes Research Foundation International (JDRF), New York, NY, USA

Hitoshi Katsuta Department of Medical Science and Technology, Graduate School of Medical Sciences, Kyushu University, Fukuoka, Japan

Hironori Kurisaki Department of Medical Science and Technology, Graduate School of Medical Sciences, Kyushu University, Fukuoka, Japan

Katharina Lind The Center for Infectious Medicine, Department of Medicine HS, Karolinska Institutet, Stockholm, Sweden

Sabina Resic Lindehammer Department of Clinical Sciences, Lund University/CRC, Skåne University Hospital SUS, Malmö, Sweden

Åke Lernmark Department of Clinical Sciences, Lund University/CRC, Skåne University Hospital SUS, Malmö, Sweden

R. David G. Leslie Blizzard Institute of Cell and Molecular Science, Queen Mary College, University of London, London, UK

Zhijun Liu Department of Medicine, University of Massachusetts Medical School, Worcester, MA, USA

Giuseppe Maccari Laboratory of Microbiology, Department of Experimental Medicine, University of Insubria Medical School, Varese, Italy

Kazuya Matsuda Department of Veterinary Pathology, School of Veterinary Medicine, Rakuno Gakuen University, Ebetsu, Hokkaido, Japan

John P. Mordes Professor of Eudoeriuology and Metabolism, Department of Medicine, University of Massachusetts Medical School, Worcester, MA, USA

N.G. Morgan Peninsula College of Medicine and Dentistry, University of Exeter, Exeter, Devon, UK

Seiho Nagafuchi Department of Medical Science and Technology, Graduate School of Medical Sciences, Kyushu University, Fukuoka, Japan

Bo Niklasson Department of Medical Cell Biology, Uppsala University, Uppsala, Sweden,

Apodemus AB, Stockholm, Sweden

Maarit Oikarinen School of Medicine, University of Tampere, Tampere, Finland

Sami Oikarinen School of Medicine, University of Tampere, Tampere, Finland

Takashi Onodera Department of Risk Assessment Science, Research Center for Food Safety, University of Tokyo, Bunkyo-ku, Tokyo, Japan

Mark Peakman Department of Immunobiology, King's College London School of Medicine, London, UK

Myra Poon The Children's Hospital at Westmead, Institute of Endocrinology and Diabetes, Sydney, Australia

Discipline of Paediatrics and Child Health, The University of Sydney, Sydney, Australia

Alberto Pugliese Diabetes Research Institute, University of Miami Miller School of Medicine, Miami, FL, USA

William D. Rawlinson Virology Research Laboratory, Prince of Wales Hospital, Sydney, Australia

School of Biotechnology and Biomolecular Science, University of New South Wales, Sydney, NSW, Australia

School of Medical Sciences, University of New South Wales, Sydney, NSW, Australia

Karena Riedweg Université Lille 2, CHRU, Laboratoire de Virologie/EA3610, Loos-lez-Lille, France

Marian Rewers Barbara Davis Center for Childhood Diabetes, University of Colorado School of Medicine, Aurora, CO, USA

S.J. Richardson Peninsula College of Medicine and Dentistry (PCMD), University of Plymouth, Plymouth, UK

Merja Roivainen Division of Health Protection, Department of Infectious Disease Surveillance and Control, Intestinal Viruses Unit, National Institute for Health and Welfare (THL), Helsinki, Finland

Alessandro Salvatoni Pediatric Endocrinology Unit, Department of Experimental Medicine, University of Insubria Medical School, Varese, Italy

Stellan Sandler Department of Medical Cell Biology, Uppsala University, Uppsala, Sweden

Famara Sane Université Lille 2, CHRU, Laboratoire de Virologie/EA3610, Loos-lez-Lille, France

Luis Sarmiento Department of Virology, "Pedro Kouri" Tropical Medicine Institute, Havana, Cuba

Guido Sebastiani Diabetes Unit, Department of Internal Medicine, Endocrine and Metabolic Sciences and Biochemistry, University of Siena, Siena, Italy

Umberto Di Mario ONLUS Research Foundation, Siena, Italy

Teemu Smura Division of Health Protection, Department of Infectious Disease Surveillance and Control, Intestinal Viruses Unit, National Institute for Health and Welfare (THL), Helsinki, Finland

Isabella Spagnuolo Diabetes Unit, Department of Internal Medicine, Endocrine and Metabolic Sciences and Biochemistry, University of Siena, Siena, Italy

Umberto Di Mario ONLUS Research Foundation, Siena, Italy

Teodora Staeva JDRF, New York, NY, USA

Glyn Stanway Department of Biological Sciences, University of Essex, Colchester, UK

Lars C. Stene Division of Epidemiology, Norwegian Institute of Public Health, Oslo, Norway

Hiroyuki Taniyama Department of Veterinary Pathology, School of Veterinary Medicine, Rakuno Gakuen University, Ebetsu, Hokkaido, Japan

Keith W. Taylor † (Deceased)

Antonio Toniolo Laboratory of Medical Microbiology, University of Insubria & Ospedale di Circolo, Varese, Italy

Malin Flodström Tullberg The Center for Infectious Medicine, Department of Medicine HS, Karolinska Institutet, Stockholm, Sweden

Francesco Vendrame Umberto Di Mario ONLUS Research Foundation, Siena, Italy

Diabetes Research Institute, University of Miami, Miami, FL, USA

Matthias G. von Herrath La Jolla Institute for Allergy and Immunology (LIAI), La Jolla, CA, USA

A. Willcox Peninsula Medical School, University of Exeter, Plymouth, UK

Danny Zipris Barbara Davis Center for Childhood Diabetes, University of Colorado, Aurora, CO, USA

Arie J. Zuckerman Emeritus Professor of Medical Microbiology, University College London Medical School, Royal Free Campus, London, UK

Part I
Background and Pathogenesis

Chapter 1
Historical Background: Earlier Studies on the Connexion Between Viruses and Diabetes

Keith W. Taylor

Although there are references to the possible relationships between mumps and diabetes in the mid-nineteenth century (Stang 1864), it was not until much later that Harris (1899), described in detail a likely association between the two diseases. In the case discussed by Harris, glycosuria in a young American farmer quickly followed the initial mumps attack, but full blown diabetes with ketosis developed over a 3-year period. It was assumed that the mumps produced pancreatitis which involved the islets. In the ensuing 30 years, sporadic cases where there was an association between mumps and diabetes were reported (Patrick 1924), but it was generally assumed that mumps was a rare cause of diabetes. Gundersen (1927), however, published a paper with the intriguing title "Is Diabetes of Infectious Origin?", in which it was suggested that what he termed infectious parotitis or mumps produced pancreatic disease leading to diabetes in the young some 3 years after the initial infection. His figures were based on death rates due to diabetes in Norway in the pre-insulin period. At that time diabetes in the young with ketosis was usually fatal, and death rates from the disease bore a relationship to its incidence.

It is now known that several other viruses can produce a parotitis, as well as pancreatic disease, including enteroviruses. Since methods for the accurate identification of viruses did not then exist, viruses other than the mumps virus could well have been involved on occasions.

The association of mumps with subsequent diabetes has been reported in isolated cases ever since.

K.W. Taylor, M.B.B.S., Ph.D., F.R.C.P.
† (Deceased)

K.W. Taylor, H. Hyoty, A. Toniolo, A.J. Zuckerman (eds.), *Diabetes and Viruses*,
DOI 10.1007/978-1-4614-4051-2_1, © Springer Science+Business Media New York 2013

Other Viruses and Diabetes

With improvements in virological techniques, however, in the 1950s and 1960s it became evident that diabetes might be associated with infection by a number of other viruses in addition to mumps in man as well as in animals. An outbreak of foot and mouth disease in cattle in Italy was accompanied by ketotic diabetes (Barboni and Manocchio 1962) with pancreatic lesions. In mice, strains of encephalomyocarditis virus caused diabetes with damage to the islets of Langerhans (Craighead and Mclane 1968). Parallel work also showed that coxsackieviruses might produce a pancreatitis in mice although without diabetes (Pappenheimer et al. 1951).

Afterwards, there were reports of coxsackieviruses inducing pancreatitis in new born infants (Kibrick and Benirschke 1958). It became clear that a number of other viruses in addition to mumps might be involved in pancreatic damage and perhaps the precipitation of diabetes in man.

A more general association of enteroviruses with the onset of diabetes in type 1 diabetes was first suggested by the work conducted in London and Birmingham in the UK during the late 1960s (Gamble et al. 1969). In preliminary work, it had been noticed that there was a marked seasonal incidence for type 1 diabetes, with autumnal and winter peaks (Gamble and Taylor 1969). It was suggested that the autumnal peak might be due to an enterovirus infection, and that the winter peak represented intercurrent non-specific infection which had worsened carbohydrate tolerance following earlier pancreatic damage.

This led to a much larger scale investigation to determine in detail whether enteroviral infection might be involved with the onset of type 1 diabetes.

The investigation involved 123 patients, all with diabetes of sudden onset. Most required insulin treatment and were ketotic. Those investigated covered a broad age range from early childhood to over 60 years of age, although with the 0–40 years group age predominating.

Using a neutralising antibody technique, higher titres of antibodies to coxsackieviruses were found than in controls than in those diabetics tested soon after onset. In these first studies, coxsackievirus B4 was the virus most commonly detected, even though it is now clear that several other enteroviruses seem to be involved.

Similar studies using classical immunological techniques to detect viruses were repeated by many other investigators during the next 30 years, with most, though not all, showing comparable results. Since the choice of patients investigated, the methods of handling blood and subsequent virus identification varied very widely, it is not surprising that results were not always clear cut. Some of these problems are discussed in Chaps. 13, 15, 17, 23, and 32. The use, however, of the polymerase chain reaction (PCR) has generally confirmed the original suggestions.

The seasonal incidence of type 1 diabetes in temperate countries was also confirmed (see Chap. 11). This supported the idea that an infective process could be associated with the precipitation of diabetes.

Rubella and Diabetes

At about the time of the first studies on enteroviruses and diabetes, interest was focussed on a long-term study of children with congenital rubella following severe rubella epidemics in Australia in the 1940s. In a 25-year follow-up of 50 such patients, one case of undiagnosed diabetes was reported (Forrest et al. 1967). By 1974, 8 of 45 patients had acquired diabetes, 4 of whom were on insulin (Menser et al. 1974). It was clear that foetal infection with this virus could be linked with diabetes in the long term.

Isolation of Viruses from Patients with Type 1 Diabetes

In an important study on of a single case of ketotic diabetes in a child who died, coxsackievirus B4 virus was isolated from the pancreas and the strain shown to produce diabetes in mice (Yoon et al. 1979). Lymphocytic infiltration of the islets of Langerhans in this case was observed. In this instance, Koch's postulates appear to have been fulfilled. A similar case was reported a year later (Champsaur et al. 1980), in an infant, this time involving coxsackievirus B5. Other isolated cases where severe enterovirus infection was associated with diabetes continued to be reported (Szopa et al. 1993).

In summary, therefore, not long after the middle of the last century a number of viruses were beginning to be clearly linked with the onset of diabetes in man and animals. The latter are reviewed elsewhere (Yoon 1991).

In humans, the most important viruses were mumps virus, rubella virus and especially enteroviruses. In a few instances, enteroviruses appeared to be directly causative, but doubts remained as to the proportion of patients with type 1 diabetes who were infected with these viruses before onset. Many of these doubts have been removed by the use of molecular methods for virus detection (Yeung et al. 2011).

References

Barboni E, Manocchio I (1962) Alterazioni pancreatiche in bovini con diabete mellito post aftoso. Arch Vet Ital 13:447–489

Champsaur H, Dussaix E, Samolyk D, Fabre N, Back C, Assan R (1980) Diabetes and coxsackie B5 infection. Lancet 1:251

Craighead JE, McLane MF (1968) Diabetes mellitus induction in mice by encephalomyocarditis virus. Science 162:913–914

Forrest JM, Menser MA, Harley JD (1967) A twenty five year follow up of congenital rubella. Lancet 2:1347–1350

Gamble DR, Taylor KW (1969) Seasonal incidence of diabetes mellitus. BMJ 3:631–633

Gamble DR, Kinsley ML, Fitzgerald MG, Bolton R, Taylor KW (1969) Viral antibodies in diabetes mellitus. BMJ 3:627–630

Gundersen E (1927) Is diabetes of infectious origin? J Infect Dis 41:197–202
Harris HF (1899) A case of diabetes mellitus quickly following mumps. Boston Med Surg J 140:465–469
Kibrick S, Benirschke K (1958) Severe generalised disease (encephalohepatomyelocarditis) occurring in the newborn period and due to infection with coxsackie virus group B. Pediatrics 22:857–875
Menser MA, Forrest JM, Honeyman MC (1974) Diabetes, HLA antigens and congenital rubella. Lancet 2:1508–1509
Pappenheimer AM, Kunz LJ, Richardson S (1951) Passage of coxsackievirus (Connecticut S strain) in adult mice with production of pancreatic disease. J Exp Med 94:45–64
Patrick A (1924) Acute diabetes following mumps. BMJ 2:802
Stang J (1864) Cited by Gundersen E, J Infect Dis 41:197, 1927
Szopa TM, Titchener PA, Portwood ND, Taylort KW (1993) Diabetes due to viruses—some recent developments. Diabetologia 36:687–695
Yeung WC, Wing Chi G, Rawlinson WD, Craig ME (2011) Enterovirus infection and type 1 diabetes: a systematic review and meta-analysis of observational molecular studies. BMJ 342:d35
Yoon JW (1991) The role of viruses in the pathogenesis of IDDM. Ann Med 23:437–449
Yoon JW, Austin M, Onodera T, Notkins AL (1979) Virus induced diabetes mellitus. Isolation of a virus from a child with diabetic ketoacidosis. New Engl J Med 300:1173–1178

Chapter 2
Viruses and Autoimmune Diabetes: A History

R. David G. Leslie, Lily Ho-Le, and Huriya Beyan

Certainty: The Ascent of the Gene

The identification of an association between type 1 diabetes and certain genes transformed our understanding of this and other related diseases, including thyroid disease and multiple sclerosis. That type 1 diabetes is genetically determined and was evident from family, twin and genetic studies. The frequency of type 1 diabetes is higher in siblings of diabetic patients (e.g. in UK 6% by age 30) than in the general population (0.4% by age 30) (Field 2002). Of genes implicated in the genetic susceptibility to type 1 diabetes, the most important are in the histocompatibility (HLA) region of chromosome 6 (Kumar et al. 1993); first sought by Singal but then sought successfully by Nerup and Cudworth (Nerup et al. 1974). Such HLA genes predispose to a number of autoimmune diseases including type 1 diabetes (Kumar et al. 1993; Concannon et al. 2005), as demonstrated in both population and family studies (Redondo et al. 2001; Kumar et al. 1993; Meyer and Thomson 2001). Genes encoding HLA molecules and located within the major histocompatibility complex (MHC) on the short arm of chromosome 6 are associated with type 1 diabetes. The MHC complex is a polymorphic gene complex in which multiple alleles exist for each genetic locus. The MHC is divided into class I (HLA-A, -B and -C), class II (HLA-DR, -DQ and -DP) and class III (genes for complement components). The classes I and II proteins coded by the relevant genes are transmembrane cell surface glycoproteins which are critically involved in the presentation of both self- and foreign antigens as short peptides to T-lymphocytes.

R.D.G. Leslie, M.B.B.S., M.R.C.S., M.D. (✉) • L. Ho-Le • H. Beyan
Blizzard Institute, Queen Mary College, University of London, London, UK
e-mail: r.d.g.leslie@qmul.ac.uk

K.W. Taylor, H. Hyoty, A. Toniolo, A.J. Zuckerman (eds.), *Diabetes and Viruses*,
DOI 10.1007/978-1-4614-4051-2_2, © Springer Science+Business Media New York 2013

HLA genes are highly polymorphic with a degree of coding region diversity unequalled elsewhere in the genome. Polymorphisms of certain genes probably originated in selection pressures exerted by environmental factors including epidemics, climatic change and availability of food. A non-human somatic study of the HLA DQ beta region suggests that this region has been in balanced polymorphism for ten or more million years (Meyer and Thomson 2001). To maintain the extraordinary diversity of HLA types over this time, selection pressures must have been operating; otherwise most alleles would have been lost through genetic drift. It has been proposed that infectious pathogens are the major cause of HLA diversity. The distribution of sequence variation is clustered in nucleotides which code for amino acids composing the antigen-binding groove. This implies that natural selection must have acted at this binding site to maintain structural diversity for peptide binding. By 1990, it seemed most likely that this peptide-binding site identified an autoantigen, and that the trimolecular complex of HLA, autoantigen and T-cell receptor reflected an autoimmune disease process (Nerup et al. 1974). Such an argument was supported by the impressive technical genetic achievements, using single nucleotide polymorphisms, association studies and genome-wide association studies (GWAS), most notably by Todd and his colleagues (Nejentsev et al. 2009). Allied to which was the comparatively unimpressive ability of epidemiologists to identify any key non-genetic factor.

But the limited degree to which HLA and other non-HLA could account for all the risk of type 1 diabetes, the missing heritability, remained an issue. The term heritability reflects gene expression or penetrance in a given environment. The best estimate of heritability can be obtained by determining concordance rates of twins. Both identical and non-identical twins share the same environment in childhood but only identical twins share the same genes. In the classic twin method the difference between the concordance rates for identical and non-identical twins is doubled to give an index of heritability. Higher concordance rates, for autoimmune diseases in general and type 1 diabetes in particular, in identical compared with non-identical twins are consistent with a genetic influence on these diseases (Salvetti et al. 2000). Estimates of heritability can be obtained from studies in Finland and the University of Southern California; in both the estimates are substantially less than 100% which means the disease is unlikely to be autosomal dominant (Hyttinen et al. 2003). Age-related genetic factors also influence the risk of type 1 diabetes, as the disease risk is lower in adults than in children, and the range of incidence across European countries is also reduced in older age (Kyvick et al. 2004). Survival analysis estimated that non-diabetic identical twins of probands diagnosed with type 1 diabetes under 25 years of age had, in one study, a 38% probability of developing diabetes compared with only 6% for twins of probands diagnosed later (Salvetti et al. 2000). Such a remarkably low twin concordance rate for adult-onset type 1 diabetes implies that the genetic impact in adult-onset type 1 diabetes is limited, and certainly lower than that in childhood-onset disease (Salvetti et al. 2000; Hyttinen et al. 2003; Kyvick et al. 2004). These effects were widely attributed to a stochastic effect by geneticists, but there remained the possibility that other non-genetically determined effects (such as epigenetic effects or environmental effects) might be important. HLA associations with these diseases could, after all, operate through susceptibility to certain undefined infections.

Uncertainty: Autoimmunity as a Disease Process

Autoimmune diseases are the third leading cause of morbidity and mortality in the developed world, only surpassed by cancer and heart disease. Most autoimmune diseases are thought to be complex disorders involving the interaction of non-genetic, probably environmental, factor(s) with more than one genetic factor. The evidence suggests that for the generality of human autoimmune diseases there is no specific "autoimmune gene" but instead a combination of normal genes and common polymorphisms (e.g. HLA haplotypes) which provide a genetic susceptibility to non-genetic factors with which they interact, resulting in an abnormal autoimmune response (Field 2002). The immune system is designed by nature to protect us from our environment (Janeway 2001). But activation of the immune response not only protects us from disease, as in infectious diseases, but also causes disease, as in autoimmunity. Autoimmunity is important to the fitness of the organism. Most individuals produce autoantibodies and autoreactive T-lymphocytes. However, only about 5% of any population develops an autoimmune disease. Control mechanisms must therefore operate to control the development of autoimmune diseases. These control mechanisms remove cytotoxic immune cells in various ways including: clonal deletion, clonal anergy and limiting antigen accessibility to the immune system. Antigen accessibility is limited by being processed for presentation to the immune system or by autoreactive T-lymphocytes circulating in an inert state. Breakdown in these control mechanisms could lead to disease. Diseases associated with autoimmune phenomena tend to distribute themselves within a spectrum of organ-specific diseases, such as type 1 diabetes and non-organ-specific diseases such as systemic lupus erythematosus. There may be clustering of diseases at either end of this spectrum; thus, T1DM is more common in patients with thyroiditis or adrenalitis. Autoimmunity, in the form of autoantibodies, is common after many infections and may well result from the mimicking of host proteins by antigens of the infectious agent. Autoimmune disease has long been considered as a shadow following infections. Epidemiological evidence shows that rheumatic fever follows streptococcal infection and *Trypanosoma cruzi* infection is the instigator of Chagas' disease. There is, however, little information regarding the mechanism by which such a train of events is initiated, e.g. there are no certain examples in humans in which molecular mimicry gives rise to autoimmune disease (Table 2.1).

For all that type 1 diabetes is considered an autoimmune disease, we must, therefore, acknowledge that the evidence is incomplete. Rose and Bona defined autoimmune diseases as those that show three features (Rose and Bona 1993): (a) defined

Table 2.1 List of potential environmental agents

General factors	Specific factors
Hygiene	Viruses (e.g. enteroviruses)
Parasites	Bacteria
Co-existent infections (TB or malaria)	Cow's milk (through early exposure)
	Toxins

Table 2.2 Autoantibodies as predictors of type 1 diabetes

Autoantibodies
• Can appear at an early age, even around the time of birth
• Can precede the clinical onset of diabetes by some years
• Have variable predictive value depending on the autoantigen recognised
• Have increasing positive predictive value with increasing numbers

autoantigens and autoantibodies must be present; (b) passive transfer of T-lymphocytes (specific or non-specific) must lead to disease development; (c) immunomodulation of subjects with disease must ameliorate symptoms. We know that the first of these is true and that the autoantibodies can predict the disease with a degree of certainty (Table 2.2). Autoantibodies, originally identified in type 1 diabetes by Bottazzo et al. (1974), has been detected to four major autoantigens, glutamic acid decarboxylase (GAD65) by Lernmark and colleagues, tyrosine phosphatase-like molecule (IA-2) by Notkins and colleagues, insulin autoantibodies (IAA) by Palmer and colleagues, and zinc transporter-8 autoantibodies (ZnT8) by Hutton and colleagues in about 90% of newly diagnosed patients with type 1 diabetes (Baekkeskov et al. 1982, 1990; Leslie et al. 2001). However, transfer of disease is ethically unacceptable though a single case has been described of apparent transfer of type 1 diabetes following a bone marrow transplant from a diabetic donor to a non-diabetic recipient (Lampeter et al. 1993). Further, there was rapid destruction of apparently normal islet insulin secretory cells when islets were transplanted from a non-diabetic twin to their diabetic identical co-twin, indicating that the destructive process must be outside the islet, insulin secretory cell specific and retain its cytotoxic memory (Sibley et al. 1985). The immune system is the most likely candidate for such an extra-islet effect. Finally we must remember that we are currently unable to immunomodulate this disease, let alone ameliorate symptoms, though there is some limited evidence that the disease process can be modified, at least from a Phase 2 study, by immunotherapy with alum-formulated GAD65; at the time of going to press, the preliminary results of a Phase 3 study appear disappointing. Further, subjects with newly diagnosed type 1 diabetes given cyclosporine, a modifier of T-cell activation, are more likely to show a transient improvement in metabolic control in the first 2 years post-diagnosis (Feutren et al. 1986). So only the presence of disease predictive autoantibodies and the trials showing the benefit of cyclosporine and alum-GAD65 in T1DM provide strong, but not definitive, support for it being an autoimmune disease.

The Failure of Immunomodulation and Its Implications

The aim of disease prediction is disease prevention. Type 1 diabetes could be prevented by avoiding those environmental factors which cause the disease process (primary prevention); or by modulating the destructive process before the onset of clinical diabetes (secondary prevention) or by trying to cure the disease process at the time of diagnosis (tertiary prevention). In the last decade, attention has focused on the possibility of immunomodulation as a secondary or tertiary form of prevention of type 1 diabetes.

Secondary Prevention of Type 1 Diabetes

Secondary prevention (that is after disease induction but before clinical diabetes develops) could prevent autoimmune diabetes by (a) protection of insulin secreting cells; (b) rest of insulin secreting cells and (c) immune modulation including antigen-based strategies. This field has been hindered by the extensive use of an animal model, the non-obese diabetic (NOD) mouse, which can be cured of diabetes in many different ways, but which offers little of value to modify human autoimmune diabetes. For example in both BB rats and NOD mice, insulin, a presumed key antigen, given therapeutically, delayed the development of diabetes and insulitis, but a study of oral insulin in at-risk children, based on such hypothetical immunomodulation, failed (Skyler et al. 2005). Such trial failures are not only disappointing, but they have also highlighted the problem with relying too heavily on an inconsistent animal model. Immunomodulation suddenly looked less promising as a cure option. So, attention switched back to non-genetic effects and the potential of identifying disease-related factors.

Possibility: Non-genetic Events Renascent

The decline in interest in the destructive immune change as a potential target of therapy has switched the focus away from autoimmunity and back to environmental events. Environmental factors have, indeed, been implicated in the aetiology of autoimmune diseases. These factors include, for autoimmune diseases, and possibly atopy, in general: temperate climate, increased hygiene and decreased rates of infection, vaccinations and antibiotics, drugs (methyl donors such as hydralazine), wheat consumption, iodine levels, increasing wealth; and also for type 1 diabetes specifically: gross national product, overcrowding in childhood and virus infections, early exposure to cow's milk, reduced rates or duration of breast feeding and vitamin D and nitrite consumption (Field 2002; Meyer and Thomson 2001). These factors will be discussed in more detail later.

Acknowledgments We wish to thank the funding agencies involved in our studies including the Diabetic Twin Research Trust, Juvenile Diabetes Research Foundation International, The Wellcome Trust, the Joint Research Board of St Bartholomew's and the Royal London Medical College, European Union and Eli Lilly. We also thank Dr S. Paschou for editorial assistance.

References

Baekkeskov S, Nielsen JH, Marner B, Bilde T, Ludvigsson J, Lernmark A (1982) Autoantibodies in newly diagnosed diabetic children immunoprecipitate human pancreatic islet cell proteins. Nature 298(5870):167–169

Baekkeskov S, Aanstoot HJ, Christgau S, Reetz A, Solimena M, Cascalho M, Folli F, Richter-Olesen H, De Camilli P (1990) Identification of the 64 K autoantigen in insulin-dependent diabetes as the GABA-synthesizing enzyme glutamic acid decarboxylase. Nature 347:151–156

Bottazzo GF, Florin-Christensen A, Doniach D (1974) Islet-cell antibodies in diabetes mellitus with autoimmune polyendocrine deficiencies. Lancet 2(7892):1279–1283

Concannon P, Erlich HA, Julier C, Morahan G, Nerup J, Pociot F, Todd JA, Rich SS (2005) Type 1 diabetes genetic consortium. Type 1 diabetes: evidence for susceptibility loci from four genome-wide linkage scans in 1,435 multiplex families. Diabetes 54:2992–3001

Feutren G, Papoz L, Assan R, Vialettes B, Karsenty G, Vexiau P, Du Rostu H, Rodier M, Sirmai J, Lallemand A, et al. Cyclosporin increases the rate and length of remissions in insulin-dependent diabetes of recent onset. Results of a multicentre double-blind trial. Lancet. 2:119–124

Field LL (2002) Genetic linkage and association studies of type 1 diabetes: challenges and rewards. Diabetologia 45:21–35

Hyttinen V, Kaprio J, Kinnunen L, Koskenvuo M, Tuomilehto J (2003) Genetic liability of type 1 diabetes and the onset age among 22,650 young Finnish twin pairs: a nationwide follow-up study. Diabetes 52:1052–1055

Janeway CA (2001) How the immune system works to protect the host from infection: a personal view. Proc Natl Acad Sci USA 98:7461–7468

Kumar D, Gemayel NS, Deapen D, Kapadia D, Yamashita PH, Lee M, Dwyer JH, Roy-Burman P, Bray GA, Mach TM (1993) North-American twins with IDDM: genetic, etiological, and clinical significance of disease concordance according to age, zygosity, and the interval after diagnosis in first twin. Diabetes 42:1351–1363

Kyvick KO, Nystrom L, Gorus F, Songini M, Oestman J, Castell C, Green A, Guyrus E, Ionescu-Tirgoviste C, Mckinney PA, Michalkova D, Ostrauskas R, Raymond NT (2004) The epidemiology of Type I diabetes mellitus is not the same in young adults as in children. Diabetologia 47:377–384

Lampeter EF, Homberg M, Quabeck K, Schaefer UW, Wernet P, Bertrams J, Grosse-Wilde H, Gries FA, Kolb H (1993) Transfer of insulin-dependent diabetes between HLA-identical siblings by bone marrow transplantation. Lancet 34:1243–1244

Leslie D, Lipsky P, Notkins AL (2001) Autoantibodies as predictors of disease. J Clin Invest 108(10):1417–1422

Meyer D, Thomson G (2001) How selection shapes variation of the human major histocompatibility complex: a review. Ann Hum Genet 65:1–26

Nejentsev S, Walker N, Riches D, Egholm M, Todd JA (2009) Rare variants of IFIH1, a gene implicated in antiviral responses, protect against type 1 diabetes. Science 324(5925):387–389

Nerup J, Platz P, Andersen OO, Christy M, Lyngsoe J, Poulsen JE, Ryder LP, Nielsen LS, Thomsen M, Svejgaard A (1974) HLA antigens and diabetes mellitus. Lancet 2(7885):864–866

Redondo MJ, Yu L, Hawa M, Mackenzie T, Pyke DA, Eisenbarth GS, Leslie RDG (2001) Heterogeneity of Type I diabetes: analysis of monozygotic twins in Great Britain and the United States. Diabetologia 44:354–362

Rose NR, Bona C (1993) Defining criteria for autoimmune diseases (Witebsky's postulates revisited). Immunol Today 14:426–430

Salvetti M, Ristori G, Bomprezzi R, Pozzilli P, Leslie RDG (2000) Twins: mirrors of the immune system. Immunol Today 21:342–347

Sibley RK, Sutherland DE, Goetz F, Michael AF (1985) Recurrent diabetes mellitus in the pancreas iso- and allograft. A light and electron microscopic and immunohistochemical analysis of four cases. Lab Invest 53:132–144

Skyler JS, Krischer JP, Wolfsdorf J, Cowie C, Palmer JP, Greenbaum C, Cuthbertson D, Rafkin-Mervis LE, Chase HP, Leschek E (2005) Effects of oral insulin in relatives of patients with type 1 diabetes: the diabetes prevention trial—type 1. Diabetes Care 28:1068–1076

Chapter 3
Genetics of Type 1 Diabetes

Robert Hermann and Jorma Ilonen

Abstract Type 1 diabetes has a strong genetic component but additional environmental factors have also an essential role. This is demonstrated by the multifold rise in incidence of the disease during the last decades in most industrialized countries. The most important genes are located within the HLA region, where class II genes largely define the disease risk. This region contributes roughly half of the genetic component. Outside the HLA region *INS*, *PTPN22*, and *CTLA4* were identified using candidate gene approach and subsequently recent genome-wide association studies have increased the number of risk-associated loci up to 50. Although the effects of common SNP polymorphisms in lately identified genes are small, their combined influence may still be considerable in complex pathways associated with activation of autoimmune process. The presence of rare mutants associated with disease risk has also generated enthusiasm since their identification for the first time in the *IFIH1* gene, where rare mutants and deteriorating gene function are in fact strongly protective. This gene is an intracellular receptor for enterovirus RNA, and the finding suggests that enterovirus infection-associated innate immune response is important in beta-cell damage. Several alternative gene combinations may also be important in individual cases reflecting existence of different routes to beta-cell destruction.

Type 1 diabetes is generated by interaction of genetic susceptibility and environmental factors. Main genetic contribution is located within the HLA class II region. In addition, up to 50 genes have been identified outside the HLA region. Rare mutants in some of these genes may have potent effects. Different gene combinations may be

R. Hermann, M.D., Ph.D.
Immunogenetics Laboratory, University of Turku, Turku, Finland

J. Ilonen, M.D., Ph.D. (✉)
Immunogenetics Laboratory, University of Turku, Turku, Finland

Department of Clinical Microbiology and Immunology,
University of Eastern Finland, Kuopio, Finland
e-mail: jorma.ilonen@utu.fi

K.W. Taylor, H. Hyoty, A. Toniolo, A.J. Zuckerman (eds.), *Diabetes and Viruses*,
DOI 10.1007/978-1-4614-4051-2_3,

important in individual cases of the disease. The IFIH1 gene serving as intracellular enterovirus RNA receptor represents a possible link between virus infection and type 1 diabetes.

Introduction

The most common form of type 1 diabetes is classified as autoimmune or type 1A diabetes, which is the main focus of this chapter, where we refer to it as type 1 diabetes (Craig et al. 2009). Type 1 diabetes is caused by reduced or complete loss of function of the insulin-producing beta-cells in the pancreatic islets of Langerhans. This organ failure is accompanied by profoundly reduced beta-cell mass, as beta-cells die during the disease process that can last for years before clinical symptoms appear. The molecular mechanism responsible for this tissue destruction is poorly understood, although it is clear that effector functions of the immune system are important. In type 1B diabetes immunity does not play a role, while IPEX and APECED are rare monogenic autoimmune syndromes including type 1 diabetes. These syndromes associated with lack of Foxp3 positive regulatory T-cells and defective negative selection in thymus demonstrate the importance of these mechanisms in the control of autoimmunity (Akirav et al. 2011; Michels and Gottlieb 2010).

Epidemiological and family studies have indicated that first degree relatives of patients with type 1 diabetes have an increased disease risk compared to the general population. This is due mainly to the increased transmission of the diabetes risk gene variants to the offspring across generations, although co-exposure to common environmental factors can have role as well. Moreover, monozygotic twins have a higher concordance rate (>50%) than dizygotic twins, producing a crucial evidence for the role of genetic predisposing factors (Kyvik et al. 1995). These studies have also shown that the risk of type 1 diabetes does not follow the simple Mendelian inheritance; it rather corresponds to a polygenic pattern.

Interestingly the disease risk of the general population displays a profound ethnic variation being most common in the Nordic countries and among the Caucasian populations (Soltesz et al. 2007). In contrast type 1 diabetes is rare in Oriental populations. The incidence of type 1 diabetes displays a steady increase in most European countries, also in the record high incidence Finland (Harjutsalo et al. 2008). Accumulation of disease susceptibility gene variants in the general population is not responsible for this phenomenon (Hermann et al. 2003a).

Genes within the HLA region, predominantly those that encode antigen-presenting molecules, confer the greatest part of the genetic risk of type 1 diabetes. HLA itself is a very complex polygenic disease superlocus where the major disease risk factor is the *DRB1-DQA1-DQB1* genotype defining polymorphisms of the antigen binding grooves of the HLA DR and DQ molecules. In addition to HLA, more than 40 genes have been confirmed to contribute to the genetic orchestra for type 1 diabetes (Barrett et al. 2009). These loci make only modest individual contributions to the overall disease risk at population level; most risk alleles have an odds ratio of 1.3 or less. However, it is important to note that these loci define several biological pathways, and may have a strong impact in individual subjects with different genomic patterns.

The Role of HLA

The HLA system (6p21) is the major disease locus for type 1 diabetes. Here the primary disease risk determinant is the *DQB1* gene that encodes the beta chain of the Class II DQ molecule, responsible for antigen presentation on the cell surface (Todd et al. 1987). It is one of the most polymorphic genes in the human genome. Its alleles in combination with the neighboring *DQA1* and *DRB1* gene variants form the *DR-DQ* haplotypes that can be categorized into risk, neutral, and protective groups, according to their predisposing effects for type 1 diabetes (Hermann et al. 2003b; Lambert et al. 2004) (see Table 3.1). The nomenclature for the HLA factors is complex, where gene names are labeled with letters and alleles separated by an asterisk are numbered (Marsh et al. 2010). The heterozygous combination of the two major susceptibility haplotype, *DRB1*03-DQA1*0501-DQB1*0201/DRB1*04-DQA1*0301-DQB1*0302* or DR3-DQ2/DR4-DQ8 in terms of serological specificities, represents the highest disease risk with an absolute risk of 2–8% (Hermann et al. 2003b; Lambert et al. 2004). Humans are evolutionarily positively selected for heterozygosity in HLA; it was, e.g., shown to protect from hepatitis C virus persistence (Hraber et al. 2007). However, in the case of type 1 diabetes the increased number of *DQ* heterodimers appears to accelerate autoimmunity when additional permissive genetic or environmental factors are present. Subjects

Table 3.1 Association of most common HLA Class II haplotypes with type 1 diabetes

HLA haplotype name	Disease risk	Notes
Risk haplotypes		
DRB1*04-DQA1*03-DQB1*0302	Strong risk	DRB1*0401, *0402, *0404, and *0405 subtypes associated with variable risk, *0403/6 with protection
DRB1*0408-DQA1*03-DQB1*0304	Strong risk	A rare haplotype
DRB1*03-DQA1*05-DQB1*02	Risk	Stronger effect in Southern Europe
DRB1*0405-DQA1*03-DQB1*02	Risk	Predominantly in Mediterraneans
DRB1*09-DQA1*03-DQB1*0303	Risk	Predominant
DRB1*0405-DQA1*03-DQB1*0401	Risk	Risk factors in Orientals
DRB1*0802-DQA1*03-DQB1*0302		
DRB1*07-DQA1*03-DQB1*02	Risk	Predominantly in blacks
Protective haplotypes		
DRB1*15-DQA1*01-DQB1*0602	Strong protective	
DRB1*07-DQA1*0201-DQB1*0303	Strong protective	
DRB1*14-DQA1*01-DQB1*0503	Strong protective	
DRB1*13-DQA1*01-DQB1*0603	Protective	
DRB1*11/12/13-DQA1*05-DQB1*0301	Protective	
DRB1*0403/06-DQA1*0301-QB1*0302	Protective	

Table 3.2 The most important non-HLA genes of type 1 diabetes

Gene name	Function	Disease variant
Insulin gene	Insulin synthesis	Promoter region VNTR class –23 A/T, +1140 A/C
Appears to control insulin expression in thymus, which affects insulin-specific autoimmunity. Class I VNTR and linked polymorphisms are associated with insulin autoantibodies		
CTLA4 gene	T-cell membrane protein Controls T-cell activation and regulatory T-cells	CT60—rs3087243 +49 A/G, AT(n) microsatellite
Strongly associated with autoimmune thyroid disease, its contribution to T1D predisposition is less pronounced		
PTPN22	Protein tyrosine phosphatase Regulates T-cell activation	Arg620Trp—rs2476601
Gain of function variant induces reduced T-cell activation, decreased IL2 secretion, that leads to activation of autoreactive T-cells		
IFIH1	Enterovirus RNA receptor	Ala946Thr—rs1990760 rs2111485
Binds long dsRNA and activates interferon regulatory factor 7 that leads to production of type 1 interferons (Fig. 3.1)		
IL2RA (CD25)	IL-2 cytokine receptor	rs11594656 intronic rs12251307 intergenic rs12722495 intronic rs2104286 intronic
Complex disease locus with multiple disease variants. T1D-predisposing alleles lower transcription of IL2RA in memory CD4 T-cells		

homozygous for any of these two haplotypes carry a considerably lower risk of type 1 diabetes—see Table 3.2. Importantly, there is also a set of HLA Class II haplotypes that confer resistance to the disease.

To establish a disease risk in a person it is necessary to take into account both HLA haplotypes. Susceptibility haplotypes behave in a recessive manner while protective haplotypes are rather dominant. In the presence of a susceptibility and a protective haplotype, the disease risk is reduced to neutral or even to the protective range. HLA Class II alleles can also be used to identify subjects at risk in the general population (Hermann et al. 2004) and in newborn screening and disease prevention programs focusing on high-risk cohorts (Hagopian et al. 2011; Kupila et al. 2001).

The HLA-encoded susceptibility haplotypes for type 1 diabetes display differences in other ethnic groups. For example, in the Japanese the *DQB1*0302* allele is associated with *DRB1*0802*, and the other main risk haplotypes for type 1 diabetes are the *DRB1*0405-DQB1*0401* and the *DRB1*0901-DQB1*0303*. It is important to emphasize further that the fulminant form of type 1 diabetes is a special subphenotype, with distinct genetic susceptibility factors (Kawabata et al. 2009).

It is known that certain HLA Class II haplotypes are associated with particular β-cell-specific autoantibody patterns. IAA and IA-2 autoantibodies are associated with the HLA *DRB1*04-DQB1*0302* haplotype, whereas GAD65 autoantibodies are found more often in patients with *HLA DRB1*03-DQA1*0501-DQB1*0201* haplotype (Vandewalle et al. 1993; Ziegler et al. 1991).

Risk modifier loci exist in the HLA Class I region as well (Nejentsev et al. 2007). Recently in the Finnish Diabetes Prediction and Prevention Study we also identified Class I effects on the progression to clinical disease after established autoimmunity (Lipponen et al. 2010). A protective effect of the *A*03* allele was detected, whereas the *B*39* allele had a promoting effect on the appearance of the clinical disease, especially in children with the high-risk HLA DR3/DR4 genotype.

Non-HLA Genes for Type 1 Diabetes

More than 40 gene variants display association with type 1 diabetes outside HLA (Barrett et al. 2009). Only those genes where substantial amount of functional information is available on their role on type 1 diabetes are discussed (Table 3.2).

The Insulin Gene Locus (INS)

At the *INS* locus the type 1 susceptibility variant maps near the promoter region (*−23A/T*, *+1140A/C*, and the 5′ VNTR at 11p15.5) (Barratt et al. 2004; Bell et al. 1984). Functional analyses suggested that the type 1 diabetes susceptibility VNTR variants (Class I alleles) are associated with lower insulin expression in the thymus, thereby creating a defect in the deletion on insulin-specific autoreactive T-cells (Pugliese et al. 1997; Vafiadis et al. 1997). The INS locus defines the "INS-dependent" disease pathway where insulin autoimmunity plays a central role in the initiation of the disease process, as the appearance of GADA (glutamic acid decarboxylase autoantibodies) and IA-2A (islet antigen two autoantibodies) follows IAA, probably by antigen spreading (Hermann et al. 2005). This theory fits well with results of previous animal studies (Nakayama et al. 2005). In IAA-negative individuals the initiation of beta-cell autoimmunity is likely controlled by factors other than the INS locus. This indicates that prevention strategies focusing on tolerance to insulin can be successful only in subjects, where the INS locus-dependent pathway is functional. This is supported by the data from the DPT-I trial that indicated a detectable effect of the intervention drug (oral insulin) in patients initially IAA positive (Skyler et al. 2005). Identification of further factors in these distinct pathways is crucial for prevention studies.

The CTLA4 Locus

The protein encoded by the *CTLA4* gene appears to have an essential role in regulatory T-cell function (Wing et al. 2008). The fine mapping of the causative genetic variant in the *CTLA4* region appears to be extremely difficult (Nistico et al. 1996).

The contribution of this locus to type 1 diabetes susceptibility displays marked *trans*-ethnic variation. Ueda et al. (2003) identified the *CT60* SNP as the potential causative disease marker; however, a series of additional SNPs display disease association in different ethnic groups. The effect of *CT60* was weaker than that of the +49A/G SNP, e.g., in the Finnish population (Douroudis et al. 2009). These data indicate that other population-specific disease variants may exist at the *CTLA4* locus, and the effect of this locus on for type 1 diabetes susceptibility is much stronger in populations in Southern Europe than in the Nordic countries. The reduction of the amount of soluble CTLA4 molecule could be a factor that predisposes to type 1 diabetes, but this issue remains unclear.

The PTPN22 Gene

The *PTPN22* gene (1p13) encodes a lymphoid protein tyrosine phosphatase (the protein product called LYP) that is expressed in lymphocytes and is acting through binding intracellular kinases. LYP is involved directly in setting thresholds for T-cell receptor signaling (Cloutier and Veillette 1999). The *Arg620Trp* polymorphism of this gene is associated with type 1 diabetes in many Caucasian populations, and this variant is a common predisposing factor for other autoimmune diseases as well (Bottini et al. 2006; Kyogoku et al. 2004). The 620Trp variant seems to be a gain of functional mutation that results in reduced intracellular Ca-mediated signaling, decreased IL2 secretion, and lower T-cell activation (Aarnisalo et al. 2008; Vang et al. 2005).

It was found that the *PTPN22* gene plays a critical role in controlling emergence of β-cell-specific autoantibodies and progression of the immune process to clinical disease (Hermann et al. 2006). The Trp/Trp genotype was associated with increased appearance of insulin autoantibodies, and it had an additive effect with the *insulin* locus and the *HLA-DRB1* gene on emergence of the autoantibodies. The *PTPN22* effect on to type 1 diabetes susceptibility was stronger in males and subjects with low-risk HLA genotypes.

The *Arg620Trp* variant is a Caucasian mutation that is virtually absent in Blacks and Orientals. In the Japanese population, a second disease variant in the promoter region was found (Kawasaki et al. 2006). Additionally, in a series of rheumatoid arthritis patients two additional SNPs were independently associated with the disease (Carlton et al. 2005).

The IFIH1 Gene

Consistent with the epidemiologic reports on the association of type 1 diabetes and virus infections, a common polymorphism resulting in an alanine to threonine change in exon 15, and few additional rare variants of the interferon-induced helicase gene

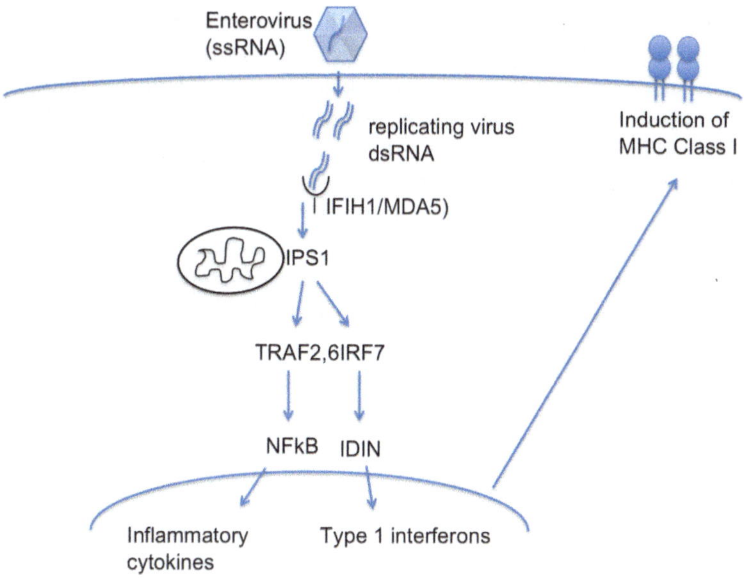

Fig. 3.1 IFIH1 and the downstream signal mechanism control type 1 interferon and cytokine responses. IFIH1 binds long dsRNA and activates downstream signaling. One arm turns on IRF7 (interferon regulatory factor 7) and then the interferon regulatory factor 7 (IRF73)-driven inflammatory network (IDIN) that controls the production of type 1 interferons. On the other arm it activates NFkB and triggers production of inflammatory cytokines

(IFIH1, 2q24.3) display a clear association with type 1 diabetes (Nejentsev et al. 2009; Smyth et al. 2006). IFIH1 is a cytosolic viral recognition receptor controlling production of type 1 interferons. These might among other effects enhance $CD8^+$ cytotoxic T lymphocyte activity against the pancreatic beta-cells by inducing expression of class I HLA molecules on them. A diverse range of genes in the interferon regulatory factor 7 (IRF73)-driven inflammatory network appear to define virus responses and seem to be involved in type 1 diabetes pathogenesis. These findings have opened up a new avenue in type 1 diabetes genetics research combining genetics with environmental factors (Heinig et al. 2010; Schoggins et al. 2011) (Fig. 3.1).

Disease Pathways and Genetic Heterogeneity

Beta-cell autoimmunity can be measured unequivocally by the presence of islet cell autoantibodies. These autoantibodies display distinct patterns of association with certain HLA haplotypes, *insulin gene* promoter variation, *PTPN22*, and a few other polymorphisms. Moreover, various islet cell autoantibodies are associated with different age of appearance, and confer different predictive values for the disease.

Based on these patterns, it is possible to build a few gene-immune states that correspond to distinct phenotypic characteristics of the disease. This is important, as it has a relevance to genetic screening and risk definitions in clinical settings. Insulin-specific autoimmunity, for instance, is defined by the presence of the *Class I INS VNTR* variant plus the HLA DR4-DQ8 haplotype that progresses to clinical disease more likely in the presence of the *PTPN22 620Trp/Trp* genotype. This pathway is rather common among young children. It was found that the effect of *PTPN22* is more pronounced among infants exposed to cow-milk-based formula before the age of 6 months (Lempainen et al. 2009). Another pathway, indicated by a few studies, is defined by the presence of the HLA-DR3-DQ2 haplotype, appearance of GAD65 autoantibodies, especially in males during or after puberty. These fingerprints call for a comprehensive approach to explore gene–environment effects to understand further disease mechanism, and their existence emphasize the importance of tailored therapeutic approaches (Åkerblom et al. 2005; Ludvigsson et al. 2011).

Genes and Environment

Genes, as the core coding elements of biological properties that define disease and health, are subject to evolutionary forces, such as selection, drift, and penetrance. It has been observed in the Finnish population that the penetrance of HLA on the phenotype of type 1 diabetes gradually changed during the last few decades. The frequency of risk HLA Class II genotypes has decreased among newly diagnosed patients, while the proportion of newly diagnosed subjects carrying neutral or protective genotypes has increased considerably (Hermann et al. 2003a). This has since been confirmed in several populations and shows strikingly a genetic predisposing locus to change penetrance across time in humans (Fig. 3.2). In addition, it points to an enhanced role of environmental factors on the increasing incidence of the disease. Third, it predicts a dramatic change in the predisposing genetic profile for a polygenic disease, unseen before. It is clear that better understanding of environmental factors is needed to define various disease pathways and interpret the role of more than 50 non-HLA genetic loci uncovered by latest GWA results.

An increasing environmental pressure increases penetrance of low-risk disease genes. Among the environmental factors viruses, especially enteroviruses represent a group of strong candidates as potentially causative agents. Enteroviral RNA is more common among patients with type 1 diabetes cases than controls either at the diagnosis or during the period before beta-cell autoantibodies appear. As it was discussed above, *IFIH1* gene and the downstream signal mechanism may provide a clue about virus-induced disease mechanisms. These findings are consistent with earlier reports on the pathogenesis of encephalomyocarditis virus-induced diabetes in mice, where the genetic factors of the host play a role (Kang and Yoon 1993). Once we consider the effect of HLA, the major locus of type 1 diabetes, on virus-induced diabetes, there is very little information published. In the Finnish DIPP study, enterovirus RNA was more often found in children with high-risk DR3/DR4 genotype

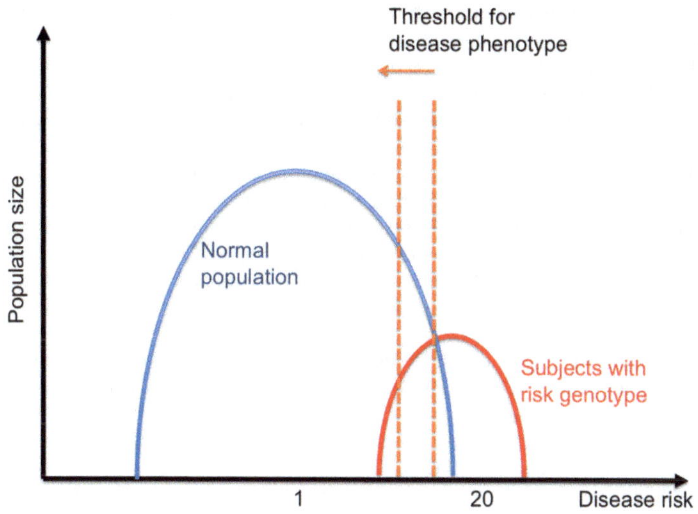

Fig. 3.2 An increasing environmental pressure increases penetrance of low-risk disease genes in type 1 diabetes worldwide. This leads to steep increase in disease incidence as the proportion of cases with low-risk HLA genotypes rises among newly diagnosed patients with type 1 diabetes

compared to those with DR4-DQ8 haplotype as the only risk factor (Oikarinen et al. 2011), but Craig et al. (2003) showed a reduced frequency of the enterovirus RNA among patients with type 1 diabetes carrying the HLA DR3-DQ2 haplotype.

References

Aarnisalo J, Treszl A, Svec P, Marttila J, Oling V, Simell O, Knip M, Korner A, Madacsy L, Vasarhelyi B, Ilonen J, Hermann R (2008) Reduced CD4$^+$ T cell activation in children with type 1 diabetes carrying the PTPN22/Lyp 620Trp variant. J Autoimmun 31:13–21

Åkerblom HK, Virtanen SM, Ilonen J, Savilahti E, Vaarala O, Reunanen A, Teramo K, Hamalainen AM, Paronen J, Groups NTS (2005) Dietary manipulation of beta cell autoimmunity in infants at increased risk of type 1 diabetes: a pilot study. Diabetologia 48:829–837

Akirav EM, Ruddle NH, Herold KC (2011) The role of AIRE in human autoimmune disease. Nat Rev Endocrinol 7:25–33

Barratt BJ, Payne F, Lowe CE, Hermann R, Healy BC, Harold D, Concannon P, Gharani N, McCarthy MI, Olavesen MG, McCormack R, Guja C, Ionescu-Tirgoviste C, Undlien DE, Ronningen KS, Gillespie KM, Tuomilehto-Wolf E, Tuomilehto J, Bennett ST, Clayton DG, Cordell HJ, Todd JA (2004) Remapping the insulin gene/IDDM2 locus in type 1 diabetes. Diabetes 53:1884–1889

Barrett JC, Clayton DG, Concannon P, Akolkar B, Cooper JD, Erlich HA, Julier C, Morahan G, Nerup J, Nierras C, Plagnol V, Pociot F, Schuilenburg H, Smyth DJ, Stevens H, Todd JA, Walker NM, Rich SS (2009) Genome-wide association study and meta-analysis find that over 40 loci affect risk of type 1 diabetes. Nat Genet 41:703–707

Bell GI, Horita S, Karam JH (1984) A polymorphic locus near the human insulin gene is associated with insulin-dependent diabetes mellitus. Diabetes 33:176–183

Bottini N, Vang T, Cucca F, Mustelin T (2006) Role of PTPN22 in type 1 diabetes and other autoimmune diseases. Semin Immunol 18:207–213

Carlton VE, Hu X, Chokkalingam AP, Schrodi SJ, Brandon R, Alexander HC, Chang M, Catanese JJ, Leong DU, Ardlie KG, Kastner DL, Seldin MF, Criswell LA, Gregersen PK, Beasley E, Thomson G, Amos CI, Begovich AB (2005) PTPN22 genetic variation: evidence for multiple variants associated with rheumatoid arthritis. Am J Hum Genet 77:567–581

Cloutier JF, Veillette A (1999) Cooperative inhibition of T-cell antigen receptor signaling by a complex between a kinase and a phosphatase. J Exp Med 189(1):111–121

Craig ME, Hattersley A, Donaghue KC (2009) Definition, epidemiology and classification of diabetes in children and adolescents. Pediatr Diabetes 10(Suppl 12):3–12

Craig ME, Howard NJ, Silink M, Rawlinson WD (2003) Reduced frequency of HLA DRB1*03-DQB1*02 in children with type 1 diabetes associated with enterovirus RNA. J Infect Dis 187:1562–1570

Douroudis K, Laine AP, Heinonen M, Hermann R, Lipponen K, Veijola R, Simell O, Knip M, Uibo R, Ilonen J, Kisand K (2009) Association of CTLA4 but not ICOS polymorphisms with type 1 diabetes in two populations with different disease rates. Hum Immunol 70:536–539

Hagopian WA, Erlich H, Lernmark A, Rewers M, Ziegler AG, Simell O, Akolkar B, Vogt R Jr, Blair A, Ilonen J, Krischer J, She J (2011) The Environmental Determinants of Diabetes in the Young (TEDDY): genetic criteria and international diabetes risk screening of 421 000 infants. Pediatr Diabetes 12:733–743

Harjutsalo V, Sjoberg L, Tuomilehto J (2008) Time trends in the incidence of type 1 diabetes in Finnish children: a cohort study. Lancet 371:1777–1782

Heinig M, Petretto E, Wallace C, Bottolo L, Rotival M, Lu H, Li Y, Sarwar R, Langley SR, Bauerfeind A, Hummel O, Lee YA, Paskas S, Rintisch C, Saar K, Cooper J, Buchan R, Gray EE, Cyster JG, Erdmann J, Hengstenberg C, Maouche S, Ouwehand WH, Rice CM, Samani NJ, Schunkert H, Goodall AH, Schulz H, Roider HG, Vingron M, Blankenberg S, Munzel T, Zeller T, Szymczak S, Ziegler A, Tiret L, Smyth DJ, Pravenec M, Aitman TJ, Cambien F, Clayton D, Todd JA, Hubner N, Cook SA (2010) A trans-acting locus regulates an anti-viral expression network and type 1 diabetes risk. Nature 467:460–464

Hermann R, Bartsocas CS, Soltesz G, Vazeou A, Paschou P, Bozas E, Malamitsi-Puchner A, Simell O, Knip M, Ilonen J (2004) Genetic screening for individuals at high risk for type 1 diabetes in the general population using HLA Class II alleles as disease markers. A comparison between three European populations with variable rates of disease incidence. Diabetes Metab Res Rev 20:322–329

Hermann R, Knip M, Veijola R, Simell O, Laine AP, Akerblom HK, Groop PH, Forsblom C, Pettersson-Fernholm K, Ilonen J (2003a) Temporal changes in the frequencies of HLA genotypes in patients with Type 1 diabetes—indication of an increased environmental pressure? Diabetologia 46:420–425

Hermann R, Laine AP, Veijola R, Vahlberg T, Simell S, Lahde J, Simell O, Knip M, Ilonen J (2005) The effect of HLA class II, insulin and CTLA4 gene regions on the development of humoral beta cell autoimmunity. Diabetologia 48(9):1766–1775

Hermann R, Lipponen K, Kiviniemi M, Kakko T, Veijola R, Simell O, Knip M, Ilonen J (2006) PTPN22 lymphoid tyrosine phosphatase Arg620Trp variant regulates insulin autoimmunity and progression to type 1 diabetes. Diabetologia 49:1198–1208

Hermann R, Turpeinen H, Laine AP, Veijola R, Knip M, Simell O, Sipila I, Akerblom HK, Ilonen J (2003b) HLA DR-DQ-encoded genetic determinants of childhood-onset type 1 diabetes in Finland: an analysis of 622 nuclear families. Tissue Antigens 62:162–169

Hraber P, Kuiken C, Yusim K (2007) Evidence for human leukocyte antigen heterozygote advantage against hepatitis C virus infection. Hepatology 46:1713–1721

Kang Y, Yoon JW (1993) A genetically determined host factor controlling susceptibility to encephalomyocarditis virus-induced diabetes in mice. J Gen Virol 74:1207–1213

Kawabata Y, Ikegami H, Awata T, Imagawa A, Maruyama T, Kawasaki E, Tanaka S, Shimada A, Osawa H, Kobayashi T, Hanafusa T, Tokunaga K, Makino H (2009) Differential association of HLA with three subtypes of type 1 diabetes: fulminant, slowly progressive and acute-onset. Diabetologia 52:2513–2521

Kawasaki E, Awata T, Ikegami H, Kobayashi T, Maruyama T, Nakanishi K, Shimada A, Uga M, Kurihara S, Kawabata Y, Tanaka S, Kanazawa Y, Lee I, Eguchi K (2006) Systematic search for single nucleotide polymorphisms in a lymphoid tyrosine phosphatase gene (PTPN22): association between a promoter polymorphism and type 1 diabetes in Asian populations. Am J Med Genet A 140:586–593

Kupila A, Muona P, Simell T, Arvilommi P, Savolainen H, Hamalainen AM, Korhonen S, Kimpimaki T, Sjoroos M, Ilonen J, Knip M, Simell O (2001) Feasibility of genetic and immunological prediction of type I diabetes in a population-based birth cohort. Diabetologia 44:290–297

Kyogoku C, Langefeld CD, Ortmann WA, Lee A, Selby S, Carlton VE, Chang M, Ramos P, Baechler EC, Batliwalla FM, Novitzke J, Williams AH, Gillett C, Rodine P, Graham RR, Ardlie KG, Gaffney PM, Moser KL, Petri M, Begovich AB, Gregersen PK, Behrens TW (2004) Genetic association of the R620W polymorphism of protein tyrosine phosphatase PTPN22 with human SLE. Am J Hum Genet 75:504–507

Kyvik KO, Green A, Beck-Nielsen H (1995) Concordance rates of insulin dependent diabetes mellitus: a population based study of young Danish twins. BMJ 311:913–917

Lambert AP, Gillespie KM, Thomson G, Cordell HJ, Todd JA, Gale EA, Bingley PJ (2004) Absolute risk of childhood-onset type 1 diabetes defined by human leukocyte antigen class II genotype: a population-based study in the United Kingdom. J Clin Endocrinol Metab 89:4037–4043

Lempainen J, Vaarala O, Makela M, Veijola R, Simell O, Knip M, Hermann R, Ilonen J (2009) Interplay between PTPN22 C1858T polymorphism and cow's milk formula exposure in type 1 diabetes. J Autoimmun 33:155–164

Lipponen K, Gombos Z, Kiviniemi M, Siljander H, Lempainen J, Hermann R, Veijola R, Simell O, Knip M, Ilonen J (2010) Effect of HLA class I and class II alleles on progression from autoantibody positivity to overt type 1 diabetes in children with risk-associated class II genotypes. Diabetes 59:3253–3256

Ludvigsson J, Hjorth M, Cheramy M, Axelsson S, Pihl M, Forsander G, Nilsson NO, Samuelsson BO, Wood T, Aman J, Ortqvist E, Casas R (2011) Extended evaluation of the safety and efficacy of GAD treatment of children and adolescents with recent-onset type 1 diabetes: a randomised controlled trial. Diabetologia 54:634–640

Marsh SG, Albert ED, Bodmer WF, Bontrop RE, Dupont B, Erlich HA, Fernandez-Vina M, Geraghty DE, Holdsworth R, Hurley CK, Lau M, Lee KW, Mach B, Maiers M, Mayr WR, Muller CR, Parham P, Petersdorf EW, Sasazuki T, Strominger JL, Svejgaard A, Terasaki PI, Tiercy JM, Trowsdale J (2010) Nomenclature for factors of the HLA system, 2010. Tissue Antigens 75:291–455

Michels AW, Gottlieb PA (2010) Autoimmune polyglandular syndromes. Nat Rev Endocrinol 6:270–277

Nakayama M, Abiru N, Moriyama H, Babaya N, Liu E, Miao D, Yu L, Wegmann DR, Hutton JC, Elliott JF, Eisenbarth GS (2005) Prime role for an insulin epitope in the development of type 1 diabetes in NOD mice. Nature 435:220–223

Nejentsev S, Howson JM, Walker NM, Szeszko J, Field SF, Stevens HE, Reynolds P, Hardy M, King E, Masters J, Hulme J, Maier LM, Smyth D, Bailey R, Cooper JD, Ribas G, Campbell RD, Clayton DG, Todd JA (2007) Localization of type 1 diabetes susceptibility to the MHC class I genes HLA-B and HLA-A. Nature 450:887–892

Nejentsev S, Walker N, Riches D, Egholm M, Todd JA (2009) Rare variants of IFIH1, a gene implicated in antiviral responses, protect against type 1 diabetes. Science 324:387–389

Nistico L, Buzzetti R, Pritchard LE, Van der Auwera B, Giovannini C, Bosi E, Larrad MT, Rios MS, Chow CC, Cockram CS, Jacobs K, Mijovic C, Bain SC, Barnett AH, Vandewalle CL, Schuit F, Gorus FK, Tosi R, Pozzilli P, Todd JA (1996) The CTLA-4 gene region of chromosome 2q33 is linked to, and associated with, type 1 diabetes. Belgian Diabetes Registry. Hum Mol Genet 5:1075–1080

Oikarinen S, Martiskainen M, Tauriainen S, Huhtala H, Ilonen J, Veijola R, Simell O, Knip M, Hyoty H (2011) Enterovirus RNA in blood is linked to the development of type 1 diabetes. Diabetes 60:276–279

Pugliese A, Zeller M, Fernandez A Jr, Zalcberg LJ, Bartlett RJ, Ricordi C, Pietropaolo M, Eisenbarth GS, Bennett ST, Patel DD (1997) The insulin gene is transcribed in the human thymus and transcription levels correlated with allelic variation at the INS VNTR-IDDM2 susceptibility locus for type 1 diabetes. Nat Genet 15:293–297

Schoggins JW, Wilson SJ, Panis M, Murphy MY, Jones CT, Bieniasz P, Rice CM (2011) A diverse range of gene products are effectors of the type I interferon antiviral response. Nature 472:481–485

Skyler JS, Krischer JP, Wolfsdorf J, Cowie C, Palmer JP, Greenbaum C, Cuthbertson D, Rafkin-Mervis LE, Chase HP, Leschek E (2005) Effects of oral insulin in relatives of patients with type 1 diabetes: The Diabetes Prevention Trial—Type 1. Diabetes Care 28:1068–1076

Smyth DJ, Cooper JD, Bailey R, Field S, Burren O, Smink LJ, Guja C, Ionescu-Tirgoviste C, Widmer B, Dunger DB, Savage DA, Walker NM, Clayton DG, Todd JA (2006) A genome-wide association study of nonsynonymous SNPs identifies a type 1 diabetes locus in the interferon-induced helicase (IFIH1) region. Nat Genet 38:617–619

Soltesz G, Patterson CC, Dahlquist G (2007) Worldwide childhood type 1 diabetes incidence—what can we learn from epidemiology? Pediatr Diabetes 8(Suppl 6):6–14

Todd JA, Bell JI, McDevitt HO (1987) HLA-DQ beta gene contributes to susceptibility and resistance to insulin-dependent diabetes mellitus. Nature 329(6140):599–604

Ueda H, Howson JM, Esposito L, Heward J, Snook H, Chamberlain G, Rainbow DB, Hunter KM, Smith AN, Di Genova G, Herr MH, Dahlman I, Payne F, Smyth D, Lowe C, Twells RC, Howlett S, Healy B, Nutland S, Rance HE, Everett V, Smink LJ, Lam AC, Cordell HJ, Walker NM, Bordin C, Hulme J, Motzo C, Cucca F, Hess JF, Metzker ML, Rogers J, Gregory S, Allahabadia A, Nithiyananthan R, Tuomilehto-Wolf E, Tuomilehto J, Bingley P, Gillespie KM, Undlien DE, Ronningen KS, Guja C, Ionescu-Tirgoviste C, Savage DA, Maxwell AP, Carson DJ, Patterson CC, Franklyn JA, Clayton DG, Peterson LB, Wicker LS, Todd JA, Gough SC (2003) Association of the T-cell regulatory gene CTLA4 with susceptibility to autoimmune disease. Nature 423:506–511

Vafiadis P, Bennett ST, Todd JA, Nadeau J, Grabs R, Goodyer CG, Wickramasinghe S, Colle E, Polychronakos C (1997) Insulin expression in human thymus is modulated by INS VNTR alleles at the IDDM2 locus. Nat Genet 15(3):289–292

Vandewalle CL, Decraene T, Schuit FC, De Leeuw IH, Pipeleers DG, Gorus FK (1993) Insulin autoantibodies and high titre islet cell antibodies are preferentially associated with the HLA DQA1*0301-DQB1*0302 haplotype at clinical type 1 (insulin-dependent) diabetes mellitus before age 10 years, but not at onset between age 10 and 40 years. The Belgian Diabetes Registry. Diabetologia 36:1155–1162

Vang T, Congia M, Macis MD, Musumeci L, Orru V, Zavattari P, Nika K, Tautz L, Tasken K, Cucca F, Mustelin T, Bottini N (2005) Autoimmune-associated lymphoid tyrosine phosphatase is a gain-of-function variant. Nat Genet 37:1317–1319

Wing K, Onishi Y, Prieto-Martin P, Yamaguchi T, Miyara M, Fehervari Z, Nomura T, Sakaguchi S (2008) CTLA-4 control over Foxp3+ regulatory T cell function. Science 322:271–275

Ziegler AG, Standl E, Albert E, Mehnert H (1991) HLA-associated insulin autoantibody formation in newly diagnosed type I diabetic patients. Diabetes 40:1146–1149

Chapter 4
Non-Genetic Factors in the Pathogenesis of Type 1 Diabetes

Serena Wai-Yan Chiu, R. David G. Leslie, and Huriya Beyan

Abstract Type 1 diabetes is an autoimmune disease characterised by immune-cell-mediated destruction of the pancreatic islet cells leading to insufficient insulin production and consequent clinical manifestations of hyperglycaemia. Many genetic variants have been identified through GWAS to detect common variants in alleles that are disease-associated; some of these variants are associated with protection from virus infection by interferon-releasing factors. Migration studies support a role for environmental factors causing a change in disease incidence. Extensive epidemiological, histological and immunological data have indicated a role for viruses in the pathogenesis of type 1 diabetes although it has proven difficult to find a causal relationship. Increasing wealth and industrialisation in developed countries may also contribute to the rising incidence of type 1 diabetes. Disproportionate maternal influences on risk of type 1 diabetes suggest that critical disease-inducing environmental events operate very early, even in utero. Early infant diet can affect the appearance of diabetes-associated autoantibodies. Disproportionate maternal influences on risk of type 1 diabetes suggest that critical disease-inducing events operate very early, even in utero. Identification of relevant disease-causing non-genetic effects, especially if they are environmental in origin, could point the way towards disease modulation or even prevention.

Non-Genetic Factors Involved in the Pathogenesis of Type 1 Diabetes

Type 1 diabetes is an autoimmune disease characterised by the immune-cell-mediated destruction of the pancreatic islet cells leading to insufficient insulin production and consequent clinical manifestations of hyperglycaemia. The aetiology of type 1 diabetes,

S. Wai-Yan Chiu • R.D.G. Leslie, M.B.B.S., M.R.C.S., M.D., F.R.C.P. (✉) • H. Beyan
Blizzard Institute, Queen Mary College, University of London, London, UK
e-mail: r.d.g.leslie@qmul.ac.uk

K.W. Taylor, H. Hyoty, A. Toniolo, A.J. Zuckerman (eds.), *Diabetes and Viruses*,
DOI 10.1007/978-1-4614-4051-2_4, © Springer Science+Business Media New York 2013

like other autoimmune diseases, can be thought of as a complex interaction between genes and the environment, although many contributing factors are yet to be elucidated. It is apparent that genes cannot be acting alone; however, it is still unclear how non-genetic events may lead to disease.

Evidence for Non-Genetic Effects

Many genetic variants have been identified through genome-wide association studies (GWAS) to detect common variants in alleles that are disease-associated. However, these variants seem to only have a small effect on disease risk, so only explain a small part of heritability, or familial clustering, which implies a role for non-genetic factors in disease development (Manolio et al. 2009). Certain human leukocyte antigen (HLA) alleles, particularly those located on the HLA class II region on chromosome 6, are known to be associated with type 1 diabetes susceptibility, e.g., HLA DRB1*03, *04; DQB1*0302 genotypes increase the risk of type 1 diabetes, while the HLA DQB1*0602 genotype confers a degree of protection (Atkinson and Eisenbarth 2001; Field 2002). However, less than 10% of people with the HLA susceptibility genes develop clinical diabetes, meaning that there must be other non-genetic factors crucial to the progression to type 1 diabetes (Knip et al. 2005). Gene associations point toward a gene–environment interaction and support a role for viruses. Associations between viral infections and type 1 diabetes, along with the discovery of a type 1 diabetes risk gene, IFIH1 (interferon induced with helicase C domain 1), have strengthened the view that viral infections may contribute to the pathogenesis of type 1 diabetes (Downes et al. 2010; Heinig et al. 2010; Todd 2010).

The incidence of autoimmune diseases has increased, particularly over the last three decades (Bach 2002). The evidence of non-genetic effect in type 1 diabetes comes from the study of populations, migrants and twins.

Populations

Population studies are of limited value in identifying the impact of non-genetic factors since it is difficult to segregate genetic from environmental influences. However, changes in disease incidence within a genetically stable population are important when disease incidence rises rapidly (Bach 2002). Such changes have been most striking in children diagnosed under 5 years of age, as in Switzerland where the incidence rose from 4.5/100,000 in 1965 to 10.5/100,000 in 2000 (Schoenle et al. 2001). In addition, a study in Sweden showed that the incidence of type 1 diabetes increased by almost double in the period from 1978 to 2000 but then remained stable from 2005 to 2007 (Berhan et al. 2011). These patterns of increasing incidence suggest that pathogenesis of type 1 diabetes may be non-genetically influenced.

Migrants

Migration studies also support a role for environmental factors causing a change in disease incidence (Serrano-Rios et al. 1999; Bach 2002). The incidence of type 1 diabetes in Asian children who migrated to Britain was much higher (11.7/100,000 per year in 1988–1990) than in their native Karachi (1/100,000 per year) (Bodansky et al. 1992; Staines et al. 1997). In a study of South Asian children in Leicestershire, UK, the incidence of type 1 diabetes was similar to those of white or other ethnic groups, and increased compared to the incidence in Asia (Raymond et al. 2001).

Twins

Monozygotic (MZ) twin studies are also of importance, as studying these genetically identical individuals will emphasise the effect of non-genetic events in influencing a phenotype, since any changes between MZ twins will be due to either environmental or stochastic events. Higher concordance rates for autoimmune diseases in identical MZ twins compared to non-identical dizygotic (DZ) twins is consistent with a genetic influence on these diseases (Hyttinen et al. 2003; Kyvik et al. 1995; Kumar et al. 1993). However, these studies generally involved small numbers, with limited follow-up and without biochemical tests. To partially resolve these problems, two clinic-based studies in the UK and the USA ascertained twins discordant for type 1 diabetes (Redondo et al. 2001). The combined analyses provided a powerful database to determine the proportion of MZ twins initially non-diabetic who subsequently developed diabetes, including the rate at which they did so and the factors that influenced this rate. The results indicate a high concordance rate in young age-at-onset diabetic twins; the rate falling substantially with age, implicating an increasing non-genetic effect with advancing age at diagnosis.

Pattern of Non-Genetic Effect

The temporal pattern of development of diabetes-associated autoantibodies such as those to glutamic acid decarboxylase (GADA), insulin (IAA), zinc transporter 8 (ZnT8A) and IA-2 (IA-2A) are indicative of a role for non-genetic factors in the development of type 1 diabetes (Ziegler and Nepom 2010). The appearance of these autoantibodies occurs in either neonatal life or later during puberty, with the characteristics of the antibodies differing at these distinct stages. In children who develop diabetes before 10 years of age, islet cell autoantibodies most commonly appear around 1–2 years of age, but are unlikely to develop in the first 6 months of life (Ziegler and Nepom 2010; Achenbach et al. 2005). At this stage, the type and affinity of these antibodies (IgG1 and high titre) are associated with particular HLA genotypes, and the first antibodies that do appear are often IAA, but quickly spread to include GADA, IA-2A and ZnT8A. In comparison, the later wave of autoimmunity occurring during puberty is typically less strongly related to HLA genotype, generally involves IAA or GADA on their own and at lower titre, and autoimmunity does not spread to involve other antigens, i.e., only one autoantibody as a rule can be detected. The dissimilar characteristics of the two peaks point to non-genetic factors in the triggering of islet autoimmunity.

Spectrum of Autoimmune Diabetes

Type 1 diabetes has been seen broadly as a form of diabetes requiring insulin therapy. However, the severity of metabolic features, both before and at the time of diagnosis of type 1 diabetes, is wide-ranging. The diabetes disease process is increasingly seen as a spectrum, with childhood-onset insulin-dependent diabetes at one end, and adult-onset non-insulin requiring diabetes at the other. Although it has proven difficult to define forms of the disease, the presence of diabetes-associated autoantibodies does serve to exclude autoimmune diabetes patients from being designated under "type 2 diabetes". Autoantibodies are associated with both type 1 diabetes and latent autoimmune diabetes of adult-onset (LADA). The definition of LADA has been set by the Immunology of Diabetes Society and Action as: (1) age 30–70 years at diagnosis; (2) at least 6 months of non-insulin requiring diabetes, and (3) the presence of diabetes-associated autoantibodies (www.actionlada.org). LADA, taken as an entity, is clearly distinct from type 2 diabetes. That distinction is demographic (LADA patients tend to be younger and less obese), clinical (metabolic syndrome is less frequent), genetic (LADA is HLA-associated), immunological (LADA patients have autoantibodies and T-cell changes) and metabolic (insulin secretory capacity and insulin resistance are both lower in LADA patients). There are two schools of thought: one, which sees LADA as part of a spectrum extending across the clinical range of autoimmune diabetes, the second, which sees it as a distinct form of autoimmune diabetes. Evidence and opinion favour the former (Leslie 2010; Brooks-Worrell and Palmer 2011).

Environmental Factors

Numerous environmental factors have been implicated in the aetiology of autoimmune diseases: for example, temperate climate, increased hygiene and decreased rates of infection, vaccinations and antibiotics, and increasing wealth. In addition to these, some factors are specific to type 1 diabetes pathogenesis of type 1 diabetes, such as overcrowding in childhood, virus infections, early exposure to cows' milk, reduced rates or duration of breastfeeding, and vitamin D and nitrite consumption (Bach 2002; Leslie and Castelli 2004; Fava et al. 1994; Hyoty and Taylor 2002; Clements et al. 1995; Cooper et al. 2011). Environmental factors could act by either triggering an already established degree of autoimmunity, or causing the destructive inflammatory response, or both, which then sets off a chain of events culminating in clinical diabetes.

The Virus Hypothesis

Extensive epidemiological, histological and immunological data have indicated a role for viruses in the pathogenesis of type 1 diabetes although it has proven difficult to find a causal relationship (Hyoty and Taylor 2002). Details are provided elsewhere in this book. The current evidence, in our opinion, strongly supports a viral origin of the disease. Specifically, evidence for a genetic association with an anti-viral innate immune response network within macrophages provides persuasive evidence for a viral effect (Heinig et al. 2010). There is evidence that two viral factors operate to influence the rate of disease progression subjects with autoantibodies: firstly, a critical gene in this anti-viral network, the interferon helicase IFIH1, has such an effect, and secondly, enterovirus infections also appear to influence the rate of progression to type 1 diabetes (Winkler et al. 2011; Oikarinen et al. 2011).

The Accelerator Hypothesis

One theory explaining the growing incidence of diabetes is the accelerator hypothesis. This hypothesis proposes that type 1 and type 2 diabetes are part of a spectrum and are caused by the same disease process, only differing in the time taken to progress to onset of clinical symptoms; this discrepancy in time frame being determined by genetics (Wilkin 2009). It is suggested that patients with type 1 diabetes have a genotype that confers increased susceptibility to the environmental factors causing β-cell stress, therefore leading to an accelerated progression to clinical diabetes compared to type 2 diabetes. Insulin resistance, being related to weight gain, plays an increasingly important role as the prevalence of childhood obesity rises, which explains the rising incidence of not only type 2 diabetes, but type 1 diabetes as well. The up-regulation of the β-cells in response to insulin resistance can cause an extreme immune reaction leading to destruction of the islets, through β-cell apoptosis (Wilkin 2001). Since genetic background determines β-cell reserve and the response to an environmental insult, the interplay between genetic and environmental factors establishes the differences between type 1 and type 2 diabetes, in particular, the dissimilar timing of the onset of clinical diabetes.

The Overload Hypothesis

Increasing wealth and industrialisation in developed countries may also contribute to the increasing incidence of type 1 diabetes. The trend of rapid weight gain in infancy over the past few decades, for example as a result of better living standards, may add to this idea, as it has been shown to correspond to an increased risk of type 1 diabetes (Hypponen et al. 2000, 1999). Also, in areas of lower socioeconomic conditions there is a decreased incidence of autoimmune diseases (Bach 2002). However, this may not be a reflection of differences in lifestyle as a result of wealth, but may be associated with the apparent reciprocal relationship seen between infection and autoimmunity, where factors such as crowded housing could increase the transmission of infections.

Improving lifestyles are mirrored in the pattern of increasing birth weights and accelerated childhood growth, both of which have implications for the development of type 1 diabetes. In support of the overload hypothesis, low compared to high birth weight by gestational age has been shown to correspond to a lower risk of type 1 diabetes (Patterson et al. 2001). Furthermore, studies have demonstrated that children who had increased linear growth, weight or BMI were more likely to progress to diabetes later in life (Hypponen et al. 2000, 1999; Blom et al. 1992; Johansson et al. 1994). Exposure to westernised heat-processed food could also affect the incidence of type 1 diabetes, and certainly there is a close relationship between T1D incidence and the gross national product of European countries (Patterson et al. 2001).

The Hygiene Hypothesis

The hygiene hypothesis is linked with increasing socioeconomic standards, which, along with the introduction of antibiotics and vaccinations, may contribute to the decreasing rate of infections (Bach 2002). A reduced exposure to infections in childhood could lead to autoimmunity later in life, which could be explained by the relative inactivity of the immune system causing increased immune responsiveness. In support of this, a higher degree of social mixing in childhood, such as in daycares or in larger families, is associated with a decreased incidence of type 1 diabetes (Bach 2002).

Gut Immune System

The gut immune system, as a major T-cell organ, plays an important role in the regulation of immune responses and has the potential to be altered by non-genetic factors such as diet. It appears that some antigens, such as cows' milk protein, when introduced within a certain period of time during infancy, affect the development of autoimmune responses later in life. Autoimmunity, in this case, may be due to increased gut permeability in the first 2 months of life, which could lead to greater immune cell infiltration (Kuitunen et al. 1994). A study showed that NOD mice deficient in MyD88 were protected from diabetes development, as the absence of MyD88 caused gut overgrowth of a certain class of bacteria, which then prevented diabetes (Wen et al. 2008). MyD88-deficient mice raised in germ-free conditions lost that protection. The results of this study indicate that gut microbiota are a critical factor to diabetes prevention in these mice.

The Weaning Diet Hypothesis

Early infant diet has been shown to affect the appearance of diabetes-associated autoantibodies. For example, the early introduction of cows' milk protein into an infant's diet has been indicated in the development of autoimmunity in type 1 diabetes (Vaarala et al. 1995). Such early dietary introduction seems to evoke a stronger immune response than if the cows' milk was introduced later. It has been proposed that this overreaction may be due to the immaturity of the gut immune system causing overt immune responses to harmless antigens. A recent study demonstrated that infants weaned on hydrolysed formula compared to cows' milk were less likely to develop diabetes-associated autoantibodies, suggesting that there may be an antigen in cows' milk that is a triggering factor of type 1 diabetes (Knip et al. 2010).

Maternal-Related Events Influence Diabetes Risk

Disproportionate maternal influences on risk of type 1 diabetes suggest that critical disease-inducing environmental events operate very early, even in utero. Children of diabetic mothers are less likely to develop type 1 diabetes than children of diabetic fathers, and the risk in mothers is less than the expected risk based on their HLA make-up (Warram et al. 1984; Bleich et al. 1993). The mean risk of diabetes in offspring of diabetic mothers and fathers in one study was 1.3% and 6.1%, respectively. This low disease risk being confined to offspring of mothers who had become diabetic after the age of 8 years, perhaps due to reduced transmission of genetic susceptibility (Warram et al. 1984).

Epigenetics or Stochastic Events

Finally, epigenetic effects could be relevant to the development of type 1 diabetes. The term epigenetics refers to alterations in gene expression without modification of the DNA sequence. Epigenetic modifications, including DNA methylation and histone modifications, affect the phenotype without altering the genotype, and such changes may be markers for disease. Environmental stimuli or stochastic events may act to alter the epigenetic state of an individual, so may contribute to the non-genetic influence involved in the pathogenesis of type 1 diabetes, and may be the missing link between genes and the environment. Epigenetic modifications may act either at the level of DNA through the methylation of cytosine in CpG dinucleotides, through post-translational changes to histones such as methylation, acetylation or phosphorylation, or through microRNA activation (Litherland 2008; Miao et al. 2008; Rakyan et al. 2011).

Concluding Remarks

In summary, there is persuasive evidence that type 1 diabetes is due, in part, to non-genetic factors. Several factors have been implicated, but evidence is confined to the genetic association with anti-viral mechanisms, and the change in the production of autoantibodies following dietary modification (Heinig et al. 2010; Knip et al. 2010). The timing of environmental events is key, perhaps occurring in two distinct waves in childhood. The potent predictive power of type 1 diabetes-associated autoantibodies, if induced by non-genetic factors, suggests that relatively few events are required before the disease process is set on a path to clinical diabetes. Some of the conflicting evidence and hypotheses could be resolved by implicating at least two or more distinct non-genetic events operating in tandem.

References

Achenbach P, Bonifacio E, Koczwara K, Ziegler AG (2005) Natural history of type 1 diabetes. Diabetes 54(Suppl 2):S25–S31

Atkinson MA, Eisenbarth GS (2001) Type 1 diabetes: new perspectives on disease pathogenesis and treatment. Lancet 358:221–229

Bach JF (2002) The effect of infections on susceptibility to autoimmune and allergic diseases. N Engl J Med 347:911–920

Berhan Y, Waernbaum I, Lind T, Mollsten A, Dahlquist G (2011) Thirty years of prospective nationwide incidence of childhood type 1 diabetes: the accelerating increase by times tends to level off in Sweden. Diabetes 60:577–581

Bleich D, Polak M, Eisenbarth GS, Jackson RA (1993) Decreased risk of type 1 diabetes in offspring of mothers who acquire diabetes during adrenarchy. Diabetes 42:1433–1439

Blom L, Persson LA, Dahlquist G (1992) A high linear growth is associated with an increased risk of childhood diabetes mellitus. Diabetologia 35:528–533

Bodansky HJ, Staines A, Stephenson C, Haigh D, Cartwright R (1992) Evidence for an environmental effect in the aetiology of insulin dependent diabetes in a transmigratory population. BMJ 304:1020–1022

Brooks-Worrell B, Palmer JP (2011) Is diabetes mellitus a continuous spectrum? Clin Chem 57:158–161

Clements GB, Galbraith DN, Taylor KW (1995) Coxsackie B virus infection and onset of childhood diabetes. Lancet 346:221–223

Cooper JD, Smyth DJ, Walker NM, Stevens H, Burren OS, Wallace C, Greissl C, Ramos-Lopez E, Hyppönen E, Dunger DB, Spector TD, Ouwehand WH, Wang TJ, Badenhoop K, Todd JA (2011) Inherited variation in vitamin d genes is associated with predisposition to autoimmune disease type 1 diabetes. Diabetes 60:1624–1631

Downes K, Pekalski M, Angus KL, Hardy M, Nutland S, Smyth DJ, Walker NM, Wallace C, Todd JA (2010) Reduced expression of IFIH1 is protective for type 1 diabetes. PLoS One 5, pii: e12646

Fava D, Leslie RD, Pozzilli P (1994) Relationship between dairy product consumption and incidence of IDDM in childhood in Italy. Diabetes Care 17:1488–1490

Field LL (2002) Genetic linkage and association studies of type I diabetes: challenges and rewards. Diabetologia 45:21–35

Heinig M, Petretto E, Wallace C, Bottolo L, Rotival M, Lu H, Li Y, Sarwar R, Langley SR, Bauerfeind A, Hummel O, Lee YA, Paskas S, Rintisch C, Saar K, Cooper J, Buchan R, Gray EE, Cyster JG, Erdmann J, Hengstenberg C, Maouche S, Ouwehand WH, Rice CM, Samani

NJ, Schunkert H, Goodall AH, Schulz H, Roider HG, Vingron M, Blankenberg S, Munzel T, Zeller T, Szymczak S, Ziegler A, Tiret L, Smyth DJ, Pravenec M, Aitman TJ, Cambien F, Clayton D, Todd JA, Hubner N, Cook SA, Cardiogenics Consortium (2010) A trans-acting locus regulates an anti-viral expression network and type 1 diabetes risk. Nature 467:460–464

Hyoty H, Taylor KW (2002) The role of viruses in human diabetes. Diabetologia 45:1353–1361

Hypponen E, Kenward MG, Virtanen SM, Piitulainen A, Virta-Autio P, Tuomilehto J, Knip M, Akerblom HK (1999) Infant feeding, early weight gain, and risk of type 1 diabetes. Childhood Diabetes in Finland (DiMe) Study Group. Diabetes Care 22:1961–1965

Hypponen E, Virtanen SM, Kenward MG, Knip M, Akerblom HK, Childhood Diabetes in Finland Study Group (2000) Obesity, increased linear growth, and risk of type 1 diabetes in children. Diabetes Care 23:1755–1760

Hyttinen V, Kaprio J, Kinnunen L, Koskenvuo M, Tuomilehto J (2003) Genetic liability of type 1 diabetes and the onset age among 22,650 young Finnish twin pairs: a nationwide follow-up study. Diabetes 52:1052–1055

Johansson C, Samuelsson U, Ludvigsson J (1994) A high weight gain early in life is associated with an increased risk of type 1 (insulin-dependent) diabetes mellitus. Diabetologia 37:91–94

Knip M, Veijola R, Virtanen SM, Hyoty H, Vaarala O, Akerblom HK (2005) Environmental triggers and determinants of type 1 diabetes. Diabetes 54(Suppl 2):S125–S136

Knip M, Virtanen SM, Seppa K, Ilonen J, Savilahti E, Vaarala O, Reunanen A, Teramo K, Hamalainen AM, Paronen J, Dosch HM, Hakulinen T, Akerblom HK, Finnish TRIGR Study Group (2010) Dietary intervention in infancy and later signs of beta-cell autoimmunity. N Engl J Med 363:1900–1908

Kuitunen M, Savilahti E, Sarnesto A (1994) Human alpha-lactalbumin and bovine beta-lactoglobulin absorption in infants. Allergy 49:354–360

Kumar D, Gemayel NS, Deapen D, Kapadia D, Yamashita PH, Lee M, Dwyer JH, Roy-Burman P, Bray GA, Mack TM (1993) North-American twins with IDDM. Genetic, etiological, and clinical significance of disease concordance according to age, zygosity, and the interval after diagnosis in first twin. Diabetes 42:1351–1363

Kyvik KO, Green A, Beck-Nielsen H (1995) Concordance rates of insulin dependent diabetes mellitus: a population based study of young Danish twins. BMJ 311:913–917

Leslie RD (2010) Predicting adult-onset autoimmune diabetes: clarity from complexity. Diabetes 59:330–331

Leslie RDG, Castelli MD (2004) Age-dependent influences on the origins of autoimmune diabetes: evidence and implications. Diabetes 53:3033–3040

Litherland SA (2008) Immunopathogenic interaction of environmental triggers and genetic susceptibility in diabetes: is epigenetics the missing link? Diabetes 57:3184–3186

Manolio TA, Collins FS, Cox NJ, Goldstein DB, Hindorff LA, Hunter DJ, McCarthy MI, Ramos EM, Cardon LR, Chakravarti A, Cho JH, Guttmacher AE, Kong A, Kruglyak L, Mardis E, Rotimi CN, Slatkin M, Valle D, Whittemore AS, Boehnke M, Clark AG, Eichler EE, Gibson G, Haines JL, Mackay TFC, McCarroll SA, Visscher PM (2009) Finding the missing heritability of complex disease. Nature 461:747–753

Miao F, Smith DD, Zhang L, Min A, Feng W, Natarajan R (2008) Lymphocytes from patients with type 1 diabetes display a distinct profile of chromatin histone H3 lysine 9 dimethylation: an epigenetic study in diabetes. Diabetes 57:3189–3198

Oikarinen S, Martiskainen M, Tauriainen S, Huhtala H, Ilonen J, Veijola R, Simell O, Knip M, Hyoty H (2011) Enterovirus RNA in blood is linked to the development of type 1 diabetes. Diabetes 60:276–279

Patterson CC, Dahlquist G, Soltesz G, Green A, EURODIAB ACE Study Group (2001) Is childhood-onset type I diabetes a wealth-related disease? An ecological analysis of European incidence rates. Diabetologia 44(Suppl 3):B9–B16

Rakyan VK, Beyan H, Down TA, Hawa MI, Maslau S, Aden D, Daunay A, Busato F, Mein CA, Manfras B, Dias KR, Bell CG, Tost J, Boehm BO, Beck S, Leslie RD (2011) Identification of type 1 diabetes-associated DNA methylation variable positions that precede disease diagnosis. PLoS Genet 7(9):e1002300

Raymond NT, Jones JR, Swift PGF, Davies MJ, Lawrence IG, McNally PG, Burden ML, Gregory R, Burden AC, Botha JL (2001) Comparative incidence of Type I diabetes in children aged under 15 years from South Asian and White or Other ethnic backgrounds in Leicestershire, UK, 1989 to 1998. Diabetologia 44(Suppl 3):B32–B36

Redondo MJ, Yu L, Hawa M, Mackenzie T, Pyke DA, Eisenbarth GS, Leslie RDG (2001) Heterogeneity of type 1 diabetes: analysis of monozygotic twins in Great Britain and the United States. Diabetologia 44:354–362

Schoenle EJ, Lang-Muritano M, Gschwend S, Laimbacher J, Mullis PE, Torresani T, Biason-Lauber A, Molinari L (2001) Epidemiology of type 1 diabetes mellitus in Switzerland: steep rise in incidence in under 5 year old children in the past decade. Diabetologia 44:286–289

Serrano-Rios M, Goday A, Larrad TM (1999) Migrant populations and the incidence of type 1 diabetes mellitus: an overview of the literature with a focus on the Spanish-heritage countries in Latin America. Diabetes Metab Res Rev 15:113–132

Staines A, Hanif S, Ahmed S, McKinney PA, Shera S, Bodansky HJ (1997) Incidence of insulin dependent diabetes mellitus in Karachi, Pakistan. Arch Dis Child 76:121–123

Todd JA (2010) Etiology of type 1 diabetes. Immunity 32:457–467

Vaarala O, Saukkonen T, Savilahti E, Klemola T, Akerblom HK (1995) Development of immune response to cow's milk proteins in infants receiving cow's milk or hydrolyzed formula. J Allergy Clin Immunol 96:917–923

Warram JH, Krolewski AS, Gottlieb MS, Kahn CR (1984) Differences in risk of insulin-dependent diabetes in offspring of diabetic mothers and diabetic fathers. N Engl J Med 311:149–152

Wen L, Ley RE, Volchkov PY, Stranges PB, Avanesyan L, Stonebraker AC, Hu C, Wong FS, Szot GL, Bluestone JA, Gordon JI, Chervonsky AV (2008) Innate immunity and intestinal microbiota in the development of type 1 diabetes. Nature 455:1109–1113

Wilkin TJ (2001) The accelerator hypothesis: weight gain as the missing link between type I and type II diabetes. Diabetologia 44:913–922

Wilkin TJ (2009) The accelerator hypothesis: a review of the evidence for insulin resistance as the basis for type I as well as type II diabetes. Int J Obes 33:716–726

Winkler C, Lauber C, Adler K, Grallert H, Illig T, Ziegler AG, Bonifacio E (2011) An interferon-induced helicase (IFIH1) gene polymorphism associates with different rates of progression from autoimmunity to type 1 diabetes. Diabetes 60:685–690

Ziegler AG, Nepom GT (2010) Prediction and pathogenesis in type 1 diabetes. Immunity 32:468–478

Part II
Studies in Animals

Chapter 5
Encephalomyocarditis Virus

Seiho Nagafuchi, Hironori Kurisaki, and Hitoshi Katsuta

Abstract EMC virus has provided the most useful animal model for virus-induced type 1 diabetes. Development of diabetes depends on many factors including virus strain, challenge dose, host factors such as genetic background, sex, immunoprotective function, inflammatory responses with macrophages, cytokines, chemokines, and chemical mediators. Autoimmunity induction is not likely to be a factor in this model, though a hit-and-run event cannot be excluded. Most importantly, the difference between the diabetogenic strain D (EMC-D) and the non-diabetogenic strain B (EMC-B) virus depends on only one amino acid change due to single point mutation of "A" to "G" at position 3155 (Thr-776 to Ala-776), suggesting that possible acquisition of diabetogenicity may occur often among environmental "non-diabetogenic" viruses. Although susceptibility to the EMC-D virus-induced diabetes depends on the genetic background of mice, the genetic determinants of the host remain to be elucidated.

Clarification of the pathogenesis of EMC virus-induced diabetes will not only promote a better understanding of the mechanisms of virus-induced diabetes in general, but will also contribute to the development of new protective strategies against viral diabetes.

Introduction

Accumulating evidence has suggested a viral origin of type 1 diabetes development. Historical study has indicated the presence of certain viruses including coxsackie virus, cyotomegalovirus, varicella-zoster virus, and rubella virus in patients with fatal

S. Nagafuchi, M.D., Ph.D. (✉) • H. Kurisaki • H. Katsuta
Department of Medical Science and Technology, Graduate School of Medical Sciences, Kyushu University, Fukuoka 812-8582, Japan
e-mail: nagafu_s@med.kyushu-u.ac.jp

K.W. Taylor, H. Hyoty, A. Toniolo, A.J. Zuckerman (eds.), *Diabetes and Viruses*,
DOI 10.1007/978-1-4614-4051-2_5, © Springer Science+Business Media New York 2013

viral infections associated with pancreatic islet cell damage (Jenson et al. 1980), suggesting that many viruses have potent diabetes inducers. Recent advances in the field have focused on the enteroviral infection as the most possible candidate virus to induce type 1 diabetes associated with immunopathologic reaction (Clements et al. 1995; Hanafusa and Imagawa 2008; Tauriainen et al. 2011; Richardson et al. 2011).

Encephalomyocarditis (EMC) virus has provided the most useful animal model for virus-induced type 1 diabetes (Jun and Yoon 2001). Development of diabetes depends on many factors including virus strain, challenge dose, host factors such as sex, immunoprotective function, inflammatory responses with macrophages, cytokines, chemokines and chemical mediators (Jun and Yoon 2001). Autoimmunity induction is not likely to operate in this model, though a hit-and-run event cannot be excluded. The clarification of the pathogenesis of EMC virus-induced diabetes will not only promote a better understanding of the mechanisms of virus-induced diabetes, but also enhance the protection strategy against virus-induced diabetes. The history and pathogenesis of EMC virus-induced diabetes are described, and the future aspects of the significance of this animal model of virus-induced diabetes are discussed.

The EMC Virus

Encephalomyocarditis virus belongs to the Picornaviridae family, as enteroviruses including coxsackie virus, genus cardiovirus, unenveloped, icosahedral structure (Fig. 5.1a) consisting of four capsid proteins (VP1 ~ 4), surrounding a core of ssRNA, moderately resistant to acidic pH (Racaniello 2007). The virion contains one molecule of positive sense, ssRNA, about 7.8 kb in size, containing a single

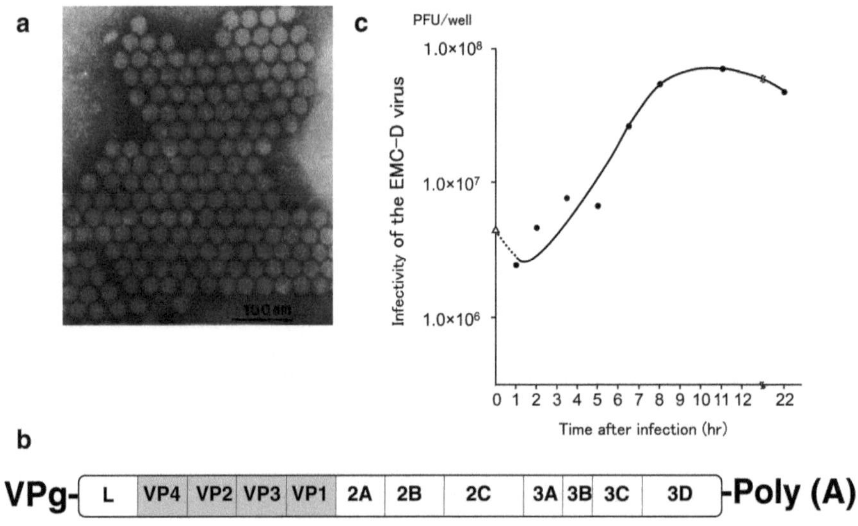

Fig. 5.1 Characteristics of encephalomyocarditis. (**a**) Negative stain of EMC-D virus. Courtesy of Dr. Yuji Ueki and Emeritus Professor Kazunobu Amako, Kyushu University. (**b**) Genomic structure of EMC virus. (**c**) Growth curve of EMC-D virus in mouse embryonic fibroblasts

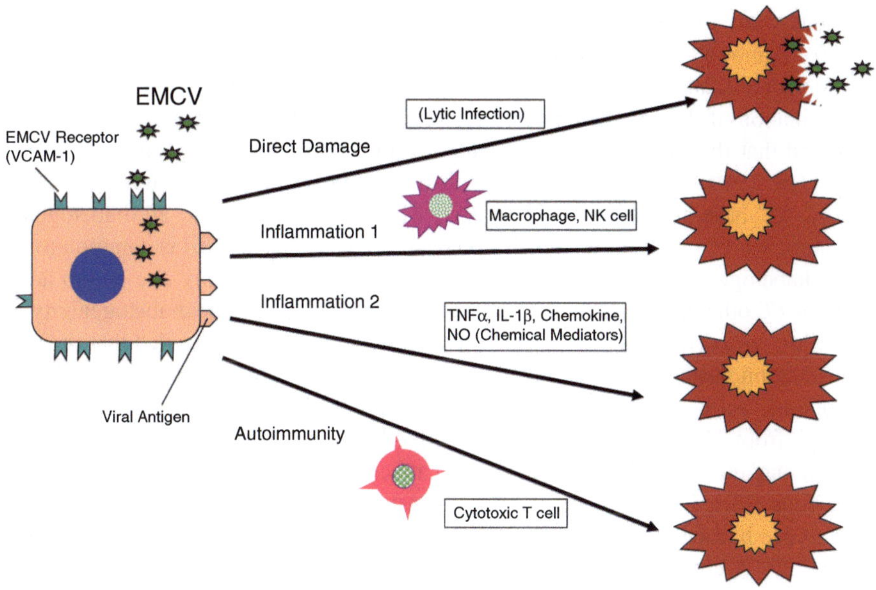

Fig. 5.2 Mechanisms of pancreatic β-cell damage due to EMCV infection

open reading frame (ORF) (Fig. 5.1b) (Racaniello 2007). The virus can be grown in a tissue culture well with a one step replication time of about 8 h (Fig. 5.1c), and can infect rodents, usually producing systemic infection representing encephalitis and myocarditis. The virus rarely infects humans. Craighead and McLane (1968) first found that the M variant of the EMC virus certainly induced diabetes in several susceptible strains of male mice. Later, Yoon et al. (1980) isolated the highly diabetogenic D variant of EMC virus and the non-diabetogenic B strain of the EMC virus, by the plaque clone purification method. EMC-D virus produces diabetes in over 90% of infected susceptible strain of mice, while EMC-B virus did not induce diabetes in any strain of mice (Yoon et al. 1980). The susceptibility depends on the strain of mice and sex, namely only male mice are susceptible to the virus (Ross et al. 1975; Yoon et al. 1980; Huber et al. 1985). These findings accelerated the study of viral genetic factors to enhance the induction of diabetes in susceptible animals as well as research on pathogenesis (Fig. 5.2).

Pathogenesis of EMC Virus-Induced Diabetes

EMC Virus

The differing diabetogenicity among EMC viruses has been noted to be dependent on the genetic variation. First, the M variant of EMC virus was obtained as highly diabetogenic, and later the diabetogenic EMC-D virus and the non-diabetogenic B variant were isolated, respectively (Craighead and MacLane 1968; Yoon et al. 1980). Although EMC-D virus and EMC-B virus could not be distinguished by either neutralization assay or competitive radioimmunoassay (Yoon et al. 1980), examination of the complete nucleotide sequences of the genomes of both variants showed that they were different in only 14 nucleotide positions (Eun et al. 1988; Bae et al. 1989). Further molecular analysis by generating mutant viruses revealed that a “G” base at position 3155 ([GCC] Ala-776) is common to all diabetogenic variants, while an “A” base at the same position ([ACC] Thr-776) is common to all non-diabetogenic variants (Bae and Yoon 1993). Therefore, only one amino acid, alanine (776th amino acid on the polyprotein), is essential for the diabetogenicity of the EMC virus (Bae and Yoon 1993). These beautiful studies revealed that the single point mutation of “A” to “G” at position 3155 (Thr-776 to Ala-776) are critical to operate as the diabetogenic EMC virus (Jun et al. 1997). It was found that a change from Thr-776 to Ala-776 reduced the hydrophilicity of the region by 37%, which may increase the efficiency of viral attachment to pancreatic beta cells (Kang and Yoon 1993; Jun et al. 1997, 1998), suggesting that the significance of the genetic difference had been supposed to influence the effectiveness of the attachment for beta cells. A challenge dose is not critical for inducing diabetes; however, it has been indicated that a high dose (10^5) PFU challenge destroys directly the pancreatic beta cells, while low dose (10^2) challenge will induce inflammatory responses which may damage beta cells (Yoon et al. 1980; Huber et al. 1985).

Protection

Innate Immunity

Since the EMC virus-induced diabetes develops within 3 days after infection, innate immunity, such as macrophages, interferons, and early inflammatory responses, is likely to be most important for determining the outcome after EMC virus infection (Yoon et al. 1980). Recent advances in the immunology of innate immunity found the significance of pattern recognition receptors (PRRs) directed against pathogen-associated molecular patterns (PAMPs) (Takeuchi and Akira 2009). They include toll-like receptors (TLR), and intracellular helicase such as melanocyte differentiation antigen (MDA) 5, same as interferon induced with the helicase C domain 1 (IFIH1), and retinoic acid inducible gene (RIG) I, for picornavirus and paramyxovirus. It was reported that polymorphism of the IFIH1 gene is associated with type 1 diabetes (Smyth et al. 2006), although it remains uncertain whether this may be associated with the viral infection. In EMC virus infection, TLR 3, 7, 8, and MDA-5 function as receptors, mediating the signal transduction pathway, inducing cellular activation including interferon production (Takeuchi and Akira 2009). McCartney et al. (2011) reported that MDA5 and TLR3 are both required to prevent diabetes in mice infected with EMC-D virus. Infection of *Tlr3*$^{-/-}$ mice caused diabetes due to impaired IFN-I responses and virus-induced β-cell damage rather than T-cell-mediated autoimmunity (McCartney et al. 2011). Mice lacking just one copy of MDA5 developed

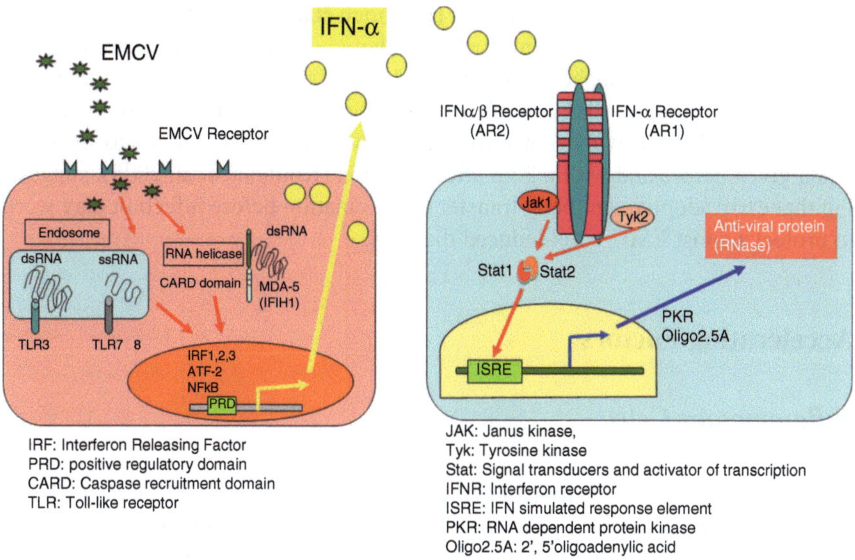

Fig. 5.3 Interferon production following EMCV infection and induction of anti-viral substances by interferon receptor signaling pathway

transient hyperglycemia when infected with EMCV-D, whereas homozygous *MDA5*$^{-/-}$ mice developed severe cardiac pathology (McCartney et al. 2011). TLR3 and MDA5 controlled EMC-D virus infection and diabetes by acting in hematopoietic and stromal cells, respectively, inducing IFN-I responses at kinetically distinct time points (McCartney et al. 2011). They conclude that optimal functioning of viral sensors and prompt IFN-I responses are required to prevent diabetes in this animal model, suggesting the significance of PRR-dependent innate immunity activation (McCartney et al. 2011).

Regarding the role of interferon, conflicting data in this model have been reported (Kaptur et al. 1989; Hirasawa et al. 1995). One study reported that interferon may worsen the EMC-D virus-induced diabetes (Kaptur et al. 1989), and others point to the significance of protective role of interferon (Hirasawa et al. 1995; McCartney et al. 2011). Possibly the challenge dose of EMC-D virus may alter the protective and/or pathogenic role of interferon in this model (Fig. 5.3).

Adaptive Immunity

Because EMC virus-induced diabetes develops as early as 3–5 days after infection, acquired and/or adaptive immunity did not likely play an important role (Yoon et al. 1980; Kounoue et al. 2008). Susceptibility to EMC-D virus-induced diabetes is not controlled by the MHC type. T-cell-deficient nude mice, B-cell-deficient (muMT) mice, or both T-cell- and B-cell-deficient Rag-2 knockout mice could resist against EMC-D virus-induced diabetes (Kounoue et al. 2008). In addition, passive transfer of lymphocytes from mice made diabetic with EMC-D virus into normal mice failed to produce diabetes (Yoon et al. 1985). Taken together, it was suggested that adaptive immunity did not affect the outcome of virus-induced diabetes. On the other hand, adoptively transferred antibody to the EMC virus was effective when given before and within 36 h after infection (Kounoue et al. 2008), suggesting that early adoptive antibody transfer or vaccination before infection may work to protect against EMC virus-induced diabetes.

Accelerating Factors

Inflammatory Cells

EMC virus belongs to the group of cytolytic viruses, and therefore a large challenge dose with the virus destroys pancreatic β-cells extensively enough to lead to diabetes (Yoon et al. 1980; Jun and Yoon 2001). At lower doses of infection, pancreatic β-cell damage is rather minimal. However, induced inflammatory response including infiltrated macrophages and produced cytokines and chemical mediators may damage further the β-cells to a reduced level of insulin secretion, leading to hyperglycemia (Baek and Yoon 1990, 1991; Hirasawa et al. 1997). Indeed, histopathologic study of EMCV-induced animals developing insulitis with infiltration of macrophages to the islets showed that they were associated with the extensive destruction of pancreatic β-cells (Baek and Yoon 1990, 1991).

Cytokines, Chemokines, and Chemical Mediators

Interleukin-1 and tumor necrosis factor (TNF) α were suggested to function as key mediators of pancreatic beta-cell destruction, inducing DNA fragmentation (Hirasawa et al. 1997; Rabinovitch et al. 1994). Nitric oxide may work as a damaging factor to worsen the deterioration of pancreatic β-cell function (Fehsel et al. 1993). Infiltrated macrophages may be responsible to produce those cytotoxic mediators (Hirasawa et al. 1997). Recently, it was reported that the CXCR3 ligand CXCL10 was produced by enterovirus-infected pancreatic β-cells, attracting cytotoxic T cells and macrophages, expressing CXCR3, associated with the induction of insulitis, leading to β-cell damage (Tanaka et al. 2009).

Other Chemicals

Streptozoin and alloxan are well-known diabetogenic substances, inducing DNA strand breads and poly ADP ribose synthetase, which lead to the lack of ATP in pancreatic β-cells, resulting in extensive β-cell damage (Yamamoto et al. 1981).

Some possible chemicals exist in the environment such as streptozoin, which is a compound of nitrourea and glucose; the streptoxocin-like substance may be generated reacting with nitrosamines in food and water, and glucose. In addition, alloxan can be derived from uric acid in the purine metabolism pathway by oxidation with superoxide substance (Santos et al. 1999). The produced alloxan would possibly work as a β-cell-specific cytotoxic chemical. Although these possible chemicals to induce β-cell damage have not been proved to operate in the environment, it should be noticed that they may work as a risk factor in addition to viral infection.

Accumulation of Environmental Insults

An interesting report has established the significance of the accumulation of environmental insults such as viruses and chemicals (Toniolo et al. 1980). The concept may also be applicable to type 2 diabetes, where accumulation of risk factors such as diet, obesity, aging, genetic risk, and little exercise together lead to impaired insulin action, resulting in the development of diabetes. Viral infection may contribute to the development of type 2 diabetes when viral infection alone is not sufficient to induce diabetes, but the damage to β-cells may reduce β-cell function to some extent. In this sense, viral infection may serve as another risk factor for the development of type 2 diabetes, although direct evidence is lacking.

Autoimmunity

The EMC virus is a cytolytic virus but does not cause persistent infection, and therefore infected cells are not likely to be attacked by cytotoxic T cells similar to the autoimmune reaction. However, infiltration of T cells with a restricted T-cell receptor repertoire has been reported, suggesting a pathogen and/or autoantigen-directed reaction (Kawagishi et al. 2003). Indeed, lysed cells after infection with the EMC virus would release self antigens and thus may possibly trigger autoimmunity to pancreatic β-cells as well as virus antigen-directed protective immune response (Flodström et al. 2002; Christen et al. 2010). The hit-and-run theory is hard to confirm and/or disclaim with evidence for or against paradigm. In addition, interferon production induced by viral infection may play a role in the development of autoreactivity to pancreatic β-cells (Fig. 5.4) (Devendra and Eisenbarth 2004). Recent advances in controlling organ-specific autoimmune diseases often associated with autoimmune type 1 diabetes, the significance of AIRE being in the thymus and Treg in the peripheral immunoregulation system have been extensively described in addition to microbial environment engagement (Nagamine et al. 1997; Kogawa et al.2002; Eisenbarth and Gottlieb 2004; Sakaguchi 2005). Programmed death factor, suppressor of cytokine signaling (SOCS), and B lymphocytes may also contribute to the prevention of autoimmunity to pancreatic β-cells (Chervonsky 2010; Yoshimura, et al. 2007; Pescovitz et al. 2009; Nagafuchi et al. 2010).

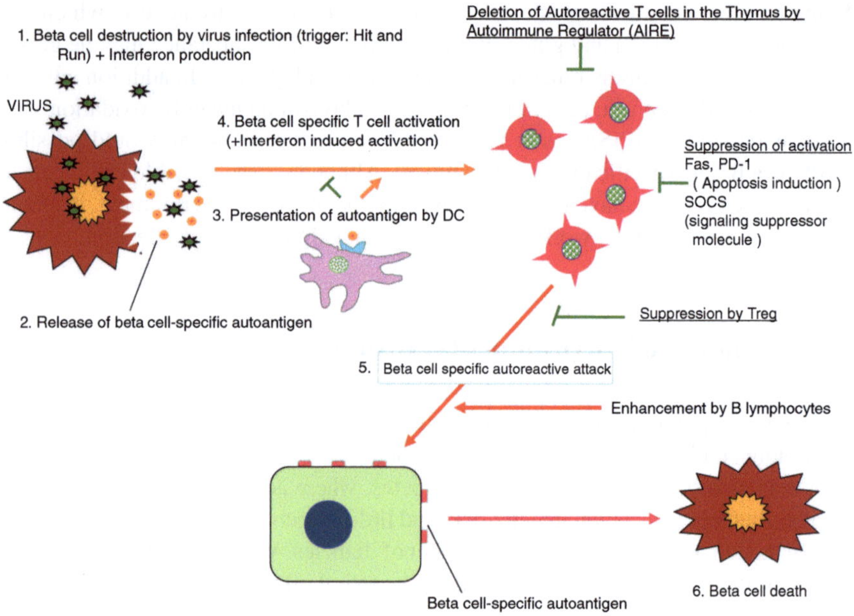

Fig. 5.4 Hypothesis: possible induction of autoreactivity to pancreatic β cells triggered by viral infection

Long-Term Complications

Susceptible mice infected with diabetogenic EMC-D virus develop hyperglycemia and develop characteristic diabetic complications similar to those in humans such as glomerulosclerosis, retinal vessel involvement, and decrease in bone formation and mineralization. The diabetic mice with glomerulosclerosis after 6 months' duration revealed the two- to threefold increase in thickness of the glomerular basement membrane. Thickening of Bowman's capsule and the renal mesangial matrix has also been observed in EMC-M virus-infected DBA/2 mice 2–6 months after infection (Yoon et al. 1982).

Host Factors

Even diabetogenic EMC-D virus induced diabetes only in a few strains of mice such as SJL/J, SWR, DBA/1J, DBA/2J, while others such as C57BL/6, CBA/J, AKR, C3H/HeJ. A/J mice are all resistant to EMC-D virus-induced diabetes (Ross et al. 1975; Yoon et al. 1980; Huber et al. 1985). Onodera and others have reported that F1 hybrid mice between susceptible SWR and resistant C57BL/6 mice were resistant to virus-induced diabetes, while the next F1 and SWR mating showed that 50% of those mice exhibited the susceptibility to the virus, thus indicating that a single autosomal recessive gene, which is inherited in a Mendelian fashion, controls the susceptibility to the virus (Onodera et al. 1978). It was indicated that the susceptibility gene may modulate the expression of virus receptors on β-cells in susceptible mice (Kang and Yoon 1993); however the exact controlling gene has remain to be identified.

Prevention

Several preventive strategies have been reported to be effective. As described in the pathogenesis section, Nitrate oxide inhibitors and chemokine suppression may reduce the diabetogenic effect of EMC virus infection. Since the antibody is very effective to prevent the EMC virus infection, vaccination and every early phase of the disease may be effective in preventing the EMC virus-induced diabetes (Kounoue et al. 2008). Immunostimulants such as BCG, *Corynebacterium parvum*, and PolyIC treatment have been shown to be effective in preventing EMD-D virus-induced diabetes (Kounoue et al. 1987; Choi et al. 2000).

Perspectives

EMC virus has contributed greatly to better understanding of the pathogenesis of virus-induced diabetes. The data described above show the development of diabetes even under the diabetogenic virus challenge, with the outcome being influenced by many factors, such as genetic background, sex, protective immunity, inflammatory cells, cytotoxic mediators, and also perhaps the regenerative activity of pancreatic β-cells (Hover and Sauter 2010). The elevation of blood glucose level may be due to the accumulation of such "risk factors," leading to the development of virus-induced diabetes. Moreover, since even susceptible strains of inbred mice develop virus-induced diabetes in a rather variable but not homogeneous fashion, the "stochastic process" in this model and/or human virus-induced diabetes should therefore be recognized. In order to acquire a better understanding of the pathogenesis of virus-induced diabetes, the assay system for the diabetogenicity of the virus together with clarification of host genetic risk factors should be exploited, as these may lead to the identification of the "diabetogenic" virus. Those studies will hopefully lead to the development of a vaccination strategy to the "diabetogenic" virus, which will in turn not only prevent the development of the virus-induced type 1 diabetes but may also contribute to reduce the risk for the development of type 2 diabetes.

Acknowledgments The authors thank Professor Takehiko Sasazuki for his helpful advice. We also thank Dr. Chiri Nagatsuka and Mrs Arisa Moroishi for their help in the preparation of the manuscript. Financial support by Ministry of Health, Labour and Welfare of Japan is acknowledged.

References

Bae YS, Eun HM, Yoon JW (1989) Genomic differences between the diabetogenic and nondiabetogenic variants of encephalomyocarditis virus. Virology 170:282–287

Bae Y, Yoon JW (1993) Determination of diabetogenicity attributable to a single amino acid Ala776, on the polyprotein of encephalomyocarditis virus. Diabetes 42:435–443

Baek HS, Yoon JW (1990) Role of macrophages in the pathogenesis of encephalomyocarditis virus-induced diabetes in mice. J Virol 64:5708–5715

Baek HS, Yoon JW (1991) Direct involvement of macrophages in destruction of β-cells leading to development of diabetes in virus-infected mice. Diabetes 40:1586–1597

Choi BK, Cho SH, Bai GH, Kim SJ, Hyun BH, Choe YK, Bae YS (2000) Prevention of encephalomyocarditis virus-induced diabetes by live recombinant mycobacterium bovis Bacillus Calmette-Guerin in susceptible mice. Diabetes 49:1459–1467

Chervonsky AV (2010) Influence of microbial environment on autoimmunity. Nat Immunol 11:28–33

Christen U, Hintermann E, Holdener M, von Herrath MG (2010) Viral triggers for autoimmunity: is the 'glass of molecular mimicry' half full or half empty? J Autoimmun 34:38–44

Clements GB, Galbraith DN, Taylor KW (1995) Coxsackie B virus infection and onset of child hood diabetes. Lancet 346:221–223

Craighead JE, McLane MF (1968) Diabetes mellitus induction in mice by encephalomyocarditis virus. Science 162:913–914

Devendra D, Eisenbarth GS (2004) Interferon alpha—a potential link in the pathogenesis of viral-induced type 1 diabetes and autoimmunity. Clin Immunol 111:225–233

Eisenbarth GS, Gottlieb PA (2004) Autoimmune polyendocrine syndromes. N Engl J Med 350:2068–2079

Eun HM, Bae Y, Yoon JW (1988) Amino acid differences in capsid protein, VP1, between diabetogenic and nondiabetogenic variants of encephalomyocarditis virus. Virology 163:369–373

Fehsel AK, Jalowy A, Qi S, Burkart V, Hartmann B, Kolb H (1993) Islet cell DNA is a target of inflammatory attack by nitric oxide. Diabetes 42:496–500

Flodström M, Maday A, Balakrishna D, Cleary MM, Yoshimura A, Sarvetnick N (2002) Target cell defense prevents the development of diabetes after viral infection. Nat Immunol 3:373–382

Hanafusa T, Imagawa A (2008) Insulitis in human type 1 diabetes. Ann NY Acad Sci 1150:297–299

Hirasawa K, Ogiso Y, Takeda M, Lee MJ, Itagaki S, Doi K (1995) Protective effects of macrophage-derived interferon against encephalomyocarditis virus-induced diabetes mellitus in mice. Lab Anim Sci 45:652–656

Hirasawa K, Jun HS, Maeda K, Kawaguchi Y, Itagaki S, Mikami T, Baek HS, Doi K, Yoon JW (1997) Possible role of macrophage-derived soluble mediators in the pathogenesis of encephalomyocarditis virus-induced diabetes in mice. J Virol 71:4024–4031

Hover D, Sauter P (2010) Pathogenesis of type 1 diabetes mellitus: interplay between enterovirus and host. Nat Rev Endocrinol 6:279–289

Huber SA, Babu PG, Craighead JE (1985) Genetic influences on the immunologic pathogenesis of encephalomyocarditis (EMC) virus-induced diabetes mellitus. Diabetes 34:1186–1190

Jenson AB, Rosenberg HS, Notkins AL (1980) Pancreatic islet-cell damage in children with fatal viral infections. Lancet 316:354–358

Jun HS, Kang Y, Notkins AL, Yoon JW (1997) Gain or loss of diabetogenicity result in from a single point mutation in recombinant encephalomyocarditis virus. J Virol 71:9782–9785

Jun HS, Kang Y, Yoon HS, Kim KH, Notkins AL, Yoon JW (1998) Determination of encephalomyocarditis viral diabetogenicity by a putative binding site of the viral capsid protein. Diabetes 47:576–582

Jun HS, Yoon JW (2001) The role of viruses in type I diabetes: two distinct cellular and molecular pathogenic mechanisms of virus-induced diabetes in animals. Diabetologia 44:271–285

Kang Y, Yoon JW (1993) A genetically determined host factor controlling susceptibility to encephalomyocarditis virus-induce diabetes in mice. J Gen Virol 74:1207–1213

Kaptur PE, Thomas DC, Giron DJ (1989) The role of interferon in the protection of ICR Swiss male mice by the nondiabetogenic variant of encephalomyocarditis virus against virus-induced diabetes mellitus. J Interferon Res 9:671–678

Kawagishi A, Kubosaki A, Takeyama N, Sakudo A, Saeki K, Matsumoto Y, Hayashi T, Onodera T (2003) Analysis of T-cell receptor Vβ gene from infiltrating T cells in insulitis and myocarditis in encephalomyocarditis virus-infected BALB/C mice. Biochem Biophys Res Commun 310:791–795

Kogawa K, Kudoh J, Nagafuchi S, Ohga S, Katsuta H, Ishibashi H, Harada M, Hara T, Shimizu N (2002) Distinct clinical phenotype and immunoreactivity in Japanese siblings with autoimmune polyglandular syndrome type 1 (APS-1) associates with compound heterozygous novel AIRE gene mutations. Clin Immunol 103:277–283

Kounoue E, Nagafuchi S, Nakamura M, Nakano S, Koga T, Nakayama M, Mitsuyama M, Niho Y, Takaki R (1987) Encephalomyocarditis (EMC) virus-induced diabetes mellitus prevented by Corynebacterium parvum in mice. Experientia 43:430–431

Kounoue E, Izumi K, Ogawa S, Kondo S, Katsuta H, Akashi T, Niho Y, Harada M, Tamiya S, Kurisaki H, Nagafuchi S (2008) The significance of T cells, B cells, antibodies, and macrophages against encephalomyocarditis (EMC)-D virus-induced diabetes in mice. Arch Virol 153:1223–1231

McCartney SA, Vermi W, Lonardi S, Rossini C, Otero K, Calderon B, Gilfillan S, Diamond MS, Unanue ER, Colonna M (2011) RNA sensor-induced type 1 IFN prevents diabetes caused by a β-cell tropic virus in mice. J Clin Invest 121:1497–1507

Nagafuchi S, Katsuta H, Anzai K (2010) Rituximab, B-lymphocyte depletion, and beta-cell function. N Engl J Med 362:761

Nagamine K, Peterson P, Scott HD, Kudoh J, Minoshima S, Heino M, Krohn KJE, Lalioti MD, Mullis PE, Antonarakis SE, Kawasaki K, Asakawa S, Ito F, Shimizu N (1997) Positional cloning of the APECD gene. Nat Genet 17:393–398

Onodera AT, Yoon JW, Brown ES, Notkins AL (1978) Evidence for a single locus controlling susceptibility to virus-induced diabetes mellitus. Nature 274:693–696

Pescovitz MD, Greenbaum CJ, Krause-Steinrauf H, Becker DJ, Gitelman SE, Goland R, Gottlieb PA, Marks JB, McGee PF, Moran AM, Raskin P, Rodriguez H, Schatz DA, Wherrett D, Wilson DM, Lachin JM, Skyler JS (2009) Rituximab, B-lymphocyte depletion, and preservation of beta-cell function. N Engl J Med 361:2143–2152

Rabinovitch A, Suarez-Pinzon WL, Shi Y, Morgan AR, Bleackley RC (1994) DNA fragmentation is an early event in cytokine-induced islet beta-cell destruction. Diabetologia 37:733–738

Racaniello VR (2007) Picornaviridae: the viruses and their replication. In: Knipe DM, Howley PM, Griffin DE, Lamb RA, Straus SE, Martin MA, Roizman B (eds) Fields virology, 5th edn. Lippincott Williams & Wilkins, Philadelphia, pp 795–838

Richardson AJ, Willcox A, Bone AJ, Morgan N, Foulis AK (2011) Immunopathology of the human pancreas in type-I diabetes. Semin Immunopathol 33:9–21

Ross ME, Onodera T, Brown KS, Notkins AL (1975) Virus-induced diabetes mellitus. IV. Genetic and environmental factors influencing the development of diabetes after infection with the M variant of encephalomyocarditis virus. Diabetes 25:190–197

Sakaguchi S (2005) Naturally arising Foxp3-expressiong $CD25^+$ $CD4^+$ regulatory T cells in immunological tolerance to self and non-self. Nat Immunol 6:345–398

Santos CX, Anjos EI, Augusto O (1999) Uric acid oxidation by peroxynitrate: multiple reaction, free radical formation, and amplification of lipid oxidation. Arch Biochem Biophys 372:285–294

Smyth DJ, Cooper JD, Bailey R, Field S, Burren O, Smink LJ, Guja C, Ionescu-Tirgoviste C, Widmer B, Dunger DB, Savage DA, Walker NM, Clayton DG, Todd JA (2006) A genome-wide association study of nonsynonymous SNPs identifies a type 1 diabetes locus in the interferon-induced helicase (IFIH1) region. Nat Genet 38:617–619

Takeuchi O, Akira S (2009) Innate immunity to virus infection. Immunol Rev 227:75–86

Tanaka S, Nishida Y, Aida K, Maruyama T, Shimada A, Suzuki M, Shimura H, Takizawa S, Takahashi M, Akiyama D, Arai-Yamashita S, Furuya F, Kawaguchi A, Kaneshige M, Katoh R, Endo T, Kobayashi T (2009) Enterovirus infection CXC chemokine ligand 10 (CXCL10), and CXCR3 circuit. Diabetes 58:2285–2291

Tauriainen S, Oikarinen S, Oikarinen M, Hyöty H (2011) Enteroviruses in the pathogenesis of type 1 diabetes. Semin Immunopathol 33:45–55

Toniolo A, Onodear T, Yoon JW, Notkins AL (1980) Induction of cumulative environmental insults from viruses and chemicals. Nature 288:383–385

Yamamoto H, Uchigata Y, Okamoto H (1981) Streptozocin and alloxan induced DNA strand breaks and poly ADP-ribose synthetase in pancreatic islets. Nature 294:284–286

Yoon JW, McLintock PR, Onodera T, Notkins AL (1980) Virus-induced diabetes mellitus. XVIII. Inhibition by a nondiabetogenic variant of encephalomyocariditis virus. J Exp Med 152:878–892

Yoon JW, Rodrigues MM, Currier C, Notkins AL (1982) Long-term complications of virus-induced diabetes mellitus in mice. Nature 296:566–569

Yoon JW, McClintock PR, Bachurski CJ, Longstreth JD, Notkins AL (1985) Virus-induced diabetes mellitus. No evidence for immune mechanisms in the destruction of β-cell by the D-variant of encephalomyocarditis virus. Diabetes 34:922–923

Yoshimura A, Naka T, Kubo M (2007) SOCS proteins, cytokine signaling and immune regulation. Nat Rev Immunol 7:454–465

Chapter 6
Enteroviruses in the Mouse Model of Type 1 Diabetes

Nora M. Chapman

Introduction

Findings of pancreatitis in mice after infection with the human enteroviruses coxsackievirus B (CVB) suggested a relationship to the onset of type 1 diabetes (Coleman et al. 1973), a correlation that had been suggested by studies that had variably found a serologic relationship of CVB4 to recent onset diabetic patients (Gamble et al. 1969). Although other enteroviruses may well be involved in induction of pancreatitis and type 1 diabetes (Tracy et al. 2010), the ability of CVBs to use the murine homolog of the coxsackievirus and adenovirus receptor CAR (Bergelson et al. 1997; Carson et al. 1997; Tomko et al. 1997; Bergelson et al. 1998) makes CVB-induced murine pancreatitis and diabetes a model for the human disease.

Properties of Diabetogenic CVBs

A CVB4 isolate from a human diabetic patient was capable of inducing insulitis and diabetes in SJL mice (Yoon et al. 1979), but, in general, most diabetogenic strains of CVB have had one or more passages in mice or murine pancreas or islets in culture (Al-Hello et al. 2005; Yoon et al. 1978a, b; Webb et al. 1976). As these enteroviruses evolve rapidly in selective cultures, passage in murine pancreatic cells is likely to increase the extent to which these viruses can infect and induce diabetes in

N.M. Chapman, Ph.D. (✉)
Department of Pathology & Microbiology, University of Nebraska Medical Center, 986495 Nebraska Medical Center, Omaha, NE 68198-6495, USA
e-mail: nchapman@unmc.edu

K.W. Taylor, H. Hyoty, A. Toniolo, A.J. Zuckerman (eds.), *Diabetes and Viruses*, DOI 10.1007/978-1-4614-4051-2_6, © Springer Science+Business Media New York 2013 mice. Sequence analysis of pancreotropic and diabetogenic strains has identified sites of variation in the 5′ nontranslated region (5′NTR), the capsid, and the nonstructural proteins (Al-Hello et al. 2005; Kang et al. 1994; Caggana et al. 1993; Titchener et al. 1994; Yin et al. 2002). Chimeras of the 5′NTR of CVB3 strains demonstrate that attenuating determinants are present in this region of the CVB3/GA strain (Chapman et al. 1994) for replication in a murine β cell line, MIN6, and for replication in the murine pancreas in vivo (Kanno et al. 2006). Variations in the capsid proteins may alter sites which control interaction with the CAR receptor or with a co-receptor, the decay accelerating factor (DAF) which may play a role in CVB virus entry (Coyne and Bergelson 2006) and in the immune response to the virus (Huber and Rincon 2008; Huber et al. 2006). Some identified sites in CVB4 VP2 and VP3 (Kang et al. 1994) align close to sites shown to be important for DAF binding in CVB3 (Pan et al. 2011), although the relatively nonpancreovirulent CVB3/GA does not differ from the pancreovirulent CVB3/28 at these sites (Chapman et al. 1994). As the extent of replication in the pancreas is related to the extent of acceleration of diabetes (Kanno et al. 2006), several of these variations may be due to selection to match the murine receptor(s) and host cell factors. Selection by growth in pancreatic islets or pancreas is likely to generate strains capable of a high rate of replication in the pancreas in vivo, but as most CVB strains have some degree of pancreovirulence in mice (Tracy et al. 2000), a virulent CVB may cause murine pancreatitis, but not diabetes, without passage in mice.

Typically, inoculation of mice with a CVB will result in much more extensive pathology of the acinar tissue than the islets (Harrison et al. 1972). Inoculation of 6-week-old SJL mice at dosages of 10^5 PFU of either the standard CVB4 serotype strain, JVB, or the E2 diabetogenic strain resulted in extensive acinar cell death (Yap et al. 2003). Infection of Swiss Albino mice with the prototypic CVB3, Nancy, was able to generate infection of the pancreas with pathology of the acinar tissue (Bopegamage et al. 2005). Survival of the islets during infection has been attributed to the relative expression of CAR (Mena et al. 2000). Studies have demonstrated CAR expression in islets of infected mice (Drescher et al. 2004) as well as viral RNA (Yap et al. 2003). Although very low level expression of CAR may limit virus infection in cultures, almost indetectable levels of CAR still allow virus replication (Carson et al. 2007), but the low level expression is likely to limit the degree of infection of the islets. Components of the innate immune response provide more antiviral protection to the murine exocrine tissue than the islets. Islets of mouse strains knockout of RNase L and the double-stranded RNA-activated protein kinase, PKR, are more resistant to infection than the acinar tissue in vivo, despite increased mortality due to CVB4 infection (Flodström-Tullberg et al. 2005). Knockouts of interferon α and β receptors, melanoma differentiation-associated protein-5 (MDA-5) and its signaling adaptor, mitochondrial antiviral signaling (MAVS) did not enhance infection of pancreatic islets after CVB4 infection (Hühn et al. 2010; Wang et al. 2010), although there was more extensive pathology of the exocrine pancreas. As MDA-5 is degraded during the course of enterovirus cell infection (Barral et al. 2007) and another viral sensor, retinoic acid-induced gene 1 (RIG-I), does not affect susceptibility to picornavirus infection (Kato et al. 2006), CVBs are likely to have evolved means of avoiding reduction of virus replication through the innate immune response to some degree.

Diabetogenic virus infections resulted in reduced neogenesis of islets postinfection indicating that there may be a lasting effect of the infection with the diabetogenic viruses (Yap et al. 2003). In studies in which RT-PCR was employed to detect viral RNA, persistence of CVB RNA is noted after loss of detection of virus by cytopathic assays (Bopegamage et al. 2005; Yap et al. 2003). In the heart, CVBs can persist in the form of a defective virus (Kim et al. 2005). The defect results in reduced levels of positive strand RNA which results in reduced levels of virus replication and cytolysis (Kim et al. 2005). These defective viruses tend to be selected in quiescent cells in culture (Kim et al. 2008) or in adult hearts (Kim et al. 2005; Chapman et al. 2008). The presence of viral antigens in islets without obvious cellular necrosis suggests the selection of defective virus in islet cells. As these genomes produce viral proteins (although at a reduced rate), the potential for alterations of function of these cells remains despite their reduced replication rate. Part of the apparent resistance to CVB infection of the islets may be due to the resistance of nondividing cells to replication of CVBs (Feuer et al. 2002; Chapman and Kim 2008), as most of the islet cells in a non-regenerating islet in vivo are quiescent (Salpeter et al. 2010). Regeneration or neogenesis of islets necessarily involves dividing cells which are more susceptible to viral infection and, consequently, can be eliminated in a pancreas with a persisting infection.

Nonobese diabetic (NOD) female mice (Atkinson and Leiter 1999; Kikutani and Makino 1992) develop spontaneously autoimmunity to pancreatic antigens, insulitis, and diabetes by 12 weeks of age. As in other mice, inoculation of NOD mice prior to 8 weeks of age results in less extensive infection of the islets than the exocrine tissue even with diabetogenic virus strains (Serreze et al. 2000; Tracy et al. 2002; Drescher et al. 2004). Infection of NOD mice with diabetogenic and nondiabetogenic CVBs at the stage in which insulitis is beginning to be manifested (8–12 weeks of age) results in the infection of islets and accelerated development of diabetes (Serreze et al. 2000; Drescher et al. 2004). Increasing dosage of a less virulent CVB can increase the extent of conversion to diabetes indicating the extent of replication in the pancreas correlates with induction of diabetes (Kanno et al. 2006). Knockouts of interleukin-4 (IL-4) do not alter the conversion to diabetes in NOD mice by CVB4 infection, whereas loss of interferon-γ (IFN-γ) does delay the onset of diabetes (Serreze et al. 2005). As CVBs can induce IFN-γ (Nair et al. 2010), the finding that higher levels of replication correlate with accelerated onset of diabetes may increase the exposure of islets to IFN-γ. As transforming growth factor-β (TGF-β) can reduce the expression of the CVB receptor (Lacher et al. 2006; Shi et al. 2010), one effect of the expression of TGF-β by beta cells may be to lower the expression of the receptor necessary for infection of beta cells (Richer et al. 2008; Peng et al. 2004). In CVB3-induced myocarditis, adoptive transfer of T regulatory cells increased TGF-β expression, decreased CAR expression, and lowered CVB3 replication in the heart (Shi et al. 2010) indicating that one protective effect of T regulatory cells may be lowering CVB replication by reducing the expression of the CVB receptor.

On the other hand, infection of NOD mice at 3–4 weeks of age with CVBs (an age at which non-NOD mice are more susceptible to pancreatitis) results in delayed onset and decreased incidence of diabetes from uninfected NOD mice (Tracy et al. 2002; Serreze et al. 2000; Filippi et al. 2009). It is known that infection or treatment with a number of agents will decrease or delay the onset of diabetes in the NOD mouse (Atkinson and Leiter 1999). Neither IL-4 nor IFN-γ are required for the delay or decrease in the development of type 1 diabetes in the NOD mouse due to CVB4 infection prior to insulitis (Serreze et al. 2005). Activation or supplementation of invariant natural killer cells (iNKT cells) in NOD mice leads to later onset or less conversion to diabetes in the NOD mouse (Lehuen et al. 1998; Naumov et al. 2001; Sharif et al. 2001), an effect requiring $CD4^+$ $CD25^+$ T regulatory cells (Ly et al. 2006). As CVB infections in the context of virus-induced expression of tumor necrosis factor-α (TNF-α) have been shown to upregulate CD1d (Huber and Sartini 2005), it is possible that the very active CVB infection of the exocrine pancreas produces an environment in which iNKT cell activation is more likely. CVB3-infected murine dendritic cells do not produce cytopathic virus but are stimulated to produce interferons, interleukins, and chemokines (Weinzierl et al. 2008). A.BY/SnJ mice susceptible to CVB3-induced chronic myocarditis produce dendritic cells (DC) which, upon infection, have similar levels of positive and negative CVB3 RNA, produce less IL-10 than those from C57BL/6 mice (resistant to CVB3-induced chronic myocarditis), and have a later peak of IL-6 and TNF-α (Weinzierl et al. 2008). It is possible that CVB-infected DCs may interact with iNKT differently resulting in changes to the level to which pancreatic antigen-specific T regulatory cells are generated via murine DCs.

As discussed above, there is evidence of persistent infection of the pancreas after inoculation of mice with diabetogenic CVBs (Yap et al. 2003; Bopegamage et al. 2005). In the heart, CVBs can persist in the form of a defective virus (Kim et al. 2005). As this slower replicating defective virus is capable of persisting without rapid cytolysis, DCs and other cells infected with defective viruses may persist so that an environment for the activation of T regulatory cells specific for pancreatic antigens may be long-lasting.

Is this relevant for human disease? CVBs are human pathogens which target the pancreas and produce pathogenic immune responses. As these viruses have also been shown to have an ability to persist after an acute infection in a form capable of replicating and producing viral proteins, but with reduced cytolysis, they are good candidates to alter the complex immune responses involved in autoimmune pathogenesis. The ability to study these human viruses in a murine model of type 1 diabetes allows an analysis of mechanisms for this disease. Enterovirus infections are common in the human population, but their frequency may be decreasing in the populations in which type 1 diabetes is increasing (Viskari et al. 2005; Tracy et al. 2010). This may have the resulting double defect of decreasing regulation of pancreatic autoimmunity in those prone to this disease, and increasing the chance that when an enterovirus infection occurs it will be at an age in which autoimmunity has made islets susceptible to infection.

References

Al-Hello H, Davydova B, Smura T, Kaialainen S, Ylipaasto P, Saario E, Hovi T, Rieder E, Roivainen M (2005) Phenotypic and genetic changes in coxsackievirus B5 following repeated passage in mouse pancreas in vivo. J Med Virol 75:566–574

Atkinson MA, Leiter EH (1999) The NOD mouse model of type 1 diabetes: as good as it gets? Nat Med 5:601–604

Barral PM, Morrison JM, Drahos J, Gupta P, Sarkar D, Fisher PB, Racaniello VR (2007) MDA-5 is cleaved in poliovirus-infected cells. J Virol 81:3677–3684

Bergelson JM, Cunningham JA, Droguett G, Kurt-Jones EA, Krithivas A, Hong JS, Horwitz MS, Crowell RL, Finberg RW (1997) Isolation of a common receptor for Coxsackie B viruses and adenoviruses 2 and 5. Science 275:1320–1323

Bergelson JM, Krithivas A, Celi L, Droguett G, Horwitz MS, Wickham T, Crowell RL, Finberg RW (1998) The murine CAR homolog is a receptor for coxsackie B viruses and adenoviruses. J Virol 72:415–419

Bopegamage S, Kovacova J, Vargova A, Motusova J, Petrovicova A, Benkovicova M, Gomolcak P, Bakkers J, van Kuppeveld F, Melchers WJG, Galama JM (2005) Coxsackie B virus infection of mice: inoculation by the oral route protects the pancreas from damage, but not from infection. J Gen Virol 86:3271–3280

Caggana M, Chan P, Ramsingh A (1993) Identification of a single amino acid residue in the capsid protein VP1 of coxsackievirus B4 that determines the virulent phenotype. J Virol 67: 4797–4803

Carson SD, Kim KS, Pirruccello SJ, Tracy S, Chapman NM (2007) Endogenous low-level expression of the coxsackievirus and adenovirus receptor enables coxsackievirus B3 infection of RD cells. J Gen Virol 88:3031–3038

Carson SD, Chapman NM, Tracy SM (1997) Purification of the putative coxsackievirus B receptor from HeLa cells. Biochem Biophys Res Commun 233:325–328

Chapman NM, Kim KS (2008) Persistent coxsackievirus infection: enterovirus persistence in chronic myocarditis and dilated cardiomyopathy. Curr Top Microbiol Immunol 323:275–292

Chapman NM, Kim KS, Drescher KM, Oka K, Tracy S (2008) 5′ terminal deletions in the genome of a coxsackievirus B2 strain occurred naturally in human heart. Virology 375:480–491

Chapman NM, Tu Z, Tracy S, Gauntt CJ (1994) An infectious cDNA copy of the genome of a non-cardiovirulent coxsackievirus B3 strain: its complete sequence analysis and comparison to the genomes of cardiovirulent coxsackieviruses. Arch Virol 135:115–130

Coleman TJ, Gamble DR, Taylor KW (1973) Diabetes in mice after coxsackie B4 virus infection. Br Med J 3:25–27

Coyne CB, Bergelson JM (2006) Virus-induced Abl and Fyn kinase signals permit coxsackievirus entry through epithelial tight junctions. Cell 124:119–131

Drescher KM, Kono K, Bopegamage S, Carson SD, Tracy S (2004) Coxsackievirus B3 infection and type 1 diabetes development in NOD mice: insulitis determines susceptibility of pancreatic islets to virus infection. Virology 329:381–394

Feuer R, Mena I, Pagarigan R, Slifka MK, Whitton JL (2002) Cell cycle status affects coxsackievirus replication, persistence, and reactivation in vitro. J Virol 76:4430–4440

Filippi CM, Estes EA, Oldham JE, von Herrath MG (2009) Immunoregulatory mechanisms triggered by viral infections protect from type 1 diabetes in mice. J Clin Invest 119:1515–1523

Flodström-Tullberg M, Hultcrantz M, Stotland A, Maday A, Tsai D, Fine C, Williams B, Silverman R, Sarvetnick N (2005) RNase L and double-stranded RNA-dependent protein kinase exert complementary roles in islet cell defense during coxsackievirus infection. J Immunol 174:1171–1177

Gamble DR, Kinsley ML, FitzGerald MG, Bolton R, Taylor KW (1969) Viral antibodies in diabetes mellitus. Br Med J 3:627–630

Harrison AK, Bauer SP, Murphy FA (1972) Viral pancreatitis: ultrastructural pathological effects of coxsackievirus B3 infection in newborn mouse pancreas. Exp Mol Pathol 17:206–219

Huber S, Song WC, Sartini D (2006) Decay-accelerating factor (CD55) promotes CD1d expression and Vgamma4+ T-cell activation in coxsackievirus B3-induced myocarditis. Viral Immunol 19:156–166
Huber SA, Sartini D (2005) Roles of tumor necrosis factor alpha (TNF-alpha) and the p55 TNF receptor in CD1d induction and coxsackievirus B3-induced myocarditis. J Virol 79:2659–2665
Huber SA, Rincon M (2008) Coxsackievirus B3 induction of NFAT: requirement for myocarditis susceptibility. Virology 381:155–160
Hühn MH, McCartney SA, Lind K, Svedin E, Colonna M, Flodström-Tullberg M (2010) Melanoma differentiation-associated protein-5 (MDA-5) limits early viral replication but is not essential for the induction of type 1 interferons after coxsackievirus infection. Virology 401:42–48
Kang Y, Chatterjee NK, Nodwell MJ, Yoon JW (1994) Complete nucleotide sequence of a strain of coxsackie B4 virus of human origin that induces diabetes in mice and its comparison with nondiabetogenic coxsackie B4 JBV strain. J Med Virol 44:353–361
Kanno T, Kim K, Kono K, Drescher KM, Chapman NM, Tracy S (2006) Group B coxsackievirus diabetogenic phenotype correlates with replication efficiency. J Virol 80:5637–5643
Kato H, Takeuchi O, Sato S, Yoneyama M, Yamamoto M, Matsui K, Uematsu S, Jung A, Kawai T, Ishii KJ, Yamaguchi O, Otsu K, Tsujimura T, Koh C-S, Reis E, Sousa C, Matsuura Y, Fujita T, Akira S (2006) Differential roles of MDA5 and RIG-I helicases in the recognition of RNA viruses. Nature 441:101–105
Kikutani H, Makino S (1992) The murine autoimmune diabetes model: NOD and related strains. Adv Immunol 51:285–322
Kim K-S, Tracy S, Tapprich W, Bailey J, Lee C-K, Kim K, Barry WH, Chapman NM (2005) 5'-Terminal deletions occur in coxsackievirus B3 during replication in murine hearts and cardiac myocyte cultures and correlate with encapsidation of negative-strand viral RNA. J Virol 79:7024–7041
Kim KS, Chapman NM, Tracy S (2008) Replication of coxsackievirus B3 in primary cell cultures generates novel viral genome deletions. J Virol 82:2033–2037
Lacher MD, Tiirikainen MI, Saunier EF, Christian C, Anders M, Oft M, Baimain A, Akhurst RJ, Korn WM (2006) Transforming growth factor-beta receptor inhibition enhances adenoviral infectability of carcinoma cells via up-regulation of coxsackie and adenovirus receptor in conjunction with reversal of epithelial-mesenchymal transition. Cancer Res 66:1648–1657
Lehuen A, Lantz O, Beaudoin L, Laloux V, Carnaud C, Bendelac A, Bach J-F, Monteiro RC (1998) Overexpression of natural killer T cells protects Valpha14-Jalpha281 transgenic nonobese diabetic mice against diabetes. J Exp Med 188:1831–1839
Ly D, Mi QS, Hussain S, Delovitch TL (2006) Protection from type 1 diabetes by invariant NK T cells requires the activity of CD4+ CD25+ regulatory T cells. J Immunol 177:3695–3670
Mena I, Fischer C, Gebhard JR, Perry CM, Harkins S, Whitton JL (2000) Coxsackievirus infection of the pancreas: evaluation of receptor expression, pathogenesis, and immunopathology. Virology 271:276–288
Nair S, Leung KC, Rawlinson WD, Naing Z, Craig ME (2010) Enterovirus infection induces cytokine and chemokine expression in insulin-producing cells. J Med Virol 82:1950–1957
Naumov YN, Bahjat KS, Gausling R, Abraham R, Exley MA, Koezuka Y, Balk SB, Strominger JL, Clare-Salzer M, Wilson SB (2001) Activation of CD1d-restricted T cells protects NOD mice from developing diabetes by regulating dendritic cell subsets. Proc Natl Acad Sci USA 98:13838–13843
Pan J, Narayanan B, Shah S, Yoder JD, Cifuente JO, Hafenstein S, Bergelson JM (2011) Single amino acid changes in the virus capsid permit coxsackievirus B3 to bind decay-accelerating factor. J Virol 85:7436–7443
Peng Y, Laouar Y, Li MO, Green EA, Flavell RA (2004) TGF-beta regulates in vivo expansion of Foxp3-expressing CD4+ CD25+ regulatory T cells responsible for protection against diabetes. Proc Natl Acad Sci USA 101:4572–4577
Richer MJ, Straka N, Fang D, Shanina I, Horwitz MS (2008) Regulatory T-cells protect from type 1 diabetes after induction by coxsackievirus infection in the context of transforming growth factor-beta. Diabetes 57:1302–1311

Salpeter SJ, Klein AM, Huangfu D, Grimsby J, Dor Y (2010) Glucose and aging control the quiescence period that follows pancreatic beta cell replication. Development 137:3205–3213

Serreze DV, Wasserfall C, Ottendorfer EW, Stalvey M, Pierce MA, Gauntt C, O'Donnell B, Flanagan JB, Campbell-Thompson M, Ellis TM, Atkinson MA (2005) Diabetes acceleration or prevention by a coxsackievirus B4 infection: critical requirements for both interleukin-4 and gamma interferon. J Virol 79:1045–1052

Serreze DV, Ottendorfer EW, Ellis TM, Gauntt CJ, Atkinson MA (2000) Acceleration of type 1 diabetes by a coxsackievirus infection requires a preexisting critical mass of autoreactive T-cells in pancreatic islets. Diabetes 49:708–711

Sharif S, Arreaza GA, Zucker P, Mi Q-S, Sondhi J, Naidenko OV, Kronenberg M, Koezuka Y, Delovitch TL, Gombert J-M, Leite-de-Moraes M, Gouarin C, Zhu R, Hameg A, Nakayama T, Taniguchi M, Lepault F, Lehuen A, Bach J-F, Herbelin A (2001) Activation of natural killer T cells by alpha-galactosylceramide treatment prevents the onset and recurrence of autoimmune Type 1 diabetes. Nat Med 7:1057–1062

Shi Y, Fukuoka M, Li G, Liu Y, Chen M, Konviser M, Chen X, Opavsky MA, Liu PP (2010) Regulatory T cells protect mice against coxsackievirus-induced myocarditis through the transforming growth factor beta-coxsackie-adenovirus receptor pathway. Circulation 121:2624–2634

Titchener PA, Jenkins O, Szopa TM, Taylor KW, Almond JW (1994) Complete nucleotide sequence of a beta-cell tropic variant of coxsackievirus B4. J Med Virol 42:369–373

Tomko RP, Xu R, Philipson L (1997) HCAR and MCAR: the human and mouse cellular receptors for subgroup C adenoviruses and group B coxsackieviruses. Proc Natl Acad Sci USA 94:3352–3356

Tracy S, Drescher KM, Chapman NM, Kim K-S, Carson SD, Pirruccello S, Lane PH, Romero JR, Leser JS (2002) Toward testing the hypothesis that group B coxsackieviruses (CVB) trigger insulin-dependent diabetes: inoculating nonobese diabetic mice with CVB markedly lowers diabetes incidence. J Virol 76:12097–12111

Tracy S, Höfling K, Pirruccello S, Lane PH, Reyna SM, Gauntt CJ (2000) Group B coxsackievirus myocarditis and pancreatitis: connection between viral virulence phenotypes in mice. J Med Virol 62:70–81

Tracy S, Drescher KM, Jackson JD, Kim K, Kono K (2010) Enteroviruses, type 1 diabetes and hygiene: a complex relationship. Rev Med Virol 20:106–116

Viskari H, Ludvigsson J, Uibo R, Salur L, Marciulionyte D, Hermann R, Soltesz G, Füchtenbusch M, Ziegler A-G, Kondrashova A, Romanov A, Kaplan B, Laron Z, Koskela P, Vesikari T, Huhtala H, Knip M, Hyöty H (2005) Relationship between the incidence of type 1 diabetes and maternal enterovirus antibodies: time trends and geographical variation. Diabetologia 48: 1280–1287

Wang JP, Cerny A, Asher DR, Kurt Jones EA, Bronson RT, Finberg RW (2010) MDA5 and MAVS mediate type I interferon responses to coxsackie B virus. J Virol 84:254–260

Webb SR, Loria RM, Madge GE, Kibrick S (1976) Susceptibility of mice to group B coxsackie virus is influenced by the diabetic gene. J Exp Med 143:1239–1248

Weinzierl AO, Szalay G, Wolburg H, Sauter M, Rammensee H-G, Kandolf R, Stevanović S, Klingel K (2008) Effective chemokine secretion by dendritic cells and expansion of cross-presenting CD4–/CD8+ dendritic cells define a protective phenotype in the mouse model of coxsackievirus myocarditis. J Virol 82:8149–8160

Yap IS, Giddings G, Pocock E, Chantler JK (2003) Lack of islet neogenesis plays a key role in beta-cell depletion in mice infected with a diabetogenic variant of coxsackievirus B4. J Gen Virol 84:3051–3068

Yin H, Berg AK, Westman J, Hellerström C, Frisk G (2002) Complete nucleotide sequence of a coxsackievirus B-4 strain capable of establishing persistent infection in human pancreatic islet cells: effects on insulin release, proinsulin synthesis, and cell morphology. J Med Virol 68:544–557

Yoon JW, Austin M, Onodera T, Notkins AL (1979) Isolation of a virus from the pancreas of a child with diabetic ketoacidosis. N Engl J Med 300:1173–1179

Yoon JW, Onodera T, Jenson AB, Notkins AL (1978a) Virus-induced diabetes mellitus. XI. Replication of coxsackie B3 virus in human pancreatic beta cell cultures. Diabetes 27: 778–781

Yoon JW, Onodera T, Notkins AL (1978b) Virus-induced diabetes mellitus. XV. Beta cell damage and insulin-dependent hyperglycemia in mice infected with coxsackie virus B4. J Exp Med 148:1068–1080

Chapter 7
Viruses and Autoimmune Diabetes in Rats

John P. Mordes, Danny Zipris, Zhijun Liu, and Elizabeth P. Blankenhorn

Abstract The role of viral infection in the pathogenesis of type 1 diabetes in humans remains an open question. Viruses are variously thought to be causative, preventive, or irrelevant. The rat models of the disease suggest that the role of viruses in the pathogenesis of autoimmunity can be multifaceted, with effects that are dependent on viral agent, host age, genetic background, and the immunological environment of the host at the time of infection. Among inbred strains with spontaneous onset of diabetes (BBDP, LEW.1AR1-iddm), the prevalence of disease generally increases with progressive removal of viruses from their environment. In contrast, in two rat strains with genetic susceptibility but little or no spontaneous diabetes in clean environments (BBDR, LEW.1WR1), infection with viruses from several families (parvovirus, enterovirus, poxvirus, herpesvirus) can trigger the disorder. The ability of viruses to do so is limited to juvenile animals, is dependent sensitively on the state of innate immunity before infections, and is strain dependent. Maternal immunization can prevent the onset of diabetes in infected offspring.

The rat models of type 1 diabetes suggest that the role of infection in autoimmunity is complex but amenable to dissection.

J.P. Mordes, M.D. (✉) • Z. Liu
Professor of Eudoeriuology and Metabolism, Department of Medicine,
University of Massachusetts Medical School, Worcester, MA, USA
e-mail: John.Mordes@umassmed.edu

D. Zipris
Barbara Davis Center for Childhood Diabetes, University of Colorado, Aurora, CO, USA

E.P. Blankenhorn
Department of Immunology and Microbiology, Drexel University College of Medicine,
Philadelphia, PA, USA

K.W. Taylor, H. Hyoty, A. Toniolo, A.J. Zuckerman (eds.), *Diabetes and Viruses*,
DOI 10.1007/978-1-4614-4051-2_7, © Springer Science+Business Media New York 2013

Introduction

Type 1 diabetes is a T cell-mediated autoimmune disorder of unknown cause (Morran et al. 2008). It probably represents multiple disease processes, with a spectrum ranging from abrupt onset soon after birth (Gillespie et al. 2002) to latent autoimmune diabetes of adults (Naik et al. 2009). The disease is heritable, but genetic susceptibility loci are insufficient for predicting diabetes onset, as most individuals with risk alleles never develop the disorder (Pearce and Merriman 2009). Interaction of genes within these loci with the environment has been proposed as a determinant of disease penetrance (Åkerblom et al. 2002; Hawa et al. 2002), but data remain inconclusive. Viral infection is a prime environmental suspect (Yoon and Jun 2004; Filippi and Von Herrath 2008), but it is unclear whether viral infection causes type 1 diabetes [accelerator hypothesis (Zipris 2009)], prevents it [hygiene hypothesis (Schaub et al. 2006)], or is a possibly irrelevant component of a stochastic process.

Rats as Models of Type 1 Diabetes

Autoimmune diabetes is relatively common among rat strains (Ellerman and Like 2000), which are an undervalued resource for studying type 1 diabetes (Leiter 2009). At least four rat strains express "spontaneous" type 1 diabetes at rates ranging from ~3% to ~90%: BBDP (Mordes et al. 2007), KDP (Yokoi et al. 2002), LEW.1AR1-iddm (Arndt et al. 2009; Lenzen et al. 2001), and LEW.1WR1 (Mordes et al. 2005). They share the same high risk class II major histocompatibility complex (MHC) allele, designated *RT1B/Du*, but they differ at class I and at non-MHC loci. Each strain appears to traverse a distinct pathway towards beta cell destruction. BBDP rats are lymphopenic and depend on regulatory T cell (Treg) deficiency for disease (Mordes et al. 2007). KDP rats are not lymphopenic but have a mutation in *Cbl-b* that leads to co-stimulation-independent activation of autoreactive T cells (Yokoi et al. 2002). The pathway to diabetes in the two LEW congenic rats, which have no known immunological defects, is unknown. These rats develop type 1 diabetes during early adolescence and young adulthood and uniformly progress rapidly to ketoacidosis and insulin dependence. Rat pancreatic insulitis resembles that of human type 1 diabetes, with no nondestructive phase of "inflammatory" insulitis (Fig. 7.1).

Other rat strains have susceptibility to type 1 diabetes that is unmasked by perturbation of the immune system. These include the BBDR, which was derived from the BBDP and never becomes spontaneously diabetic (Ellerman and Like 2000) and the LEW.1WR1, whose rate of spontaneous type 1 diabetes is ~2.5% (Mordes et al. 2007). Both have a normal immunophenotype. One trigger that induces rat type 1 diabetes in both is Treg depletion (Mordes et al. 2001, 2005). Both also become diabetic following innate immune activation with polyinosinic:polycytidylic acid (poly I:C). Poly I:C, an analog of dsRNA, is a ligand of TLR3 and also Rig-I and

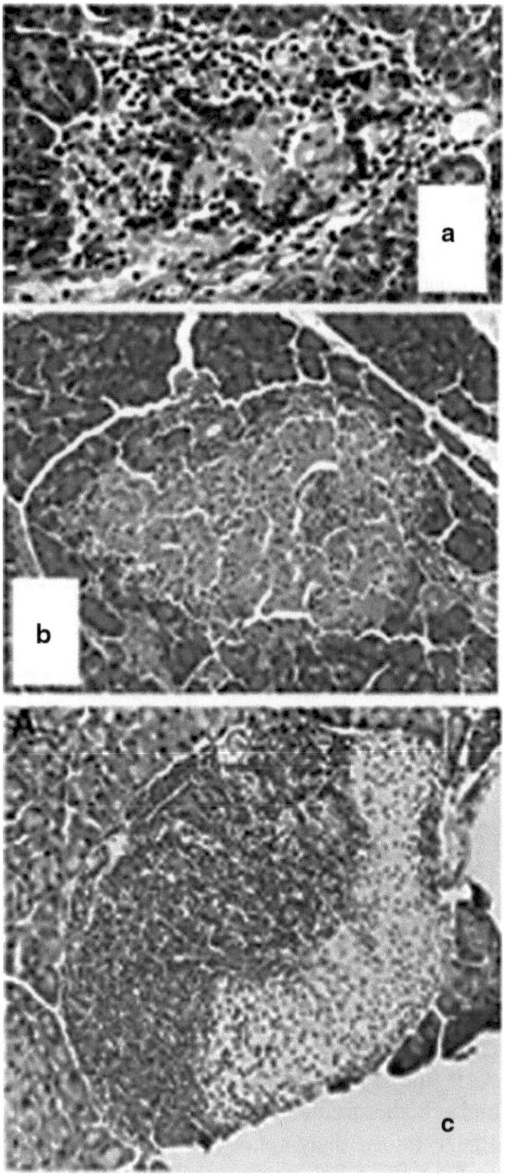

Fig. 7.1 (**a**) Insulitis in a 2-year-old girl (Gepts 1965), (**b**) early insulitis in a LEW.1WR1 rat and (**c**) early insulitis in an NOD mouse

IFIH1 (also designated MDA5). IIFIH1 has been identified as a susceptibility gene in human type 1 diabetes GWAS studies (Nejentsev et al. 2009). Poly I:C triggers diabetes in LEW.1WR1 rats at a rate close to 100%, but the TLR4 ligand LPS has no effect (Mordes et al. 2005).

Viral infection has strong effects on the penetrance of diabetes in strains with both spontaneous and induced type 1 diabetes. These effects are nonuniform and

Table 7.1 Changes in diabetes penetrance associated with viral infection

Strain	Spontaneous diabetes in VAF vivaria	Diabetes frequency with viral infection	Diabetes after viral infection following poly I:C priming
BBDP	~90%	Lower penetrance (~50%) in SPF housing[a] Prevention by LCMV No effect of deliberate KRV infection	Not known
LEW.1AR1-iddm	60–80%	Lower penetrance (~20%) in SPF housing[a]	Not known
BBDR	0%	After KRV: ~30% T1D rarely or not at all after RCMV or CoxB4	Up to 100%
LEW.1WR1	3%	After KRV or RCMV ~40%	100% of KRV-treated rats 10% of RCMV-treated rats T1D in a small percentage of CoxB4-treated rats

Some changes in penetrance of autoimmune diabetes in four strains of rats as a function of exposure to viral infection. Priming doses of poly I:C are given on three consecutive days before virus infection and by themselves do not cause diabetes. *VAF* viral antibody free, *SPF* specific pathogen free, *KRV* Kilham rat virus, *RCMV* rat cytomegalovirus, *LCMV* lymphocytic choriomeningitis virus, *CoxB4* coxsackie B4 virus, *Poly I:C* polyinosinic:polycytidylic acid

[a]Prior to rederivation into VAF vivaria; specific viral agents not characterized. For details and references, see text

suggest that the interaction of a type 1 diabetes-susceptible genome and viral infection is a complex function of genetic background, age, virus type, and state of the host when infected (Table 7.1).

Viral Infection in Spontaneous Diabetes in Rats

Spontaneous diabetes in BBDP rats occurs in more than half of animals housed in either gnotobiotic (Rossini et al. 1979) or "conventional" vivaria (Guberski 1994). As is true of NOD mice (Serreze and Leiter 2001), however, diabetes occurs earlier and at higher frequency in viral antibody free (VAF) vivaria. Deliberate infection of BBDP rats with lymphocytic choriomeningitis virus (LCMV) prevents diabetes, perhaps by deleting effector T cells (Dyrberg et al. 1988). Infection with other viruses does not affect disease frequency or age at onset (Like et al. 1991). The rate of diabetes in inbred LEW.1AR1-iddm rats was initially ~20% (Lenzen et al. 2001) but has increased to >60% in progressively cleaner environments (Jörns et al. 2005). The effect of intentional infection in these animals is unknown. Spontaneous type 1 diabetes in LEW.1WR1 rats was first detected in an inbred population after being housed for 10 years in a VAF environment (Mordes et al. 2005). Diabetes frequency was ~3% and has been constant thereafter despite (diabetic × diabetic) breeding, but as discussed below that rate can increase dramatically after viral infection. Whether an environmental factor played a role in the evolution of KDP rats [80% T1D penetrance (Yokoi et al. 2002)], from their LETL progenitors [20% penetrance (Kawano et al. 1991)], is unclear (N. Yokoi, personal communication).

Little is known about the mechanisms underlying the increase in disease frequency in BBDP and LEW.1AR1-iddm rats that occurred as vivaria became "cleaner." Nor is it known specifically which environmental changes led to the higher incidence of type 1 diabetes. From an epidemiological standpoint, however, one cannot help being struck by the consonance of these observations with the "hygiene hypothesis" to account for the increasing prevalence of human type 1 diabetes and other autoimmune diseases (Schaub et al. 2006). Clearly, rat strains (BBDP and LEW.1AR1-iddm) that progress to type 1 diabetes "on their own" harbor genetic elements whose response to the environment interferes with disease progression. Understanding the underlying biology could help in assessing the hygiene hypothesis in human type 1 diabetes, but the necessary studies have not been undertaken.

Viral Triggering of Type 1 Diabetes: Kilham Rat Virus and the BBDR Rat

Whereas little is known of the mechanisms behind the down-modulatory effect of viral infection on "spontaneous" diabetes in the BBDP and LEW.1AR1-iddm rats, much more is known about mechanisms of viral-triggering of T1D in rats strains that rarely or never become diabetic in VAF vivaria. The initial dataset was developed in the BBDR rat.

Epidemiology

BBDR rats in virus-free facilities remain free of spontaneous diabetes, but when infected with Kilham rat virus (KRV a single stranded DNA parvovirus) many become diabetic. H-1 and rat parvovirus-1 (RPV-1) also infect rats but are not diabetogenic. KRV is most homologous with H-1 (Jacoby et al. 1996); its structural proteins are ~80% homologous and their non-structural proteins are 100% homologous.

Naturally occurring KRV infection affected ~1% of one generation of BBDR rats before its eradication by cesarean rederivation of the colony (Guberski et al. 1991). Intentional infection (10^7 PFU of KRV-UMass Strain) usually induces diabetes in ~30% of BBDR rats (Guberski et al. 1991). As few as 10^3 virions induce a few BBDR rats (<5%) to express diabetes, but even KRV inocula as high as 10^8 virions do not induce T1D in >50% of animals.

KRV-triggered diabetes occurs only if animals are infected when young-post-weaning. Infection is associated with pancreatic insulitis but not with infection of islet cells or with exocrine pancreatitis (Brown et al. 1993). The ability to induce autoimmune diabetes in BBDR rats is virus specific; infection with H-1 induces no diabetes despite causing a robust cellular and humoral immune response (Zipris et al. 2003). KRV infection increases serum IL-12p40 in treated animals, and it increases IL-12p40, IP-10, and IFN-γ mRNA transcript levels, particularly in the pancreatic lymph nodes (Zipris et al. 2005). Rat cytomegalovirus (RCMV) infection only rarely leads to type 1 diabetes (Tirabassi et al. 2010). Curiously, inoculation with RCMV followed 4 days later by KRV increases the penetrance of diabetes, whereas inverting the order of infection completely prevents hyperglycemia (Tirabassi et al. 2010).

Cellular Mechanisms of Disease

The mechanism by which KRV induces diabetes is beginning to be understood. It does not involve direct infection of beta cells (Brown et al. 1993). It is also unlikely to involve molecular mimicry. That hypothesis was disconfirmed in studies in which BBDR rats were injected with viral vectors encoding KRV proteins (Chung et al. 2000). No diabetes occurred despite the generation of cellular and humoral immune responses to those proteins.

A more likely mechanism is alteration of the immunoregulatory environment. The virus infects T and B lymphocytes (McKisic et al. 1995), but without causing the severe T lymphopenia characteristic of the BBDP rat. Infected lymphocytes are phenotypically normal but have diminished proliferative and cytolytic responses (McKisic et al. 1995). KRV may trigger diabetes in the BBDR rat by changing their immunoregulatory environment, specifically by reducing the frequency of $CD4^+$ $CD25^+$ Treg cells (Brown et al. 1993). This was inferred by comparing KRV and H1. Both produce similar host immune responses, but only KRV concurrently reduces the number splenic $CD4^+$ $CD25^+$ Tregs. Supporting the possible role of altered immunoregulation is the observation that $ART2^+$ Treg depletion synergizes with KRV to induce type 1 diabetes at a higher frequency than either agent alone (Ellerman et al. 1996).

The results suggest a mechanism that links underlying genetic predisposition [based on susceptibility genes shared with the BBDP rat (Mordes et al. 2009)] to environmental perturbation. The virus-specific disruption of immune regulation transforms a "regulated predisposition" into autoimmune diabetes and suggests a pathway by which infection accelerates diabetes.

Immunological Perturbants Enhance Diabetogenicity of Viral Infection

Further insights into the molecular mechanisms of disease acceleration by infection have come from the observation that noninfectious environmental perturbants can synergize with KRV to induce diabetes. The combination of KRV infection and a brief course of poly I:C (one incapable of inducing diabetes by itself) induces diabetes in 100% of BBDR rats (Ellerman et al. 1996; Zipris et al. 2003). Other TLR ligands can also synergize with KRV infection (Zipris et al. 2005); these include heat-killed *Escherichia coli* and *Staphylococcus aureus*, which are natural TLR agonists.

TLR ligation leads to a pleiotrophic immune response and activation of antigen presenting cells that could disequilibrate an immune system genetically predisposed to autoimmunity. Consistent with this view, both macrophage depletion (Chung et al. 1997) and inhibition of macrophage-derived iNOS by aminoguanidine (Mendez et al. 2004) render BBDR rats resistant to diabetes induction in response to KRV plus poly I:C. This suggests that upregulation of macrophage-derived proinflammatory cytokines and subsequent T cell activation are important for T1D triggered by KRV infection.

The Central Role of TLR9 in KRV-Induced Diabetes

The synergy of activators of innate immunity with KRV to enhance the penetrance of rat type 1 diabetes suggested that KRV itself might act via innate immunity (Lien and Zipris 2009). Consistent with this suggestion, KRV infection strongly stimulates BBDR splenocytes to produce proinflammatory cytokines including IL-6 and IL-12p40 (Zipris et al. 2007). Genomic KRV DNA also induced BBDR splenocytes to produce IL-12p40, and KRV-induced upregulation of B lymphocytes was blocked by TLR9 antagonists including inhibitory CpG oligonucleotide and chloroquine. The latter is thought to act via endosomal acidification, preventing signaling of TLRs localized in these organelles. Finally, in vivo treatment of BBDR rats with chloroquine reduced the incidence of KRV-induced diabetes and decreased circulating levels of IL-12p40 (Zipris et al. 2007). Taken together, these studies identify a role for TLR9 signaling in KRV-induced diabetes in BBDR rats.

Proinflammatory Gene Signature in KRV-Induced Diabetes

These data suggest that proinflammatory responses induced by KRV are linked with the process leading to islet destruction (Zipris et al. 2005). This is supported by the finding that down-modulation of virus-induced inflammation in BBDR rats correlates with lower disease frequency (Zipris et al. 2007). As shown in Table 7.2 and (Wolter et al. 2009), DNA microarray analyses indicate that KRV, but not H-1, alters the global gene expression profile in pancreatic lymph nodes. KRV induces transcripts for many genes associated with inflammation, including IL-1, IL-6, IL-12, IL-23, IL-18, and type I interferons in BBDR (Wolter et al. 2009) and LEW1.WR1 (Londono et al. 2010) rats. Interestingly, the priming dose of poly I:C that leads to increased penetrance of virus-triggered diabetes does not have the same effect on gene expression as does KRV alone. We hypothesize that poly I:C enhances diabetes penetrance via activation of as yet unknown genes or via post-translational mechanisms such as modulation of T effector and/or T regulatory cell functions (Lien and Zipris 2009; Pasare and Medzhitov 2003). We have also detected transcripts for IRF-7, CXCL-10, and KRV in pancreatic islets purified from LEW1.WR1 rats on day 5

Table 7.2 Upregulation of genes in pancreatic lymph nodes from KRV-infected BBDR rats

Affymetrix identifier	Gene name	Relative expression		Fold increase	*p*-value
		Uninfected	KRV infected		
Cytokines					
1387835_at	Interleukin 1 receptor antagonist (Il1ra)	6.065	459.87	75.81	0.032
1368592_at	Interleukin 1, alpha (Il1a)	76.40	329.39	4.311	0.001
1369031_at	Interleukin 18 binding protein (Il18bp)	593.85	2391.4	4.02	0.004
1369191_at	Interleukin 6 (Il6)	99.51	398.26	4	0.001
1369565_at	Interleukin 12b (Il12b)	116.55	366.36	3.14	0.05
1398256_at	Interleukin 1, beta (Il1b)	783.73	2429.39	3.1	0.003
1369665_a_at	Interleukin 18 (Il18)	1190.13	3076.44	2.58	0.001
Interferon pathways					
1376908_at	Interferon-induced protein with tetratricopeptide repeats 3 (Ifit3)	135.10	2738.2	20.26	0.009
1370790_at	Interferon, gamma (Ifng)	10.10	137.81	13.63	0.019
1383564_at	Interferon regulatory factor 7 (Irf7)	1552.71	20100.6	12.94	0.0001
Chemokines					
1379365_at	Chemokine (C–X–C motif) ligand 11 (Cxcl11)	126.91	7374.95	58.11	0.001
1387969_at	Chemokine (C–X–C motif) ligand 10 (Cxcl10)	158.78	4968.9	31.29	0.005
1382454_at	Chemokine (C–X–C motif) ligand 9 (Cxcl9)	444.56	11532.7	31.12	0.019
TLR pathways					
1387982_at	Toll-like receptor 4 (Tlr4)	116.10	555.30	4.78	0.003
1382531_at	Toll-like receptor 7 (Tlr7)	289.15	822	2.84	0.003
1368490_at	CD14 molecule (Cd14)	634.50	1715.67	2.70	0.016
NK response					
1379293_at	Granzyme A (Gzma)	90.12	2127	23.6	0.002
1370628_at	Granzyme B (Gzmb)	117.88	2715.52	23.03	0.005
1369872_a_at	Fc fragment of IgE, low affinity II, receptor for (CD23) (FCER2)	471.36	44.68	10.54	0.002
1368455_at	natural killer cell group 7 sequence (nkg7)	141.01	931.05	6.5	0.002
1368377_at	Granzyme C (Gzmc)	52.54	317.07	6.03	0.008
1370483_at	Similar to transmembrane NK cell receptor 2B4	167.56	379.06	2.26	0.026

Groups of three BBDR rats 22–25 days of age were left untreated, or treated with 1×10^7 PFU of KRV (Zipris et al. 2005). Total RNA was isolated from pancreatic lymph nodes removed from animals 5 days after virus infection. Differential expression analysis was evaluated by ANOVA with a cutoff threshold *p* value of 0.05 and a false detection rate (FDR) of 0.1. Values shown are the fold increase in the expression level of the indicated gene in lymph nodes from KRV-infected rats relative to the expression in control uninfected animals

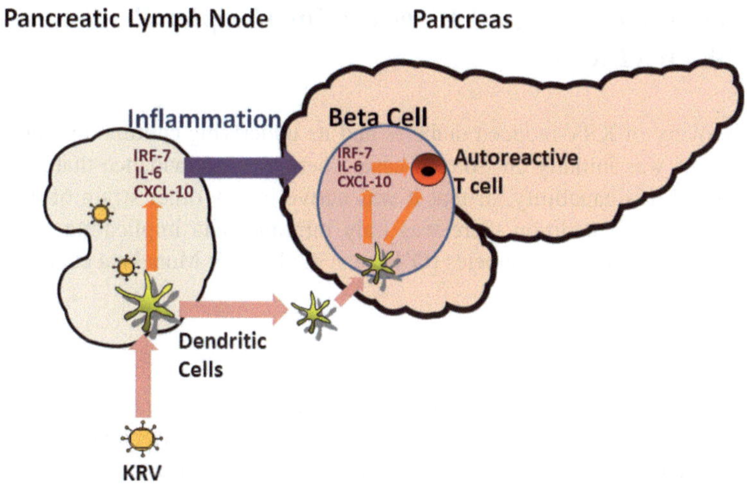

Fig. 7.2 Postulated mechanism of KRV-induced autoimmune diabetes in the rat. Infection with KRV leads to innate immune activation in pancreatic lymph nodes and beta cells shortly after virus infection. This response results in the production of proinflammatory cytokines and chemokines, i.e. IL-6 and CXCL-10, as well as IRF-7, a transcription factor associated with type I interferon production. The proinflammatory response induced in conjunction with the presentation of islet-derived antigens to T cells in the microenvironment of the pancreas and/or pancreatic lymph nodes culminate in the upregulation of anti-islet T cells and ultimately islet destruction. The relationship of the Iddm37 locus (Blankenhorn et al. 2009) to this process is currently under study

following virus infection, at which time inflammation is just detectable in the spleen and pancreatic lymph nodes (unpublished observations). As shown in Fig. 7.2, it is hypothesized that virus-induced diabetes involves upregulation of the innate immune system in islet beta cells shortly after infection (Wolter et al. 2009).

Anti-inflammatory Therapy Prevents Virus-Induced Diabetes

Further documentation of the role of inflammation in virus-triggered diabetes came from studies of dexamethasone, an anti-inflammatory/immunosuppressive glucocorticoid. Brief treatment with low-dose dexamethasone for 5 days beginning on the day of infection completely prevented diabetes (Londono et al. 2010). A short course of dexamethasone also prevented diabetes induced by poly I:C priming followed by KRV; strikingly, a single dose of dexamethasone given on the day of infection prevented disease in ~50% of the animals (Londono et al. 2010) without compromising the ability of the immune system to eliminate the virus from the host. Treatment of LEW.1WR1 rats with non-steroidal anti-inflammatory agents also prevented KRV-induced diabetes, but was less effective than dexamethasone (unpublished observations). Thus, targeting virus-induced inflammation before diabetes onset can attenuate autoimmunity directed against pancreatic islets.

Viral Triggering of Type 1 Diabetes: The Complex Responses of LEW.1WR1 Rats

The discovery of KRV-induced diabetes and its underlying cellular and molecular mechanisms was initially an isolated case. There was no evidence that any other virus shared this capability, or that it was active in any other strain of diabetes-susceptible rat. In addition there was only minimal data implicating parvoviral infection in human type 1 diabetes (O'Brayan et al. 2005; Munakata et al. 2005).

Epidemiology

To address these concerns, the Worcester group took advantage of the LEW.1WR1 rat, which develops diabetes spontaneously at a low rate, ~2.5%, and is susceptible to increased disease penetrance after Treg depletion or exposure to poly I:C (Mordes et al. 2005). It was found that infection with viruses from several families triggers autoimmune diabetes in these animals (Tirabassi et al. 2010). Both KRV and RCMV induced diabetes in up to 60% of LEW.1WR1 rats, whereas H-1, vaccinia, and Coxsackie B4 viruses did not. As is true for KRV in the BBDR rat, diabetes can be triggered by virus only in young LEW.1WR1 rats and does not involve direct infection of islets. By 45 days of age, susceptibility to viral triggering disappears. Histologically, pancreata from diabetic rats showed end-stage insulitis, but islets from identically inoculated rats that did not become diabetic showed almost no inflammation, suggesting that progression to diabetes after infection is a bimodal phenomenon.

The observation that CMV, a herpesvirus, induces diabetes in LEW.1WR1 but not BBDR rats is interesting for two reasons. First it demonstrates that the ability of virus to affect the incidence of type 1 diabetes is dependent on both host genetic background and the nature of the infecting agent. Second, CMV is a candidate pathogen in human type 1 diabetes (Pak et al. 1988), for the first time linking a human hypothesis of induction of type 1 diabetes in an animal model.

Immunological Perturbants Enhance Diabetogenicity of Viral Infection

As is the case with the BBDR rat, simultaneous inoculation of KRV and RCMV or inoculation of RCMV followed 4 days later by KRV-induced diabetes in nearly all animals, whereas inverting the sequence of infections prevented the disease (Tirabassi et al. 2010), underscoring how difficult it may be to identify an effect of infection on type 1 diabetes incidence. Pretreatment of rats with an activator of innate immunity increased the diabetogenicity of KRV but not RCMV, and was associated with a moderate rate of diabetes after either infection with vaccinia or coxsackie B4 virus (Tirabassi et al. 2010), the latter being a candidate for involvement in human type 1 diabetes (Jaidane and Hober 2008).

Genetics of Susceptibility to Virus-Triggered Type 1 Diabetes

Both BBDR and LEW.1WR1 rats share at least two major rat T1D susceptibility loci, the *RT1B/Du* class II MHC and the dominant non-MHC *Iddm14* locus (Blankenhorn et al. 2009). Both these type 1 diabetes loci are necessary for type 1 diabetes induced by Treg depletion or by virus inoculation (Blankenhorn et al. 2009). The identity of the gene underlying *Iddm14* is under study, and an excellent candidate gene is a single variable region in the T cell receptor beta-chain locus [(Mordes et al. 2009) and unpublished observations]. The susceptibility of BBDR and LEW.1WR1 rats to type 1 diabetes after infection with KRV, together with the resistance of other *RT1B/Du/Iddm14*-permissive rat strains, suggested a role for background genes.

To search for such genes, (LEW.1WR1 × WF)F2 rats were inoculated with KRV after poly I:C priming. WF rats are highly resistant to type 1 diabetes despite having the *RT1B/Du* class II MHC. The analysis revealed a new locus near the RT1 MHC, termed *Iddm37*, that is a major determinant of susceptibility to type 1 diabetes specifically in response to viral infection (Blankenhorn et al. 2009). Interestingly, one gene linked to autoimmune diabetes in mouse and human, UBD or diubiquitin, lies within this region. The UBD alleles of LEW.1WR1 and WF have been sequenced and both coding and noncoding nucleotide substitutions were found, including the insertion of a SINE element in the UBD promoter that diminishes significantly UBD expression in WF rats (Blankenhorn et al. 2009). The fact that this locus is important in type 1 diabetes triggered by viral infection but not by other triggers suggests that virus–*Iddm37* interactions may govern rat onset type 1 diabetes after infection.

Prevention of Virus-Triggered Type 1 Diabetes by Maternal Immunization

Inoculation of LEW.1WR1 dams with either KRV or the combination of KRV and RCMV prior to pregnancy can protect the progeny of these immunized animals from virus-induced diabetes in a virus-specific manner (Tirabassi et al. 2010). The outcome was interpreted as evidence of maternal immunization leading to prevention of virus-triggered autoimmunity. The suggestion that immunization can prevent at least some cases of autoimmune diabetes reanimates a longstanding theoretical prevention strategy for type 1 diabetes (Coon et al. 1999).

Conclusions

Exposure to viruses, including KRV, RCMV, vaccinia, and Coxsackievirus B4, can affect the incidence of autoimmune diabetes in genetically predisposed rats. This connection between infection and autoimmunity is complex. In the BBDP and LEW.1AR1-iddm strains, data support the “hygiene” hypothesis. Conversely, in

LEW.1WR1 and BBDR rats, data support the "accelerator" hypothesis. Unfortunately, apart from the analyses of KRV infection of BBDR rats, only limited mechanistic data are available to explain this complex set of phenomena.

Nonetheless, the observational data from the rat suggest possible explanations for some difficulty encountered in searches for evidence of viral modulation of human type 1 diabetes onset. First, rates of diabetes penetrance induced by viruses of different families can be quite low and variable. Effects may be detectable only in specific genetic backgrounds, making their detection in outbred populations difficult. In addition, the effect of infection on type 1 diabetes may depend on pre-initiation of autoimmunity (as shown in studies of poly I:C priming). On occasion, multiple infections might act in concert to precipitate diabetes (e.g., RCMV before KRV). Such animal data are consonant with the "fertile field hypothesis" (Von Herrath et al. 2003), which proposes that viral infection alone might not be able to induce human type 1 diabetes in the absence of other inflammatory factors. We suggest that additional studies in rats in which the genetic background can be held constant, the permissive cytokine milieu identified, and the immune environment manipulated may provide better understanding of viral infection in human type 1 diabetes.

Acknowledgements Supported in part by grants 7-11-BS-102 (JPM), 7-08-RA-106 (JPM) and 7-06-BS-18 (EPB) from the American Diabetes Association, 1-2006-745, 1-2007-584, and 5-2008-224 from the Juvenile Diabetes Research Foundation (DZ), and R21AI088480 (EBP) and DK32520 from the National Institutes of Health. The contents of this publication are solely the responsibility of the authors and do not necessarily represent the official views of the National Institutes of Health.

References

Åkerblom HK, Vaarala O, Hyöty H, Ilonen J, Knip M (2002) Environmental factors in the etiology of type 1 diabetes. Am J Med Genet 115:18–29

Arndt T, Wedekind D, Weiss H, Tiedge M, Lenzen S, Hedrich HJ, Jorns A (2009) Prevention of spontaneous immune-mediated diabetes development in the LEW.1AR1-iddm rat by selective CD8+ T cell transfer is associated with a cytokine shift in the pancreas-draining lymph nodes. Diabetologia 52:1381–1390

Blankenhorn EP, Cort L, Greiner DL, Guberski DL, Mordes JP (2009) Virus-induced autoimmune diabetes in the LEW.1WR1 rat requires *Iddm14* and a genetic locus proximal to the major histocompatibility complex. Diabetes 58:2930–2938

Brown DW, Welsh RM, Like AA (1993) Infection of peripancreatic lymph nodes but not islets precedes Kilham rat virus-induced diabetes in BB/Wor rats. J Virol 67:5873–5878

Chung YH, Jun H-S, Kang Y, Hirasawa K, Lee B-R, Van Rooijen N, Yoon J-W (1997) Role of macrophages and macrophage-derived cytokines in the pathogenesis of Kilham rat virus-induced autoimmune diabetes in diabetes-resistant BioBreeding rats. J Immunol 159:466–471

Chung YH, Jun HS, Son M, Bao M, Bae HY, Kang Y, Yoon JW (2000) Cellular and molecular mechanism for Kilham rat virus-induced autoimmune diabetes in DR-BB rats. J Immunol 165:2866–2876

Coon B, An LL, Whitton JL, Von Herrath MG (1999) DNA immunization to prevent autoimmune diabetes. J Clin Invest 104:189–194

Dyrberg T, Schwimmbeck PL, Oldstone MBA (1988) Inhibition of diabetes in BB rats by virus infection. J Clin Invest 81:928–931

Ellerman KE, Like AA (2000) Susceptibility to diabetes is widely distributed in normal class II^u haplotype rats. Diabetologia 43:890–898

Ellerman KE, Richards CA, Guberski DL, Shek WR, Like AA (1996) Kilham rat virus triggers T-cell-dependent autoimmune diabetes in multiple strains of rat. Diabetes 45:557–562

Filippi CM, Von Herrath MG (2008) Viral trigger for type 1 diabetes: pros and cons. Diabetes 57:2863–2871

Gepts W (1965) Pathology and anatomy of the pancreas in juvenile diabetes mellitus. Diabetes 14:619–633

Gillespie KM, Gale EAM, Bingley PJ (2002) High familial risk and genetic susceptibility in early onset childhood diabetes. Diabetes 51:210–214

Guberski DL (1994) Diabetes-prone and diabetes-resistant BB rats: animal models of spontaneous and virally induced diabetes mellitus, lymphocytic thyroiditis, and collagen-induced arthritis. Ilar News 35:29–37

Guberski DL, Thomas VA, Shek WR, Like AA, Handler ES, Rossini AA, Wallace JE, Welsh RM (1991) Induction of type 1 diabetes by Kilham's rat virus in diabetes resistant BB/Wor rats. Science 254:1010–1013

Hawa MI, Beyan H, Buckley LR, Leslie RDG (2002) Impact of genetic and non-genetic factors in type 1 diabetes. Am J Med Genet 115:8–17

Jacoby RO, Ball-Goodrich LJ, Besselsen DG, McKisic MD, Riley LK, Smith AL (1996) Rodent parvovirus infections. Lab Anim Sci 46:370–380

Jaidane H, Hober D (2008) Role of coxsackievirus B4 in the pathogenesis of type 1 diabetes. Diabetes Metab 34:537–548

Jörns A, Gunther A, Hedrich HJ, Wedekind D, Tiedge M, Lenzen S (2005) Immune cell infiltration, cytokine expression, and beta-cell apoptosis during the development of type 1 diabetes in the spontaneously diabetic LEW.1AR1/Ztm-iddm rat. Diabetes 54:2041–2052

Kawano K, Hirashima T, Mori S, Saitoh Y, Kurosumi M, Natori T (1991) New inbred strain of Long-Evans Tokushima lean rats with IDDM without lymphopenia. Diabetes 40:1375–1381

Leiter EH (2009) Type 1 diabetes genes in rats: few or many? Diabetes 58:796–797

Lenzen S, Tiedge M, Elsner M, Lortz S, Weiss H, Jorns A, Klöppel G, Wedekind D, Prokop CM, Hedrich HJ (2001) The LEW.1AR1/Ztm-*iddm* rat: a new model of spontaneous insulin-dependent diabetes mellitus. Diabetologia 44:1189–1196

Lien E, Zipris D (2009) The role of Toll-like receptor pathways in the mechanism of type 1 diabetes. Curr Mol Med 9:52–68

Like AA, Guberski DL, Butler L (1991) Influence of environmental viral agents on frequency and tempo of diabetes mellitus in BB/Wor rats. Diabetes 40:259–262

Londono P, Komura A, Hara N, Zipris D (2010) Brief dexamethasone treatment during acute infection prevents virus-induced autoimmune diabetes. Clin Immunol 135:401–411

McKisic MD, Paturzo FX, Gaertner DJ, Jacoby RO, Smith AL (1995) A nonlethal rat parvovirus infection suppresses rat T lymphocyte effector functions. J Immunol 155:3979–3986

Mendez II, Chung YH, Jun HS, Yoon JW (2004) Immunoregulatory role of nitric oxide in Kilham rat virus-induced autoimmune diabetes in DR-BB rats. J Immunol 173:1327–1335

Mordes JP, Bortell R, Groen H, Guberski DL, Rossini AA, Greiner DL (2001) Autoimmune diabetes mellitus in the BB rat. In: Sima AAF, Shafrir E (eds) Animal models of diabetes: a primer. Harwood Academic Publishers, Amsterdam, pp 1–41

Mordes JP, Guberski DL, Leif JH, Woda BA, Flanagan JF, Greiner DL, Kislauskis EH, Tirabassi RS (2005) LEW.1WR1 rats develop autoimmune diabetes spontaneously and in response to environmental perturbation. Diabetes 54:2727–2733

Mordes JP, Poussier P, Rossini AA, Blankenhorn EP, Greiner DL (2007) Rat models of type 1 diabetes: genetics, environment, and autoimmunity. In: Shafrir E (ed) Animal models of diabetes: frontiers in research. CRC Press, Boca Raton, pp 1–39

Mordes JP, Cort L, Norowski E, Leif JH, Fuller J, Lernmark Å, Greiner DL, Blankenhorn EP (2009) Analysis of the *Iddm14* rat diabetes susceptibility locus in multiple rat strains: identification of a susceptibility haplotype in the *Tcrb-V* locus. Mamm Genome 20:162–169

Morran MP, Omenn GS, Pietropaolo M (2008) Immunology and genetics of type 1 diabetes. Mt Sinai J Med 75:314–327

Munakata Y, Kodera T, Saito T, Sasaki T (2005) Rheumatoid arthritis, type 1 diabetes, and Graves' disease after acute parvovirus B19 infection. Lancet 366:780

Naik RG, Brooks-Worrell BM, Palmer JP (2009) Latent autoimmune diabetes in adults. J Clin Endocrinol Metab 94:4635–4644

Nejentsev S, Walker N, Riches D, Egholm M, Todd JA (2009) Rare variants of IFIH1, a gene implicated in antiviral responses, protect against type 1 diabetes. Science 324:387–389

O'Brayan TA, Beck MJ, Demers LM, Naides SJ (2005) Human parvovirus B19 infection in children with new onset Type 1 diabetes mellitus. Diabet Med 22:1778–1779

Pak CY, Eun HM, McArthur RG, Yoon JW (1988) Association of cytomegalovirus infection with autoimmune type 1 diabetes. Lancet 2:1–4

Pasare C, Medzhitov R (2003) Toll pathway-dependent blockade of $CD4^{+}CD25^{+}$ T cell-mediated suppression by dendritic cells. Science 299:1033–1036

Pearce SH, Merriman TR (2009) Genetics of type 1 diabetes and autoimmune thyroid disease. Endocrinol Metab Clin North Am 38:289

Rossini AA, Williams RM, Mordes JP, Appel MC, Like AA (1979) Spontaneous diabetes in the gnotobiotic BB/W rat. Diabetes 28:1031–1032

Schaub B, Lauener R, Von Mutius E (2006) The many faces of the hygiene hypothesis. J Allergy Clin Immunol 117:969–977

Serreze DV, Leiter EH (2001) Genes and pathways underlying autoimmune diabetes in NOD mice. In: Von Herrath MG (ed) Molecular pathology of insulin dependent diabetes mellitus. Karger, New York, pp 31–67

Tirabassi RS, Guberski DL, Blankenhorn EP, Leif JH, Woda BA, Liu Z, Winans DA, Greiner DL, Mordes JP (2010) Infection with viruses from several families triggers autoimmune diabetes in LEW.1WR1 rats: prevention of diabetes by maternal immunization. Diabetes 59:110–118

Von Herrath MG, Fujinami RS, Whitton JL (2003) Microorganisms and autoimmunity: making the barren field fertile? Nat Rev Microbiol 1:151–157

Wolter TR, Wong R, Sarkar SA, Zipris D (2009) DNA microarray analysis for the identification of innate immune pathways implicated in virus-induced autoimmune diabetes. Clin Immunol 132:103–115

Yokoi N, Komeda K, Wang HY, Yano H, Kitada K, Saitoh Y, Seino Y, Yasuda K, Serikawa T, Seino S (2002) *Cblb* is a major susceptibility gene for rat type 1 diabetes mellitus. Nat Genet 31:391–394

Yoon J-W, Jun H-S (2004) Role of viruses in the pathogenesis of type 1 diabetes mellitus. In: LeRoith D, Taylor SI, Olefsky JM (eds) Diabetes mellitus. A fundamental and clinical text. Lippincott Williams & Wilkins, Philadelphia, pp 575–590

Zipris D (2009) Epidemiology of type 1 diabetes and what animal models teach us about the role of viruses in disease mechanisms. Clin Immunol 131:11–23

Zipris D, Hillebrands JL, Welsh RM, Rozing J, Xie JX, Mordes JP, Greiner DL, Rossini AA (2003) Infections that induce autoimmune diabetes in BBDR rats modulate $CD4^{+}CD25^{+}$ T cell populations. J Immunol 170:3592–3602

Zipris D, Lien E, Xie JX, Greiner DL, Mordes JP, Rossini AA (2005) TLR activation synergizes with Kilham rat virus infection to induce diabetes in BBDR rats. J Immunol 174:131–142

Zipris D, Lien E, Nair A, Xie JX, Greiner DL, Mordes JP, Rossini AA (2007) TLR9-signaling pathways are involved in Kilham rat virus-induced autoimmune diabetes in the biobreeding diabetes-resistant rat. J Immunol 178:693–701

Chapter 8
Reovirus

Takashi Onodera and Toshiharu Hayashi

Abstract Initial studies examined the effect of reovirus types 1 and 3 isolated from humans. Subsequently viruses were passaged in primary β-cell cultures from suckling mice. Animals were infected with the virus. Viral infection led to reduction in pancreatic insulin levels which persisted for 2-week period and associated with abnormal levels of glucose tolerance tests. Histopathological examinations of the pancreas showed areas of focal, coagulative necrosis of some of the islets and mononuclear cell infiltrate from 5 days after infection. Electron microscopy revealed the presence of reticulogranular matrix which was characteristics of reovirus growth in the cytoplasm. Using double antibody labeling, it was possible to demonstrate the presence of reovirus antigen in insulin-containing β-cells. Mice infected with reovirus type 2, isolated from a cow, also developed diabetes as an immune-mediated syndrome. Reovirus type 2 without adaptation to pancreatic β-cells infects islet cells and some acinar cells in newborn mice. The animals develop insulitis with hypoinsulinemia and impaired glucose tolerance in mice approximately 2 weeks after infection.

Introduction

Mammalian reoviruses are nonenveloped viruses that contain a genome of ten double-stranded RNA segments (Nibert and Schiff 2001). Reoviruses replicate in the cytoplasm of host cells and produce cell death by apoptosis (DeBiasi et al. 2001;

T. Onodera, D.V.M., Ph.D. (✉)
Department of Risk Assessment Science, Research Center for Food Safety,
University of Tokyo, Yayoi 1-1-1, Bunkyo-ku, Tokyo 113-8657, Japan
e-mail: aonoder@mail.ecc.u-tokyo.ac.jp

T. Hayashi
Faculty of Agriculture, Department of Veterinary Pathology,
Yamaguchi University, Yoshida 1677-1, Yamaguchi 753-8515, Japan

K.W. Taylor, H. Hyoty, A. Toniolo, A.J. Zuckerman (eds.), *Diabetes and Viruses*,
DOI 10.1007/978-1-4614-4051-2_8,

Oberhaus et al. 1997; Tyler et al. 1995). Mammals serve as hosts for reovirus infection, but disease is restricted to very young individuals, including mild gastroenteritis (Tyler et al. 1995; Zurney et al. 2009). In vitro, human β-cells are susceptible to reovirus type 3 (Reo-3) infection (Yoon et al. 1981) and biliary atresia (the most common disorders in infants) is characterized by a progressive inflammatory sclerosing cholangitis. Increasing evidence indicates that Reo-3 and rotavirus infections can cause human infantile biliary atresia (Wilson et al. 1994). There is little evidence for the involvement of reovirus in the pathogenesis of human type 1 diabetes, but the three reovirus types provide excellent models for studying the pathogenesis of endocrine disorders in neonatal mice (Samuel 1998). Reovirus is capable of establishing persistent infection in cultured cells, but not in immunocompetent animals (Morrison et al. 1993). Mice infected with reovirus develop endocrine abnormalities, including growth hormone deficiency (Onodera et al. 1981), hypothyroidism (Onodera et al. 1990), and diabetes mellitus (Onodera et al. 1978, 1981). In this chapter, results are summarized on reovirus-induced diabetes and thyroiditis. Results on reovirus type 2-induced autoimmune diabetes in newborn mice will be also presented. The BN-77 reovirus type 2 strain used in these studies was isolated from a cow with diarrhea (Kurogi et al. 1980).

Diabetes Mellitus

The hypothesis is examined that viruses may induce type 1 diabetes in humans. Initial studies examined the effects in mice of a human isolate of Reo-3. Subsequently, Reo-3 was passaged in primary β-cell cultures from suckling mice. Animals were infected with 10^5 PFU. Viral infection led to a fall of pancreatic insulin levels persisting for 2 weeks and associated with abnormal glucose tolerance tests. Histopathological examination of the pancreas showed areas of focal coagulative necrosis in some of the islets together with mononuclear cell infiltrate starting at 5 days post-infection (Figs. 8.1 and 8.2). Electron microscopy revealed a reticulogranular matrix characteristic of reovirus assembly into the cytoplasm. Double antibody labeling showed that reovirus antigens co-localized with insulin-containing β-cells (Onodera et al. 1978).

A more extensive study examined the endocrine effects of infection with reovirus type 1 (Onodera et al. 1981). Infected mice demonstrated impaired glucose tolerance tests and a decrease in pancreatic immunoreactive insulin. Focal necrosis and disruption of pancreatic islets were also detected. Approximately 20% of cells in each islet were positive for reovirus antigens (Fig. 8.3). Electron microscopy revealed that viral particles were located primarily in β-cells, but α- and δ-cells were also infected (Fig. 8.4). Significant numbers of mice exhibited growth retardation, suggesting that the pituitary gland could also be a target for Reo-1. Examination of the pituitary by electron microscopy revealed viral particles in growth hormone-producing cells. Mice exhibiting growth retardation had on the average a more than 50% reduction of circulating plasma growth hormone levels.

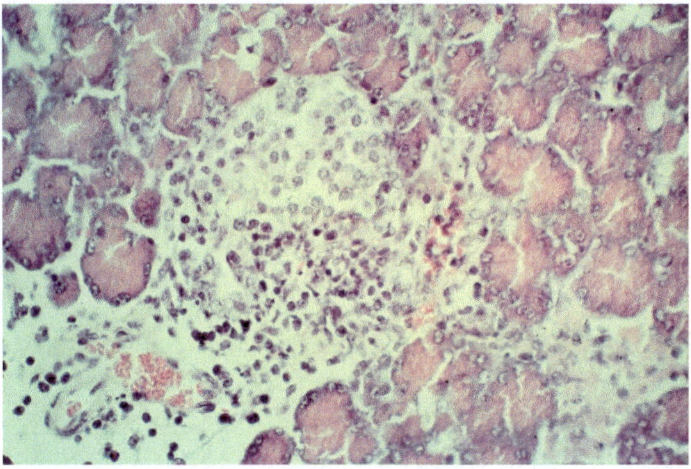

Fig. 8.1 Microscopic changes in pancreatic islets of Langerhans of mice infected with reovirus type 3. Infiltration of mononuclear round cells was observed 5 days after infection

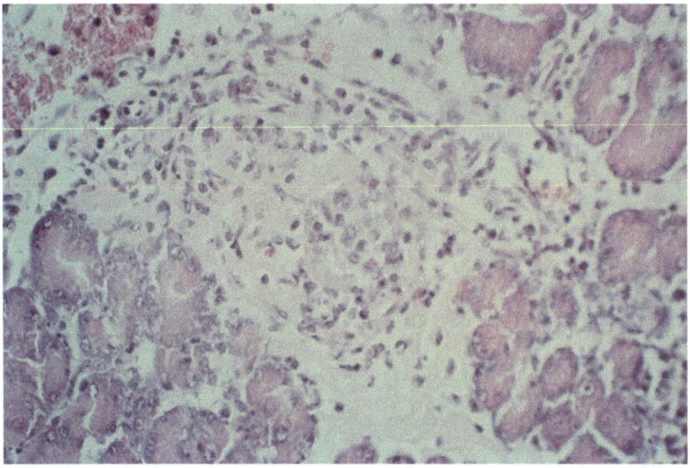

Fig. 8.2 Microscopic changes in pancreatic islets of Langerhans of mice infected with reovirus type 3. Coagulative necrosis observed 7 days after infection

Immunosuppression affects strongly the host response to reovirus infection (Onodera et al. 1982). Mice infected with reovirus type 1 were treated with rabbit anti-mouse thymocyte serum, rabbit anti-mouse lymphocyte serum, or cyclophosphamide (20 mg/kg). Upon infection, immunosuppressed mice failed to develop glucose intolerance. Development of the runting syndrome was also suppressed by the treatment of anti-lymphocyte serum or cyclophosphamide. Immunosuppression

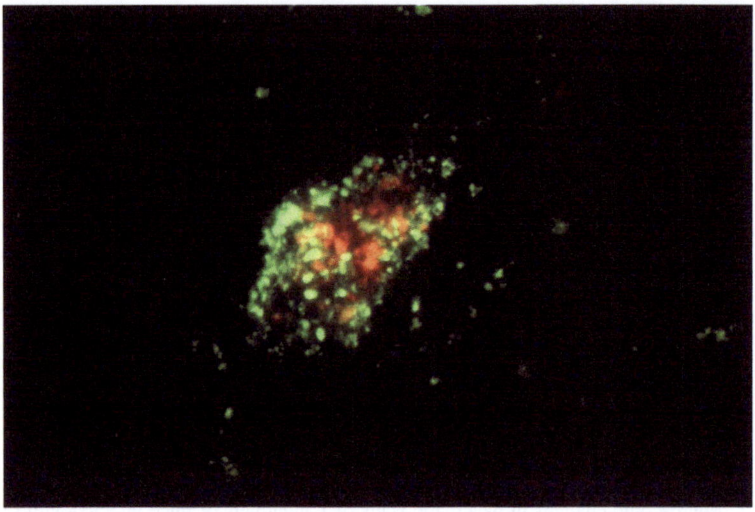

Fig. 8.3 Frozen sections from pancreas of mice 7 days after infection with reovirus type 1. Double immunofluorescence: FITC-labeled antibody to reovirus type 1 (*green*) and TRITC-labeled antibody to insulin (*red*). Viral antigen was seen in many cells throughout the islets

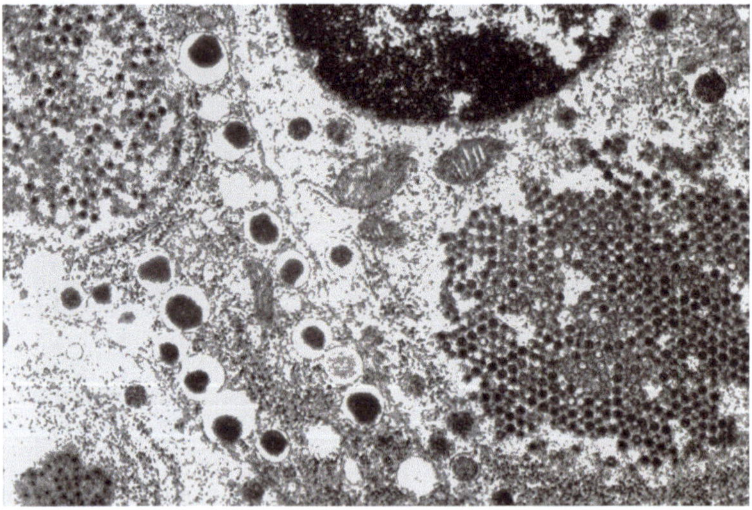

Fig. 8.4 Electron microscopy changes in islets of Langerhans of mice infected with reovirus type 1. Pancreatic beta cells with insulin granules and reticulogranular matrix containing developing reovirus particles

did not alter the pattern of viral growth in liver, heart, and pancreas and reduced, but did not eliminate the antibody response to virus.

Mice infected with reovirus type 2 also developed diabetes as an immune-mediated syndrome (Hayashi et al. 1995). Reovirus type 2 (without adaptation to pancreatic β-cells) does infect islet cells and pancreatic acinar cells of newborn mice. NC mice develop insulitis with hypoinsulinemia and glucose intolerance approximately

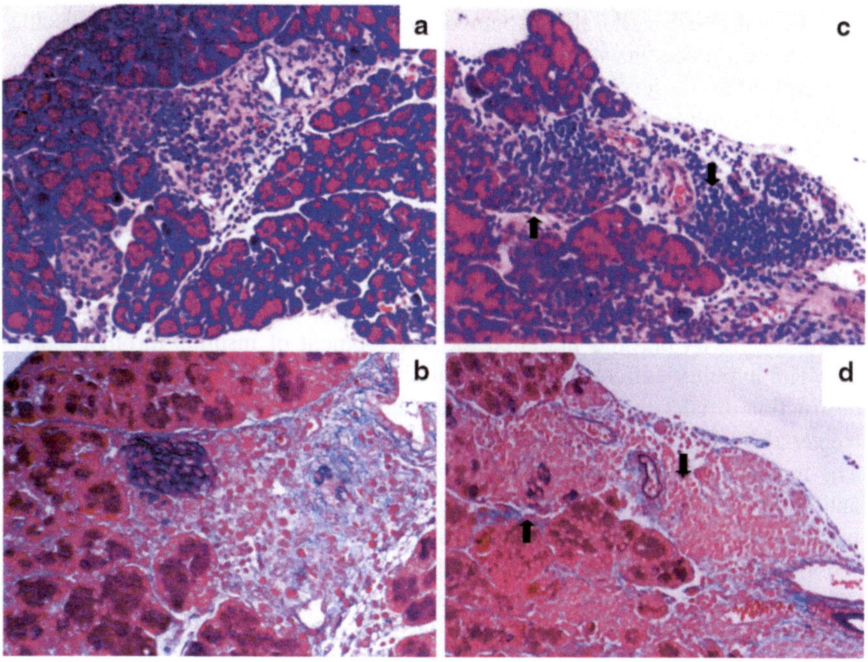

Fig. 8.5 Mild (**a** and **b**) and severe (**c** and **d**, *arrows*) insulitis in mice infected with reovirus type 2 and treated with CpG oligodeoxynucleotides 17 days after infection (Hayashi et al. 2002a, b). (**a** and **c**) Hematoxylin eosin staining. (**b** and **d**) Aldehyde fuchsin staining

2 weeks post-infection (Onodera et al. 1990). In these mice, no prominent pathologic changes were observed in organs other than the pancreas, with the exception of liver where small focal necrosis and neutrophil/lymphocyte infiltration was observed (Onodera et al. 1989). No evidence of virus growth was observed in the thymus and spleen of NC mice. From the pancreas, 10^{2-3} and 10^{4-6} PFU of virus were recovered on days 3 and 5, respectively. On day 7, virus titers were decreasing to 10^{2-3} PFU; no virus was isolated from day 14 onwards (Onodera et al. 1989). Thymic atrophy with the production of antinuclear-antibodies suggested that Reo-2 could trigger autoimmunity (Onodera et al. 1991). When autoimmunity-prone DBA/1 J mice (carrying the VAS-1 gene that is associated with autoimmunity; Holmdahl et al. 1990) were infected with reovirus type 2, mild insulitis developed in association with glucose intolerance (Hayashi et al. 2001). Shortly after infection, viral antigens and virus particles were detected in β- and α-cells (unpublished observations) without development of glucose intolerance. Virus was rapidly cleared in association with the production of interferon α/β (Samuel 1998) and of neutralizing antibody (Hayashi et al. 2002a, b). Thereafter, helper T(Th)1 dominant responses (interferon-γ, IL-12, IL-18), together with sensitized cytotoxic T (Tc) cells against β-cell antigens, could led to the destruction of β-cells (Hayashi 1997). In addition, the cellular immune response enhanced the expression of major histocompatibility complex (MHC) class I and class II antigens (Fig. 8.5) that was

elicited by IFN-α/β and/or IFN-γ (Samuel 1998). In addition, cell adhesion molecules (e.g., intercellular adhesion molecule-1, lymphocyte function associated antigen-1) participated to the development of insulitis (Hayashi et al. 1995). During the early phase of insulitis, no virus antigens or particles could be detected in islet cells. Virus isolation from pancreas was also negative. Treatment with monoclonal antibody against mouse IFN-γ inhibited the development of insulitis and the increase of blood glucose levels (Hayashi et al. 1995). Enhanced mRNA expression for IFN-γ (but not IL-4) in spleen cells from Reo-2 infected mice was detected shortly after infection. During the early phase of insulitis, cytotoxic T cells against β-cells were demonstrated (Hayashi et al. 2001). In this model, treatment with an IL-4 expressing DNA plasmid inhibited the development of insulitis through mutual Th1/Th2 inhibitory effects (Hayashi et al. 2003). The role of autoantibodies in the destruction of islet cells remains unclear (Hayashi et al. 2002a, b). Thus, insulitis triggered by Reo-2 can be induced by autoimmune mechanisms rather than by Th1-mediated virus clearance mechanisms. As reported above, in contrast with its antiviral role, virus-induced IFN-α/β appeared deleterious, acting as initiator of β-cell-specific autoimmunity (Hober and Sauter 2010). There are a number of reports on the pathogenic role of IFN-α/β in autoimmune pathology (Mathian et al. 2005) that associate IFN production with Th1 diseases such as lupus erythematosus (Crow and Kirou 2004) and type 1 diabetes in humans (Selmi et al. 2006; Hayashi 2010a, b).

The majority of mice infected with Reo-2 recovered from overt disease within 2–3 weeks. However, when DBA/1 J mice were infected with Reo-2 and treated with synthetic CpG olygodeoxynucleotides that enhance Th1 reactions (Hasegawa and Hayashi 2003), the animals developed more prolonged and severe insulitis as compared to infected but untreated mice (Hayashi et al. 2002a, b). These observations suggest that environmental factors such as bacterial infections may act as enhancing factors of insulitis. In addition, there may be no direct association of Reo-2 with the development of insulitis, since the viral association with type 1 diabetes can be multifactorial. The "one organism–one disease" paradigm that is central to Koch's postulates might not invariably apply to microbe-induced autoimmune disease as proposed originally by von Herrath et al. (2003) with the "fertile field hypothesis." A viral infection alone might not be able to cause overt autoimmunity, but might be an essential precipitator or modulator when the required predisposing environmental events have occurred. These events include, but are not restricted to, the generation of sufficient numbers of autoreactive cells systemically and the activation of antigen-presenting cells in the pancreas. More importantly, in terms of viral etiology being responsible for type 1 diabetes, it was pointed out that despite a large body of evidence describing associations between viruses and type 1 diabetes in individuals who are genetically prone, a clear identification of causative infectious agents has not been successful so far. If that were the case, viral footprints might be hard to detect systemically or in the target organ once autoimmunity has been initiated, and different infections might have to act in concert for precipitating overt autoimmunity. In such a context, it is possible that one single viral event (e.g., reovirus type 2 infection in early life) encountered before the onset of type 1 diabetes—possibly unrelated to other infectious events and not necessarily cross-reactive with self-antigens—may have the ability to trigger the final cascade leading to complete β-cell destruction and diabetes when sufficient damage is inflicted to β-cells (Christen and von Herrath 2004).

Elimination of $CD4^+CD25^+$ Treg cells in DBA/1 J mice infected with reovirus type 2 (in addition to treatment with synthetic CpG) induced severe insulitis with the production of autoantibodies against islet cells (Hayashi et al. 2006) including anti-GAD65 antibodies (unpublished data). Infection and autoimmunity were followed by overt hyperglycemia and the majority of mice died with severe diarrhea, oily hair, and emaciation. These observations suggest that exposure to bacterial components (exogenous environmental factors) and functional defects of peripheral tolerance (intrinsinc factors) may contribute to the development of type 1 diabetes in reovirus type 2 infection. In addition, it has been suggested that the impaired function, rather than the numbers of $CD4^+CD25^+$ T-cells, can be associated with type 1 diabetes in humans (Brusko et al. 2005). Originally $CD4^+CD25^+$ T cells were thought as maintaining immune tolerance (Sakaguchi et al. 1995) and controlling autoimmune diabetes in NOD mice (Salomon et al. 2000) as well as other autoimmune diseases (Toubi 2008). Subsequently, it was stressed that second signals (e.g., infectious agents) together with the depletion of $CD4^+CD25^+$ T cells were needed for development of autoimmune diseases (McHugh and Shevach 2002). Our studies seem to support this hypothesis since no growth retardation, insulitis, or clinically overt diabetes occurred in the control group treated only with a monoclonal antibody against $CD4^+CD25^+$ T cells. The experiments indicated that the mere depletion of Treg cells was not sufficient to induce autoimmune disease. Thus, the development of reovirus type 2-associated autoimmune diabetes in DBA/1 J mice can be due to the combined effect of $CD4^+CD25^+$ T cell depletion and enhanced Th1 activity consequent to CpG treatment. Tregs modulate the response to autoantigens and probably play a role in the pathogenesis of type 1 diabetes (Łuczyński et al. 2010). On the other hand, the function of Tregs in patients with type 1 diabetes remains unclear. Thus, further study is needed to clarify this point for Treg function in type 1 diabetes, since Tregs represent an active mechanism for suppressing autoreactive T cells that escape central tolerance (Łuczyński et al. 2010).

Humoral Autoimmunity

Since reovirus infections are associated with autoimmunity, we investigated the possibility that endocrine disorders observed in reovirus type 1 infected mice were mediated through the release of autoantigens from virus-infected cells. Using indirect immunofluorescence in paraformaldehyde-fixed sections, autoantibodies against cells of anterior pituitary, pancreatic islets, and gastric mucosa were detected. No autoantibodies to adrenal, thyroid, ovary, and testis cells were detected. Induction of autoantibodies was age dependent. One-hundred percent of mice infected at 3 days of age developed autoantibodies against growth hormone. Mice infected at 30 days of age failed to develop these autoantibodies.

The development of pituitary autoimmunity was specific to reovirus type 1. In fact, Reo-3 infection failed to induce pituitary autoantibodies. Using reovirus reassortants it was shown that the ability to induce autoantibodies to growth hormone-producing cells was a property associated with the virus S1 gene that codes for the hemagglutinin and determines the viral tropism. The reovirus type 1 S1 gene targets virus to the anterior pituitary as well as ependymal cells. Presumably, growth of virus in pituitary cells is required (but not sufficient) for the development of autoantibody to growth hormone and pituitary cells. However, the molecular mechanism by which reovirus induces autoantibodies remains obscure. Reo-3 is capable of replicating in β-cells, but fails to induce autoantibodies (Onodera et al. 1981).

Spleen cells from reovirus type 1-infected mice that develop autoantibodies were used to produce hybridomas secreting monoclonal autoantibodies to endocrine tissues (Haspel et al. 1983). Monoclonal antibodies reacted not only against pancreas, pituitary, stomach cells, but also with cell nuclei. Several hybridoma clones produced autoantibodies against islet cells. These antibodies reacted with cells at the periphery of islets. Only one hybridoma clone produced antibodies reactive with the central portion of the islet where β-cells are located. This monoclonal antibody had high activity against rat insulin. Similarly, some of the antibodies directed against pituitary cells reacted with epitopes of growth hormone.

Thyroiditis

When mice were infected with reovirus type 1 in the neck region, virus antigens were detected by immunofluorescence in subcutaneous tissue (Onodera and Awaya 1990; Srinivasappa et al. 1988). Although initial studies failed to show thyroid autoimmunity, reovirus type 1 does cause autoimmune thyroiditis in SJL/J and BALB/c mice infected intraperitoneally. These animals developed thyroiditis associated with autoantibodies to thyroglobulin. When mice were infected subcutaneously, focal thyroiditis with mononuclear cell infiltration was seen. However, no changes in thyroid function were observed. Treatment with synthetic thymic factor suppressed the autoantibody response. Reo-3 was unable to infect the thyroid and to induce thyroid autoimmunity. The differential ability of reovirus type 1 and Reo-3 to infect the thyroid was mapped to the viral hemagglutinin (S1 gene product), indicating that the hemagglutinin itself targets the virus to thyroid cells.

In conclusion, experiments in mice with the three reovirus types indicate that these agents can trigger a vast array of consequences ranging from organ-specific to systemic autoimmunity. In some cases, the ability of virus to induce autoimmune reactions could be mapped to the viral genes that govern viral tropism.

References

Brusko TM, Wasserfall CH, Clare-Salzler MJ, Schatz DA, Atkinson MA (2005) Functional defects and the influence of age on the frequency of CD4+CD25+ T-cells in type 1 diabetes. Diabetes 54:1407–1414

Christen U, von Herrath MG (2004) Induction, acceleration or prevention of autoimmunity by molecular mimicry. Mol Immunol 40:1113–1120

Crow MK, Kirou KA (2004) Interferon-α in systemic lupus erythematosus. Curr Opin Rheumatol 16:541–547

DeBiasi RL, Edelstein CL, Sherry B, Tyler KL (2001) Calpain inhibition protects against virus-induced apoptotic myocardial injury. J Virol 75:351–361

Hasegawa K, Hayashi T (2003) Synthetic CpG oligodeoxynucleotides accelerate the development of lupus nephritis during preactive phase in NZBxNZWF1 mice. Lupus 12:838–845

Haspel MV, Onodera T, Prabhakar BS, Horita M, Suzuki H, Notkins AL (1983) Virus-induced autoimmunity: monoclonal antibodies that react with endocrine tissues. Science 220:304–306

Hayashi T (1997) Mechanisms of development of insulitis with abnormal glucose tolerance in reovirus type 2-infected mice. Jpn J Vet Res 44:224–225

Hayashi T (2010a) Therapeuticstrategies for SLE involving cytokines: mechanism-oriented therapies especially IFN-γ targeting gene therapy. J Biomed Biotechnol. doi:10.1155/2010/461641

Hayashi T (2010b) Environmental factors and concepts in the induction of type 1 diabetes. In: Aucoin L, Prideus T (eds) Handbook of type 1 diabetes mellitus: etiology, diagnosis, and treatment. Nova Scientific, New York, pp 1–45

Hayashi T, Yamamoto S, Onodera T (1995) Prevention of reovirus type 2-induced diabetes-like syndrome in DBA/1 suckling mice by treatment with antibodies against intercellular adhesion molecule-1 and lymphocyte function associated antigen-1. Int J Exp Pathol 76:403–409

Hayashi T, Morimoto M, Iwata H, Onodera T (2001) Possible involvement of IL-12 in reovirus type-2-induced diabetes in newborn DBA/1 mice. Scand J Immunol 53:572–578

Hayashi T, Yoshinaka K, Hasegawa K, Maeda K, Onodera T (2002a) CpG oligodeoxynucleotides accelerate reovirus type 2-induced insulitis in DBA/1suckling mice. Int J Exp Pathol 83:217–223

Hayashi T, Anayama H, Hasegawa K, Iwata H, Onodera T (2002b) Systemic administration of histamine reduces reovirus type-2-induced insulitis in suckling DBA/1 mice. J Comp Pathol 128:153–160

Hayashi T, Yasutomi Y, Hasegawa K, Sasaki Y, Onodera T (2003) Interleukin-4-expressing plasmid DNA inhibits reovirus type-2-triggered autoimmune insulitis in DBA/1 J suckling mice. Int J Exp Pathol 84:101–106

Hayashi T, Hasegawa K, Sasaki Y, Onodera T (2006) Elimination of CD4+CD25+ regulatory T cells breaks down reovirus type 2-Triggered and CpG ODN-Induced prolonged mild autoimmune insulitis in DBA/1 mice. Scand J Immunol 63:116–124

Hober D, Sauter P (2010) Pathogenesis of type 1 diabetes mellitus: interplay between enterovirus and host. Nat Rev Endocrinol 6:279–289

Holmdahl R, Andersson M, Goldschmidt TJ, Gustafsson K, Jansson L, Mo JA (1990) Type II collagen autoimmunity in animals and provocations leading to arthritis. Immunol Rev 118: 193–232

Kurogi H, Inaba Y, Takahashi E, Sato K, Omori T (1980) Serological relationship between Japanese strains of reovirus of bovine origin and type 2 reovirus strains. Natl Inst Anim Health Q (Tokyo) 20:32–33

Łuczyński W, Stasiak-Barmuta A, Juchniewicz A, Wawrusiewicz-Kurylonek N, Iłendo E, Kos J, Kretowski A, Gûrska M, Chyczewski L, Bossowski A (2010) The mRNA expression of pro- and anti-inflammatory cytokines in T regulatory cells in children with type 1 diabetes. Folia Histochem Cytobiol 48:93–100

Mathian A, Weinberg A, Gallegos M, Banchereau J, Koutouzov S (2005) IFN-α induces early lethal lupus in preautoimmune (New Zealand Black x New Zealand White) F1 but not in BALB/c mice. J Immunol 74:2499–2506

McHugh RS, Shevach EM (2002) Cutting edge: depletion of CD4⁺CD25⁺ regulatory T cells is necessary, but not sufficient, for induction of organ-specific autoimmune disease. J Immunol 168:5979–5983
Morrison LA, Fields BN, Dermody TS (1993) Prolonged replication in the mouse central nervous system of reoviruses isolated from persistent infected cultures. J Virol 67:3019–3026
Nibert ML, Schiff LA (2001) Reovirus and their replication. In: Knipe DM, Howley PM (eds) Fields virology, 4th edn. Lippincott Wiliams & Wilkins, Philadelphia, p 1679
Oberhaus SM, Smith RL, Clayton GH, Dermody TS, Tyler KL (1997) Reovirus infection and tissue injury in the mouse central nervous systems are associated with apoptosis. J Virol 71:2100–2106
Onodera T, Jenson AB, Yoon J-W, Notkins AL (1978) Virus-induced diabetes mellitus: reovirus infection of pancreatic beta cells in mice. Science 201:529–531
Onodera T, Toniolo A, Ray UR, Jenson AB, Knazek RA, Notkins AL (1981) Virus-induced diabetes mellitus. XX. Polyendocrinopathy and autoimmunity. J Exp Med 153:1457–1473
Onodera T, Ray UR, Melez KA, Suzuki H, Toniolo A, Notkins AL (1982) Virus induced diabetes mellitus: autoimmunity and polyendocrine disease prevented by immunosuppression. Nature 297:66–68
Onodera T, Taniguchi T, Yoshihara K, Sato M, Homma S (1989) Pathogenicity of a reovirus type 2 in suckling mice: insulitis and lymphoid atrophy. Jpn J Vet Sci 51:1071–1074
Onodera T, Taniguchi T, Yoshihara K, Shimizu S, Satoh M, Hayashi T (1990) Reovirus type 2-induced diabetes in mice prevented by immunosuppression and thymic hormone. Diabetologia 33:192–196
Onodera T, Taniguchi T, Yoshihara K, Shimizu S, Satoh M, Hayashi T (1991) Thymic atrophy in type 2 reovirus infected mice; Immunosuppression and effects of thymic hormone. Thymus 18:95–109
OnoderaT AA (1990) Anti-thyroglobulin antibody induced with recombinant reovirus infection in BALB/c mice. Immunology 71:581–585
Sakaguchi S, Sakaguchi N, Asano M, Itoh M, Toda M (1995) Immunologic tolerance maintained by activated T cells expressing IL-2 receptor α-chains (CD25): breakdown of a single mechanism of self-tolerance causes various autoimmune diseases. J Immunol 155:1151–1164
Salomon B, Lenschow DL, Rhee L, Ashourian N, Singh B, Sharpe A, Bluestone JA (2000) B7/CD28 costimulation is essential for the homeostasis of the CD4⁺CD25⁺ immunoregulatory T cells that control autoimmune diabetes. Immunity 12:431–440
Samuel CE (1998) Reoviruses and the interferon system. Curr Top Microbiol Immunol 233:125–145
Selmi C, Lleo A, Zuin M, Podda M, Rossaro L, Gershwin EM (2006) Interferon alpha and its contribution to autoimmunity. Curr Opin Investig Drugs 7:451–456
Srinivasappa J, Garzelli C, Onodera T, Ray UR, Notkins AL (1988) Virus induced thyroiditis. Endocrinology 122:563–566
Toubi E (2008) The role of CD4⁺CD25⁺ T regulatory cells in autoimmune diseases. Clin Rev Allergy Immunol 34:338–344
Tyler KL, Squier MK, Rodgers SE, Schneider BE, Oberhaus SM, Grdina TA, Cohen JJ, Dermody TS (1995) Differences in the capacity of reovirus strains to induce apoptosis are determined by the viral attachment protein σ1. J Virol 69:6972–6979
Von Herrath MG, Fujinami RS, Whitton LJ (2003) Microorganisms and autoimmunity: making the barren field fertile? Nat Rev Microbiol 1:151–157
Wilson GA, Morrision LA, Fields BN (1994) Association of the reovirus S1 gene with serotype 3-induced-biliary atresia in mice. J Virol 68:6458–6465
Yoon J-W, Selvaggio S, Onodera T, Wheeler J, Jenson AB (1981) Infection of cultured human pancreatic cell with reovirus type 3. Diabetologia 20:462–467
Zurney J, Kobayashi T, Holm GH, Dermody TS, Sherry B (2009) Reovirus mu2 protein inhibits interferon signaling through a novel mechanism involving nuclear accumulation of interferon regulatory factor 9. J Virol 83:2178–2187

Chapter 9
Ljungan Virus and Diabetes

Martin Blixt, Stellan Sandler, and Bo Niklasson

Abstract The LV, a member of the *Parechovirus* genus and the *Picornaviridae* family, was isolated originally from a wild reservoir, the bank vole (*M. glareolus*). LV is associated with myocarditis, encephalitis, pregnancy-related diseases, and diabetes in wild rodents. The same pathological conditions can be induced in mice when examined in the laboratory. Several mouse and rat animal models used in diabetes research have been found to be carriers of the LV. The role of the viruses in human diabetes pathogenesis is conjectural. Variations in the incidence of human type 1 diabetes have been found to correlate with the fluctuations in native rodent populations in central Sweden. Increased prevalence of antibodies to LV has been detected in Swedish type 1 diabetes cases. However, LV has not been isolated and its viral RNA has not been detected in type 1 diabetes patients. Investigations of the pancreatic islets in infected animal have shown virus particles in the pancreatic islets, altered islet function, and changed islet structure. Treating infected animals with antiviral compounds results in retardation of progression of the disease.

Discovery of the Ljungan Virus

It has been proposed that lethal myocarditis and type 1 diabetes in humans may be caused by one or more infectious agents carried by rodents, based on the association between rodent density and disease incidence (Niklasson et al. 1998). This hypothesis

M. Blixt • S. Sandler
Department of Medical Cell Biology, Uppsala University,
Box 571, SE-751 23 Uppsala, Sweden

B. Niklasson, M.D., Ph.D. (✉)
Department of Medical Cell Biology, Uppsala University,
Box 571, SE-751 23 Uppsala, Sweden

Apodemus AB, Nordenflychtsvägen 64, 112 52 Stockholm, Sweden
e-mail: bo.niklasson@apodemus.se

K.W. Taylor, H. Hyoty, A. Toniolo, A.J. Zuckerman (eds.), *Diabetes and Viruses*,
DOI 10.1007/978-1-4614-4051-2_9,

initiated a search for an etiologic agent to the diseases mentioned above in small rodents. Ljungan virus (LV) is a member of the *Parechovirus* genus in the family *Picornaviridae*. It was subsequently isolated from one of its wild rodent reservoirs the bank vole (*Myodes glareolus*), near the Ljungan River in central Sweden (Niklasson et al. 1999). The LV genome possesses a deviant picornavirus-like organization. The Ljungan River LV isolate had two structurally different 2A proteins whose functions are not mutually exclusive. This feature has never previously been observed within the family of picornaviruses. Later it has been shown that LV isolates from both European and American bank voles share the same genomic feature, thus suggesting a continuous world wide presence of the virus (Johansson et al. 2002, 2003). LV has also been confirmed in older virus isolates from several species of wild rodents in different parts of the world including native rodents from the USA, England, Denmark, Germany, and Italy (Hauffe et al. 2010; Main et al. 1976; Whitney et al. 1970). Indeed LV has been found in all countries investigated suggesting a global distribution. It has been noted that voles and lemmings both in the wild and when colonized in the laboratory develop several different pathological signs and symptoms including myocarditis, diabetes, encephalitis and stereotypic behavior (Niklasson et al. 2003a; Schoenecker et al. 2000). The fact that it has not been possible to establish an LV free colony of any of these wild rodent species has hampered the possibility to prove that LV is an etiologic agent causing these conditions. However, it has been possible to induce all diseases found in the wild state in mice and guinea pigs under controlled laboratory conditions (Bo Niklasson, unpublished observation). Successful virus isolation has been achieved by intracerebral inoculation into 1-day-old suckling mice, as well as in tissue culture using Vero, GMK, and A-549 cells. In vivo propagation of LV in suckling mouse brain (SMB) yields progeny virus of high titer (Niklasson et al. 1999), albeit SMB propagation of LV is only possible in a small proportion of LV strains. LV also replicates in different cell lines of both animal and human origin. Although the virus replicates in a large number of different cell lines and primary cells, virus titers in both cells and culture supernatants are often low or very low with only discrete cytopathogenic effects detected. Moreover, the virus isolation success rate is low from both cell culture and suckling mice is very low even when the starting material contains a high copy number of LV based on quantitative LV-specific PCR (Donoso Mantke et al. 2007). When studied in laboratory mice LV has an acute phase with very high virus load in all organs if the infection occurs early in life followed by a chronic phase with very low copy number measured by PCR (Donoso Mantke et al. 2007). This chronic persistent infection is associated with a very poor antibody response. Neutralizing antibodies and antibodies measured using ELISA or indirect immunofluorescence have been low or undetectable.

Rodent Diabetes and the Ljungan Virus

The bank vole (*M. glareolus*) is a natural reservoir for the LV. However, other species of rodents, both resident in the wild and those bred in the laboratory, have also been found to be carriers of the LV. In Scandinavia the *Myodes rufocanus*, *Microtus agrestis*, *Lemmus lemmus,* and *Myopus schisticolor* are other species that have been confirmed to carry the LV using virus isolation, immunohistochemistry (IHC), or serology (Bo Niklasson et al., unpublished observations; Niklasson et al. 1999, 2003a, b). Also, in the laboratory the type 1 diabetes animal model, the BB rat, has been tested positive for LV (Niklasson et al. 2007a).

Voles and lemmings in the northern hemisphere have cyclic population variations with density peaks every third to fourth year (Hansson and Henttonen 1985, 1988; Hoernfeldt 1994). At density peaks both voles and lemmings have subgroups of individuals that display elevated blood glucose levels when tested after capture. When normoglycemic voles and lemmings are brought to the laboratory and kept in captivity some individuals eventually develop elevated blood glucose levels. Furthermore, after 4–6 weeks in captivity these animals were confirmed glucose intolerant when tested with a glucose tolerance test (Niklasson et al. 2006). These data suggest that environmental factors play a role in the development of bank vole diabetes. Stress is one factor that in the laboratory has shown to be of importance in the development of hyperglycemia in bank voles (Freimanis et al. 2003). At population density peaks it has been suggested that the experienced stress level in voles and lemmings is high and may result in increased glucose intolerance. One example of this is the gray-sided vole (*M. rufocanus*). At high population density approximately 70% of the tested gray-sided voles had elevated blood glucose levels upon capture, whereas gray-sided voles trapped during moderate population density did not reveal any elevated blood glucose upon capture nor were there any glucose intolerant animals found. However, after 5 weeks in captivity approximately 40% of the normoglycemic animals trapped at moderate population density had developed glucose intolerance (Niklasson et al. 2006).

Bank voles trapped in Zealand in Denmark, a region without major cyclic density peaks, did not show elevated blood glucose levels when measured upon capture (Niklasson et al. 2003a). Though when kept in captivity for 1 month, these bank voles had developed elevated blood glucose levels. Also, the pancreatic islets in the hyperglycemic animals had developed lesions that were not present in normoglycemic animals. These islet malformations resembled so-called hydropic degeneration (Toreson 1951) and were suggested to be a result of beta cell loss since the insulin positive area in the pancreas as examined in light microscope appeared to be reduced. Also, in these animals Picornavirus particles have been observed in the islet cells (Niklasson et al. 2003b). These observations may suggest that the LV could be involved in the destruction of the islets cells. When bank voles still normoglycemic after 1 month in captivity were infected with LV and the pancreatic islet structure was subsequently examined after 6 weeks post inoculation, islet lesions were observed in these animals similar as in bank voles that developed elevated blood glucose levels (Niklasson et al. 2003a). Taken together these data suggest that the combination of stress and LV infection may lead to reduction of beta cell mass eventually development of glucose intolerance and diabetes.

Diabetes in Colonized Bank Voles

The bank vole may turn out be a unique animal model in the field of diabetes research (Blixt et al. 2007). One special feature of the model is the diabetes type duality. The bank vole diabetes can present itself both as a type 1 and as a type 2 diabetes. Trapped bank voles kept in captivity with developed hyperglycemia display polydipsia, glucoseuria, ketonuria, ketonemia, and hyperlipidemia (Niklasson et al. 2003a; Schoenecker et al. 2000). These symptoms in combination with the presence of islet autoantibodies and a reduced beta cell mass suggest that the bank vole model has subdued to type 1 diabetes (Niklasson et al. 2003a). However, colonized bank voles kept in breeding display features resembling more of the type 2 form of the disease. Thus, the bank vole colony kept in breeding at the Astrid Fagreus laboratory in Stockholm has a diabetes incidence of approximately 20% (Blixt et al. 2007). The majority of these animals develop their glucose intolerance between 10 and 25 weeks of age. The glucose intolerant animals display hyperglycemia, hyperinsulinemia, and intra-islet lesions as described above. However, no sign of immune cell infiltration or insulitis could be seen in these islets. In addition elevated serum insulin and body weights argue in favor for a type 2 diabetic state. A possibility is that the diabetic condition first exhibits a type 2 diabetes phenotype, and later elderly animals display features of type 1 diabetes. This would in some aspects resemble human latent autoimmune diabetes in the adult (LADA) (Naik and Palmer 2003).

Association Between Ljungan Virus and Other Diseases than Diabetes in Man

LV infected bank voles in captivity develop several different pathological signs and symptoms including myocarditis, diabetes, encephalitis, and stereotypic behavior (Niklasson et al. 2003a; Schoenecker et al. 2000). Studies on laboratory mice showed that more than half of the dams infected with LV during pregnancy and exposed to stress gave birth to pups that died during the perinatal period. Malformation of the central nervous system including hydrocephaly and anencephaly was seen in some of these offspring (Samsioe et al. 2006). Suckling mice infected during the first 2 days develop severe encephalitis with hydrocephaly noted in a fraction of these animals (Bo Niklasson, unpublished observations). Progressive fatal paralysis has been seen in adult guinea pigs infected with LV (Bo Niklasson, unpublished observations).

Based on these findings the role of LV infection in humans during pregnancy has been studied. We have focused our studies on perinatal death intrauterine fetal death (IUFD) and sudden infant death syndrome (SIDS). Thus, variations in the incidence of IUFD closely tracked the fluctuations in native rodent populations. LV was detected in the brain tissue in four out of ten cases of IUFD investigated by IHC. The virus was also detected in the placenta in five of the ten IUFD cases, but none of 20 placentas from normal pregnancies (Niklasson et al. 2007b, 2009b; Samsioe et al. 2009).

Variation in the incidence of SIDS using the Swedish cause-of-death database also revealed an association with changes in the population fluctuations of native rodents. Formalin-fixed tissues from brain, heart and lung tissue were investigated from cases of SIDS or SIDS with lymphocyte infiltration of the myocardium (myocarditis) or myocarditis alone, cases using LV-specific IHC. The LV was detected in the brain, heart, and lung tissue from all three of the patient categories (Niklasson et al. 2009a).

Finally, LV was diagnosed in 1 out of 10 cases with hydrocephalus, 5 out of 9 cases with anencephaly and in 1 out of 18 trisomy 21 control cases by either IHC or RT-PCR (Niklasson et al. 2009b).

References

Blixt M, Niklasson B, Sandler S (2007) Characterization of beta-cell function of pancreatic islets isolated from bank voles developing glucose intolerance/diabetes: an animal model showing features of both type 1 and type 2 diabetes mellitus, and a possible role of the Ljungan virus. Gen Comp Endocrinol 154:41–47

Donoso Mantke O, Kallies R, Niklasson B, Nitsche A, Niedrig M (2007) A new quantitative real-time reverse transcriptase PCR assay and melting curve analysis for detection and genotyping of Ljungan virus strains. J Virol Methods 141:71–77

Freimanis T, Heller KE, Schonecker B, Bildsoe M (2003) Effects of postnatal stress on the development of type 1 diabetes in bank voles (*Clethrionomys glareolus*). Int J Exp Diabesity Res 4:21–25

Hansson I, Henttonen H (1985) Gradients in density variations of small rodents: the importance of latitude and snow cover. Oecologia 67:394–402

Hansson L, Henttonen H (1988) Rodent dynamics as community processes. Trends Ecol Evol 3:195–200

Hauffe HC, Niklasson B, Olsson T, Bianchi A, Rizzoli A, Klitz W (2010) Ljungan virus detected in bank voles (*Myodes glareolus*) and yellow-necked mice (*Apodemus flavicollis*) from Northern Italy. J Wildl Dis 46:262–266

Hoernfeldt B (1994) Delayed density dependence as a determinant of vole cycles. Ecology 75:791–806

Johansson S, Niklasson B, Maizel J, Gorbalenya AE, Lindberg AM (2002) Molecular analysis of three Ljungan virus isolates reveals a new, close-to-root lineage of the Picornaviridae with a cluster of two unrelated 2A proteins. J Virol 76:8920–8930

Johansson ES, Niklasson B, Tesh RB, Shafren DR, Travassos da Rosa AP, Lindberg AM (2003) Molecular characterization of M1146, an American isolate of Ljungan virus (LV) reveals the presence of a new LV genotype. J Gen Virol 84:837–844

Main AJ, Shope RE, Wallis RC (1976) Characterization of Whitney's Clethrionomys gapperi virus isolates from Massachusetts. J Wildl Dis 12:154–164

Naik RG, Palmer JP (2003) Latent autoimmune diabetes in adults (LADA). Rev Endocr Metab Disord 4:233–241

Niklasson B, Hornfeldt B, Lundman B (1998) Could myocarditis, insulin-dependent diabetes mellitus, and Guillain-Barre syndrome be caused by one or more infectious agents carried by rodents? Emerg Infect Dis 4:187–193

Niklasson B, Kinnunen L, Hornfeldt B, Horling J, Benemar C, Hedlund KO, Matskova L, Hyypia T, Winberg G (1999) A new picornavirus isolated from bank voles (*Clethrionomys glareolus*). Virology 255:86–93

Niklasson B, Heller KE, Schonecker B, Bildsoe M, Daniels T, Hampe CS, Widlund P, Simonson WT, Schaefer JB, Rutledge E, Bekris L, Lindberg AM, Johansson S, Ortqvist E, Persson B, Lernmark A (2003a) Development of type 1 diabetes in wild bank voles associated with islet autoantibodies and the novel ljungan virus. Int J Exp Diabesity Res 4:35–44

Niklasson B, Hornfeldt B, Nyholm E, Niedrig M, Donoso-Mantke O, Gelderblom HR, Lernmark A (2003b) Type 1 diabetes in Swedish bank voles (*Clethrionomys glareolus*): signs of disease in both colonized and wild cyclic populations at peak density. Ann N Y Acad Sci 1005: 170–175

Niklasson B, Nyholm E, Feinstein RE, Samsioe A, Hornfeldt B (2006) Diabetes and myocarditis in voles and lemmings at cyclic peak densities—induced by Ljungan virus? Oecologia 150:1–7

Niklasson B, Hultman T, Kallies R, Niedrig M, Nilsson R, Berggren PO, Juntti-Berggren L, Efendic S, Lernmark A, Klitz W (2007a) The biobreeding rat diabetes model is infected with Ljungan virus. Diabetologia 50:1559–1560

Niklasson B, Samsioe A, Papadogiannakis N, Kawecki A, Hornfeldt B, Saade GR, Klitz W (2007b) Association of zoonotic Ljungan virus with intrauterine fetal deaths. Birth Defects Res A Clin Mol Teratol 79:488–493

Niklasson B, Almqvist PR, Hornfeldt B, Klitz W (2009a) Sudden infant death syndrome and Ljungan virus. Forensic Sci Med Pathol 5:274–279

Niklasson B, Samsioe A, Papadogiannakis N, Gustafsson S, Klitz W (2009b) Zoonotic Ljungan virus associated with central nervous system malformations in terminated pregnancy. Birth Defects Res A Clin Mol Teratol 85:542–545

Samsioe A, Feinstein R, Saade G, Sjoholm A, Hornfeldt B, Fundele R, Klitz W, Niklasson B (2006) Intrauterine death, fetal malformation, and delayed pregnancy in Ljungan virus-infected mice. Birth Defects Res B Dev Reprod Toxicol 77:251–256

Samsioe A, Papadogiannakis N, Hultman T, Sjoholm A, Klitz W, Niklasson B (2009) Ljungan virus present in intrauterine fetal death diagnosed by both immunohistochemistry and PCR. Birth Defects Res A Clin Mol Teratol 85:227–229

Schoenecker B, Heller KE, Freimanis T (2000) Development of stereotypies and polydipsia in wild caught bank voles (*Clethrionomys glareolus*) and their laboratory-bred offspring. Is polydipsia a symptom of diabetes mellitus? Appl Anim Behav Sci 68:349–357

Toreson WE (1951) Glycogen infiltration (so-called hydropic degeneration) in the pancreas in human and experimental diabetes mellitus. Am J Pathol 27:327–347

Whitney E, Roz AP, Rayner GA (1970) Two viruses isolated from rodents (*Clethrionomys gapperi* and *Microtus pennsvlvanicus*) trapped in St. Lawrence County, New York. J Wildl Dis 6:48–55

Chapter 10
Virus-Related Diabetes in Cattle

Kazuya Matsuda and Hiroyuki Taniyama

Abstract Twelve cattle with IDDM presented emaciation, polyuria, polydipsia, glycosuria, persistent hyperglycemia, and decreased glucose tolerance. In chronic cases, major histopathological findings in the pancreas were a decrease in the size and number of pancreatic islets, interlobular and interacinar fibrosis, mild lymphocytic insulitis, and vacuolation of a few islets. Cells comprising atrophic islets were immunohistochemically revealed to be α- and δ-cells. In acute cases, most islets consisted of vacuolated and severely degranulated β-cells and also contained numerous necrotic cells. Lymphocytic insulitis was common. Bovine IgG-immunoreactive islet cells were seen frequently in the vacuolated islets. Atrophic and vacuolated islets had no or small numbers of granules immunoreactive to GAD, respectively. In seven cases the bovine viral diarrhea virus was sought. Noncytopathic virus was isolated from blood leukocytes and sera of all cases. Viral antigen was demonstrated immunohistochemically in the acinar cells of the exocrine pancreas and epithelial cells of the pancreatic ducts. In four cases, viral antigen was also recognized in the islet cells. The results suggest that autoimmune insulin-dependent diabetes may have been induced by persistent BDV infection, resulting in the gradual and selective destruction of β-cells in pancreatic islets.

Introduction

Insulin-dependent diabetes mellitus (IDDM) is believed to result from the destruction of insulin producing β-cells induced by cellular and/or humoral autoimmune mechanisms induced by different environmental factors, including viral infections and chemicals (Yoon 1990). Most human patients with insulin-dependent diabetes have

K. Matsuda, D.V.M., Ph.D. • H. Taniyama, D.V.M., Ph.D. (✉)
Department of Veterinary Pathology, School of Veterinary Medicine,
Rakuno Gakuen University, 582 Bunkyodai-Midorimachi, Ebetsu, Hokkaido 069-8501, Japan
e-mail: taniyama@rakuno.ac.jp

K.W. Taylor, H. Hyoty, A. Toniolo, A.J. Zuckerman (eds.), *Diabetes and Viruses*,
DOI 10.1007/978-1-4614-4051-2_10,

an underlying autoimmune reaction to β-cells that may be triggered by various viral infections in genetically predisposed individuals (Gepts and DeMey 1978). Many viruses, such as group B coxsackieviruses, mumps virus, Epstein–Barr virus, cytomegalovirus, influenza virus, rubella virus, and herpesvirus, are suspected to be diabetogenic factors and are regarded as an important cause of diabetes in human beings (Yoon 1990). Diabetes mellitus in domestic animals has been reported most commonly not only in dogs (Davison et al. 2003) and cats (von Sandersleben et al. 1983) but, sporadically, also in other species such as cattle (Christensen and Schambye 1950; Kaneko and Rhode 1964; Barboni et al. 1966; Phillips et al. 1971; Mostaghni and Ivoghli 1977; Gould 1981; Baker et al. 1983; von Tontis and Wittwer 1986), horse (Tasker et al. 1966), sheep (Baker et al. 1931; Mattheeuws et al. 1982), and pig (Biester 1925). In cattle, these reports described mainly clinical signs, treatment and pathological changes, but the causes were not clearly indicated. A possible association between bovine diabetes mellitus and bovine diarrhea virus (BDV) infection was first proposed by Tajima et al. (1992).

BDV is a small enveloped virus with a positive-strand RNA genome, classified in the family *Flaviviridae* and genus *Pestivirus*. There are two BDV biotypes designated noncytopathic and cytopathic, depending on their effect on tissue culture cells (Gillespie et al. 1960). BDV is one of the major pathogens of dairy cattle and may cause a wide range of lesions or clinical syndromes (Divers 2008). BDV causes fever, mucosal erosions, diarrhea, respiratory diseases, abortions or reproductive failure, congenital anomalies, persistent infection of fetuses and many other signs. Fetuses that are transplacentally exposed to noncytopathic BDV between approximate ages of 40 and 125 days gestation may become persistently infected (PI). If a PI animal is challenged by a heterologous cytopathic BDV, or if a noncytopathic virus harbored in an animal is converted into the cytopathic biotype by accidental insertion of host RNA, the affected animal develop fatal "mucosal disease," characterized by fever, diarrhea, weight loss, mucosal ulcerations of the gastrointestinal tract, digital lesions, and/or dermatologic lesions (Van Metre et al. 2008).

The clinical, histopathological, and immunohistochemical findings from natural cases of bovine insulin-dependent diabetes are described, and a significant association with BDV infection is reviewed (Taniyama et al. 1993, 1995, 1999a, b, c).

Case Profiles

Case profiles and clinicopathologic findings of 12 cattle with IDDM, which underwent necropsy during 16 years are reviewed. The 12 animals were divided further into two groups: slow-onset diabetes (eight cases) and acute-onset diabetes (four cases), on the basis of the histopathological findings of the pancreas (as described below). The 12 diabetic cattle included 7 Japanese Black and 5 Holstein–Friesian breeds, 8 females and 4 males from 6- to 48 months of age. All diabetic animals were persistently hyperglycemic. Upon intravenous glucose tolerance test, glucose levels returned to the preinjection values within 2 h in control cattle, but remained markedly elevated levels after 3 h in diabetic cattle. Urine tests (N-Multistix, Miles-Sankyo,

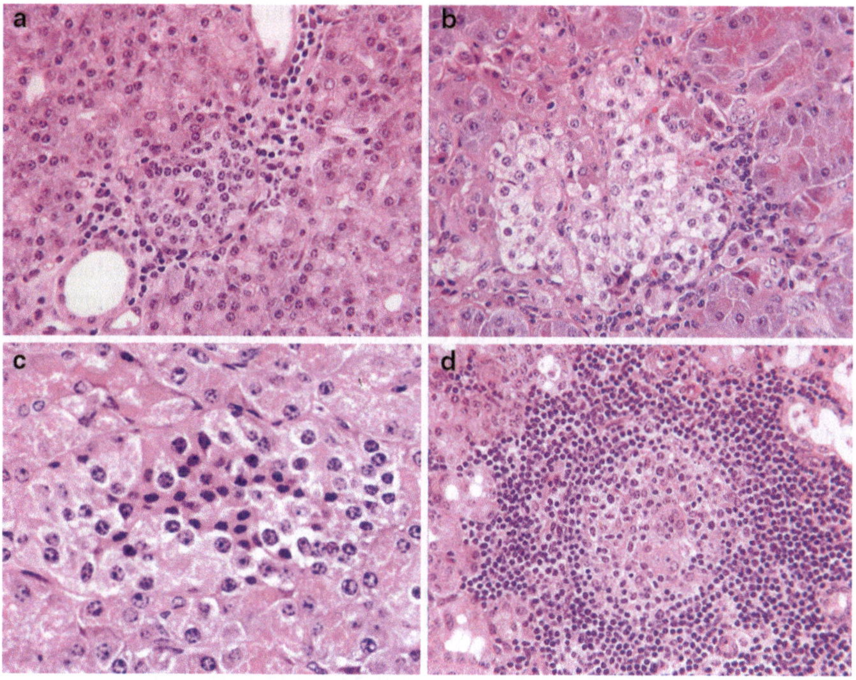

Fig. 10.1 Histological changes in pancreatic islets of cattle with chronic (**a**, **b**) and acute (**c**, **d**) IDDM (hematoxylin and eosin staining). (**a**) Atrophic islet with mild lymphocytic infiltration at the periphery. (**b**) Islet comprising vacuolated cells with lymphocytic infiltration. (**c**) Islet containing islet cells with eosinophilia, shrinkage of cytoplasm, nuclear pyknosis, and karyorrhexis. (**d**) Atrophic islet with severe lymphocytic infiltration at the periphery and in the inner part of the islet

Tokyo, Japan) showed acidic pH and glycosuria. Cattle with chronic IDDM showed anorexia, severe emaciation, polyuria, and polydipsia. Animals with acute IDDM presented emaciation, diarrhea, dehydration, and had erosions and ulcers on the mucosa of the muzzle, lips, oral cavity, and tongue. Clinicopathological manifestations (severe emaciation, polyuria, polydipsia, glycosuria, persistent hyperglycemia, and decreased glucose tolerance) agreed with the diagnosis of IDDM.

Histopathological Analysis of Pancreas in Cattle with Insulin Dependent Diabetes

Chronic Onset Insulin-Dependent Diabetes

The most characteristic changes in the pancreas were a decrease in the size and number of pancreatic islets, interlobular and interacinar fibrosis, and mild lymphocytic infiltration in a few islets (Fig. 10.1a). Occasionally, the islets disappeared completely in the pancreatic lobes. Almost all cells in the atrophied islets did not contain aldehyde fuchsin and Masson-Goldner (AFMG)-positive granules in the residual small amount of cytoplasm. Mild lymphocytic infiltration at both the periphery and the inner part of the islets was observed in some lobes (Fig. 10.1a and b). Islets with lymphocytic infiltration contained necrotic islet cells and stromal concentrically laminated basophilic bodies which were interpreted as calcification by staining with the von Kossa method. Mild fibrosis with or without lymphocytic infiltration was observed in some atrophied islets. Residual islets consisted of many cells with vacuolated cytoplasm, including a small number of AFMG-positive granules, indicating that the cells were degranulated β-cells (Fig. 10.1b). In two cases, large numbers of glycogen granules were present in their cytoplasm. Accumulation of glycogen granules was also observed in the cytoplasm of the pancreatic ductal epithelium in these cases. Using the silver impregnation method, interlobular and interacinar fibrosis was diffuse and often accompanied by atrophy of acinar cells with reduced numbers of secretory granules. Mild infiltration of lymphocytes was present around small and large pancreatic ducts, occasionally with mild or severe fibrosis. In chronic onset human insulin-dependent diabetes, histopathologic changes of the pancreas consist in the decrease of the number and size of pancreatic islets, vacuolar degeneration of residual islet cells, and interlobular and interacinar fibrosis (Gepts 1965; Gepts and DeMey 1978). In particular, a complete or almost complete absence of β-cells in islets and diffuse fibrosis in the exocrine pancreas are considered hallmarks of long-term insulin-dependent diabetes (Gepts and DeMey 1978). Thus, the major histological changes in the pancreas of diabetic cattle described here were consistent with those of human cases. Furthermore, calcification, necrosis, and fibrosis of the pancreatic islets found in selected cattle cases were also described in chronic onset human insulin-dependent diabetes (Kloppel 1984).

Acute-Onset Diabetes Mellitus

In cattle with acute-onset diabetes, the majority of pancreatic islets consisted of islet cells with vacuolated cytoplasm and small numbers of AFMG-positive granules. In addition, islets contained many islet cells with increased eosinophilia, shrinkage of cytoplasm, and karyorhexis, indicating necrosis of β-cells (Fig. 10.1c). Glycogen accumulation was not observed in the vacuolated cytoplasm of islet cells. In some cases, each pancreatic lobe had slightly atrophied islets composed of islet cells with scanty cytoplasm. Mild or severe lymphocytic infiltration at both the periphery and inner part of the islets was observed frequently in all cases (Fig. 10.1d). Most inflammatory cells consisted of small lymphocytes and, occasionally, few large mononuclear cells and plasma cells. In some lobes, mitoses were seen in the enlarged islets composed of hypertrophic cells. In the exocrine glands, focal lymphocytic infiltration was sometimes found in the interlobular connective tissues, mainly around small-sized pancreatic ducts. Interlobular and interacinar fibrosis was not observed. In acute-onset human diabetes, it is difficult to analyze pathomorphologic changes in the pancreas, because treatment for this type of disease is usually prolonged (Kloppel 1984). Thus, the histopathological properties of the pancreas in the early stages of diabetes are still unclear. However, pathological observations in acute-onset diabetes suggest that in at least some cases β-cells are destroyed by an inflammatory process which selectively affects pancreatic islets, i.e. insulitis (Doniach and Morgan 1973; Freytag and Kloppel 1973; Gepts 1965; Gepts and DeMey 1978; Junker et al. 1977; MacCuish et al. 1974). In the majority of cases, the inflammatory infiltrate is composed of lymphocytes, occasionally with macrophages and mast cells (Kloppel 1984). Recent descriptions of the histopathology of pancreas in human insulin-dependent diabetes increasingly point to lymphocytic insulitis as the most characteristic lesion in acute-onset diabetes. Hydropic or vacuolar degeneration of β-cells has also been recognized in human cases. The β-cells are swollen and show a complete degranulation and an apparently empty cytoplasm. It has been noted that hydropic or vacuolar changes in β-cells can either be the result of a harmless glycogen deposition in the cytoplasm (Toreson et al. 1964) or represent a truly degenerative lesion (ballooning degeneration) (Kremer 1947).

Immunohistochemical Analysis of Pancreas in Cattle with Insulin-Dependent Diabetes

Pancreatic Hormones

The majority of cells in the atrophied islets in both types of insulin-dependent diabetes were infrequently reactive to anti-insulin antibody. Only a few cells that possessed poorly granulated cytoplasm were faintly reactive to glucagon and/or somatostatin antibodies. However, almost all islet cells reacted strongly to chromogranin antibody. On the other hand, vacuolated islet cells in acute and chronic onset diabetes had small numbers of granules that were strongly reactive to both insulin and chromogranin antibodies. Some anti-glucagon or anti-somatostatin reactive cells were observed at the periphery of the islets. These findings demonstrated the almost complete absence of β-cells in the atrophied islets and that the α- and δ-cells with slight amounts of cytoplasm containing small number of secretory granules remained in atrophied islets. This consideration, together with the presence of vacuolated islet cells immunoreactive for insulin, suggests the selective degranulation and loss of β-cells in the pancreatic islets of bovine diabetes as described in previous reports (Bommer et al. 1981; Gepts and DeMey 1978; Kaneko and Rhode 1964; Mattheeuws et al. 1982). In acute-onset diabetic cattle, an increase in acinar islet cells and a proliferation of ductal epithelial cells showing insulin-immunoreactivity were observed. Islet regeneration or neoformation, represented by a proliferation and endocrine differentiation of ductular and acinar cells showing large numbers of insulin-positive granules in the cytoplasm is not a specific, but a prominent finding in the pancreas of acute onset human diabetes (Gepts 1965, 1981; Gepts and DeMey 1978; Kloppel 1984).

Bovine IgG

Islets containing various numbers of cells showing immunoreactivity to anti-bovine IgG antibody were observed more frequently in acute than in chronic onset diabetes (Fig. 10.2). Furthermore, many islets with lymphocytic infiltration contained small numbers of plasma cells with cytoplasm reactive to anti-bovine IgG antibody. Though not in all cases, some atrophied islets were composed of bovine IgG-immunoreactive islet cells. The deposition of IgG, as well as histological findings in the pancreatic islets, suggests that autoantibodies [e.g., islet cell antibody (ICA), complement-fixing ICA (CF-ICA), islet-cell surface antibody (ICSA)] may have been produced in diabetic animals (Bottazzo et al. 1974; Hallberg et al. 1995; Handwerger et al. 1980; Lernmark et al. 1978; Lohmann et al. 1997; MacCuish et al. 1974).

Glutamic acid Decarboxylase

Glutamic acid decarboxylase (GAD) is a major target for autoantibodies and autoreactive T cells in human insulin-dependent diabetes (Davenport et al. 1998; Kawasaki et al. 1994; Myers et al. 1995). These autoantibodies are present in the sera of most patients with insulin-dependent diabetes diagnosed recently and appear years before clinical symptoms. Therefore, they are considered early markers of type 1 diabetes (Lohmann et al. 1997; Tuomilehto et al. 1994). In pancreatic islets, GAD is contained in β-cells and forms gamma-aminobytyric acid (GABA) by decarboxylation of glutamate (Kawasaki et al. 1993). GABA is believed to be an endocrine transmitter that participates in the synthesis of insulin and the regulation of hormonal secretion of α- and δ-cells (Kawasaki et al. 1993). However, there is

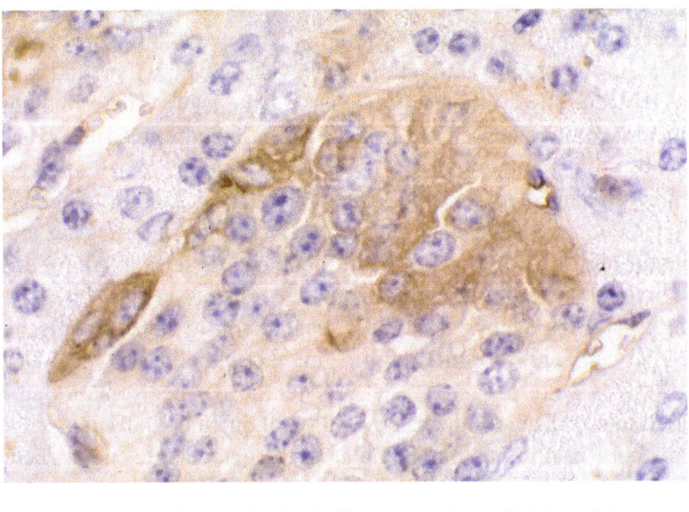

Fig. 10.2 Immunohistochemistry for bovine IgG counterstained with Mayer's hematoxylin in the pancreatic islet of a diabetic bovine. Some islet cells are immunoreactive

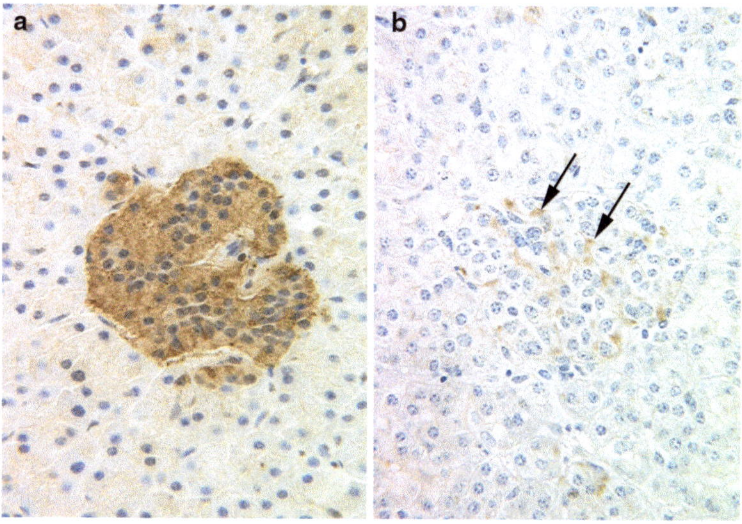

Fig. 10.3 Immunohistochemistry for GAD counterstained with Mayer's hematoxylin in pancreatic islets of non-diabetic (**a**) and diabetic (**b**) cattle. (**a**) Almost all islet cells are strongly GAD-positive. (**b**) Few islet cells are faintly positive (*arrows*)

still uncertainty about the regulation and function of GABA in islet cells of patients with insulin-dependent diabetes.

On the basis of immunohistochemical findings in control cattle, almost all islet cells were strongly positive to anti-GAD antibody, but exocrine cells and ductal epithelium failed to react with GAD antibody (Fig. 10.3a). The majority of cells in atrophied islets of diabetic cattle were not reactive to anti-GAD and anti-insulin antibodies, suggesting that the residual cells in atrophied islets did not contain GABA and had lost insulin synthesis (Fig. 10.3b). Enlarged pancreatic islets contained variable numbers of vacuolated cells located at the center of the islets. These vacuolated cells contained small numbers of granules reactive to anti-GAD and anti-insulin antibodies. This suggests that GAD is released gradually from vacuolated β-cells into the blood, possibly eliciting GAD autoantibodies in cattle, as it occurs in humans with type 1 diabetes (Baekkeskov et al. 1990; Kawasaki et al. 1994). Some pancreatic lobes had exocrine cells and proliferating ductal epithelium with insulin immunoreactivity. Epithelial cells of the pancreatic ducts in some lobes showed a positive reaction to anti-GAD antibody. These changes are regarded as an attempt to compensate for an inadequate supply of insulin from pancreatic islets.

Bovine Viral Diarrhea Virus Infection and Diabetes Mellitus

Seven cases, including animals with chronic and acute-onset insulin-dependent diabetes, were investigated for BDV infection. Isolation of BDV from the blood and mesenteric lymph nodes was attempted using postmortem specimens. Noncytopathic

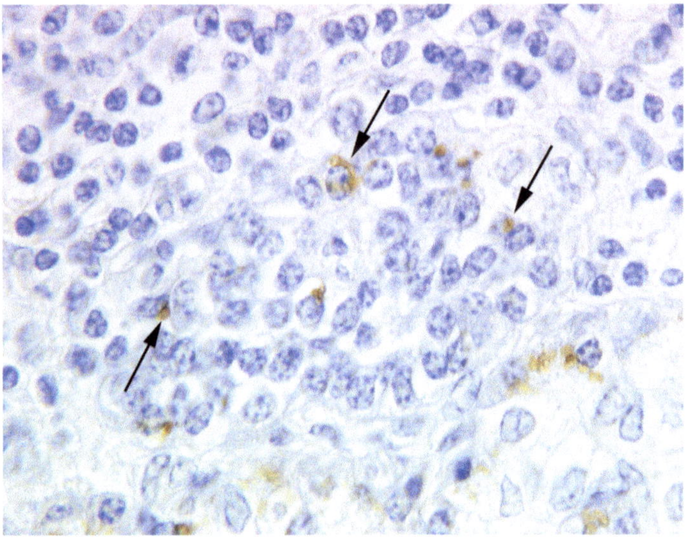

Fig. 10.4 Immunohistochemistry for BDV using primary antibody against BDV strain MD74, counterstained with Mayer's hematoxylin in a pancreatic islet of a diabetic bovine. Immunoreactivity is detected in the cytoplasm of selected cells (*arrows*)

virus was isolated from the blood leukocytes and sera of all investigated cattle and from lymph nodes of some animals. Histologically, multiple necrotic foci were observed in the mucous membranes of the upper and lower alimentary tract in the seven animals. Squamous epithelial cells adjacent to the basal layer in the upper alimentary tract showed hydropic changes, i.e., swelling or shrinkage, with various degrees of neutrophile leukocyte infiltration. The small and large intestines also showed villous atrophy and dilatation of mucosal glands, suppurative cryptitis, mild infiltration of lymphocytes and macrophages, and fibroplasia in the lamina propria. Lymphoid germinal centers were necrotic, and mature lymphocytes were depleted in Peyer's patches and lymph nodes draining the alimentary tract. Histological changes of the pancreas are described above.

Immunohistochemical detection of BDV antigen was attempted using virus-specific monoclonal antibody (TropBio, Townville, Australia). The primary antibody was manufactured by immunization Balb/c mouse with BDV strain MD74, and used for screening for the presence of pestiviruses. In control cattle, immunoreactivity for BDV was absent in all investigated organs. In the alimentary tract of diabetic cattle, BDV immunoreactivity was observed in some residual epithelial cells of the mucous membrane of the upper and lower alimentary tract in all animals. A few immunopositive lymphoid cells were also observed in the lamina propria and in lymphoid germinal centers of Peyer's patches. In the pancreas of all seven cattle, immunoreactivity with BDV antibody was detected in the acinar cells of the exocrine pancreas and the epithelial cells of pancreatic ducts. In four of these animals, similar immunoreactivity was recognized in the cytoplasm of islet cells in residual islets, or in the enlarged islets containing vacuolated cells (Fig. 10.4). BDV antigen-positive cells were more frequently seen in the enlarged islets than in the atrophied islets. Furthermore, some islets with lymphocytic infiltrates had a comparatively small number of BDV antigen-positive cells. These histological and immunohistochemical findings suggest that autoimmune insulin-dependent diabetes may have been induced by persistent BDV infection, resulting in the gradual loss of β-cells in pancreatic islets (Yoon 1990; See and Tilles 1998).

In humans, numerous studies have indicated that different viral infections can be diabetogenic (Yoon 1990). Pregnant cattle infected with BDV give birth frequently to PI calves. Such calves usually have a chronic form of bovine viral diarrhea, characterized by gradual emaciation, disturbance of alimentary or respiratory function, breakdown of immunological tolerance, and/or autoimmune disease (Kahrs et al. 1970; Kendrick 1971; Scott et al. 1973). In man, autoimmune type 1 diabetes is frequently associated with congenital or neonatal viral infection. The clinical onset is preceded by a lengthy pathological process in most cases (Menser et al. 1978; Handwerger et al. 1980; Rayfield et al. 1986). Experimentally, several viruses, including encephalomyocarditis virus, mengovirus-2 T, coxsackievirus B4, and rubella virus have been shown to induce insulin-dependent diabetes in genetically susceptible newborn mice (Craighead et al. 1974; Menser et al. 1978). The slow process of β-cell destruction may begin as an active immune response to a foreign antigen closely resembling a normal β-cell component (Menser et al. 1978; Rayfield et al. 1986). It seems possible that incorporation of host antigens by BDV may render such antigens foreign to cattle. Alternatively, persistent BDV infection may induce inflammation and favor the release of β-cell specific autoantigens.

References

Baekkeskov S, Aanstoot HJ, Christgau S, Reetz A, Solimena M, Cascalho M, Folli F, Richter-Olesen H, De Camilli P (1990) Identification of the 64 K autoantigen in insulin-dependent diabetes as the GABA-synthesizing enzyme glutamic acid decarboxylase. Nature 347:151–156

Baker LH, Reid JJ, Owen M (1931) Diabetic coma of feed-lot sheep. J Am Vet Med Assoc 79:97–99

Baker JS, Jackson HD, Sommers EL (1983) Diabetes mellitus in a four-year-old pregnant Holstein. Comp Cont Educ Pract 5:S328–S331

Barboni E, Manocchio I, Asdrubali G (1966) Observations on diabetes in cattle due to experimental epizootic aphthae (Preliminary note). Article in Italian. Nuovi Ann Ig Microbiol 17: 223–226

Biester HE (1925) Diabetes in a pig showing pancreatic lesions. J Am Vet Med Assoc 67:99–109

Bommer G, Morohoshi T, Drenk CR, Oberholzer M, Heitz PU, Klöppel G (1981) Immunzytochemische Morphometrie am endokrinen Pankreas juveniler und adulter Diabetiker. Verh Dtsch Ges Pathol 65:387

Bottazzo GF, Florin-Christensen A, Doniach D (1974) Islet-cell antibodies in diabetes mellitus with autoimmune polyendocrine deficiencies. Lancet 2:1279–1283

Christensen NO, Schambye P (1950) Om diabetes mellitus hos kvag. Nord Vet Med 2:863–900

Craighead JE, Kanich RE, Kessler JB (1974) Lesions of the islets of Langerhans in encephalomyocarditis virus-infected mice with diabetes mellitus-like disease. I. Acute lesions. Am J Pathol 74:287–300

Davenport C, Radford PM, Al-Bukhari TA, Lai M, Bottazzo GF, Todd I (1998) Heterogeneity in the occurrence of a subset of autoantibodies to glutamic acid decarboxylase in autoimmune polyendocrine patients with islet cell antibodies. Clin Exp Immunol 111:497–505
Davison LJ, Ristic JM, Herrtage ME, Ramsey IK, Catchpole B (2003) Anti-insulin antibodies in dogs with naturally occurring diabetes mellitus. Vet Immunol Immunopathol 91:53–60
Divers TJ (2008) Respiratory diseases. In: Divers TJ, Peek SF (eds) Rebhun's diseases of dairy cattle, 2nd edn. Saunders, St. Louis, pp 79–129
Doniach I, Morgan AG (1973) Islets of Langerhans in juvenile diabetes mellitus. Clin Endocrinol 2:233–248
Freytag G, Kloppel G (1973) Insulitis—a morphological review. Curr Top Pathol 58:49–90
Gepts W (1965) Pathologic anatomy of the pancreas in juvenile diabetes mellitus. Diabetes 14:619–633
Gepts W (1981) Islet changes in human diabetes. In: Cooperstein SJ, Watkins D (eds) The islets of Langerhans. Academic, New York, pp 321–356
Gepts W, DeMey J (1978) Islet cell survival determined by morphology. An immunocytochemical study of the islet of Langerhans in juvenile diabetes mellitus. Diabetes 27(Suppl 1):251–261
Gillespie JH, Baker JA, McEntee K (1960) A cytopathogenic strain of virus diarrhea virus. Cornell Vet 50:73–79
Gould AC (1981) Diabetes mellitus in cattle. Vet Rec 109:539
Hallberg A, Juhlin C, Berne C, Kampe O, Karlsson FA (1995) Islet cell antibodies: variable immunostaining of pancreatic islet cells and carcinoid tissue. J Intern Med 238:207–213
Handwerger BS, Fernandes G, Brown DM (1980) Immune and autoimmune aspects of diabetes mellitus. Hum Pathol 11:338–352
Junker K, Egeberg J, Kromann H, Nerup J (1977) An autopsy study of the islets of Langerhans. Acta Pathol Microbiol Scand A 85:699–706
Kahrs RF, Scott FW, de Lahunta A (1970) Congenital cerebellar hypoplasia and ocular defects in calves following BVD of pregnant cows. J Am Vet Med Assoc 156:1443–1450
Kaneko JJ, Rhode EA (1964) Diabetes mellitus in a cow. J Am Vet Med Assoc 144:367–373
Kawasaki E, Moriuchi R, Watanabe M, Saitoh K, Brunicardi FC, Watt PC, Yamaguchi T, Mullen Y, Akazawa S, Miyamoto T, Nagataki S (1993) Cloning and expression of large isoform of glutamic acid decarboxylase from human pancreatic islet. Biochem Biophys Res Commun 192:1353–1359
Kawasaki E, Takino H, Yano M, Uotani S, Matsumoto K, Takao Y, Yamaguchi Y, Akazawa S, Nagataki S (1994) Autoantibodies to glutamic acid decarcoxylase in patients with IDDM and autoimmune thyroid disease. Diabetes 43:80–86
Kendrick JW (1971) Bovine viral diarrhea-mucosal disease virus infection in pregnant cows. Am J Vet Res 32:533–544
Kloppel G (1984) Islet histopathology in diabetes mellitus. In: Kloppel G, Heitz PU (eds) Pancreatic pathology. Churchill Livingstone, New York, pp 154–192
Kremer HU (1947) Juvenile diabetes as a sequel to mumps. Am J Med 3:257–258
Lernmark A, Freedman ZR, Hofmann C, Rubenstein AH, Steiner DF, Jackson RL, Winter RJ, Traisman HS (1978) Islet-cell-surface antibodies in juvenile diabetes mellitus. N Engl J Med 299:375–380
Lohmann T, Seissler J, Verlohren HJ, Schroder S, Rotger J, Dahn K, Morgenthaler N, Scherbaum WA (1997) Distinct genetic and immunological features in patients with onset of IDDM before and after age 40. Diabetes Care 20:524–529
MacCuish AC, Barnes EW, Irvine WJ, Duncan LJP (1974) Antibodies to pancreatic islet cells in insulin dependent diabetes with coexistent autoimmune disease. Lancet 2:1529–1531
Mattheeuws D, Derijcke J, Rottiers R, Hoorens J (1982) Diabetes mellitus in two twin male lambs. Vet Q 4:135–138
Menser MA, Forrest JM, Bransby RD (1978) Rubella infection and diabetes mellitus. Lancet 1:57–60
Mostaghni K, Ivoghli B (1977) Diabetes mellitus in the bovine. Cornell Vet 67:24–28
Myers MA, Rabin DU, Rowley MJ (1995) Pancreatic islet cell cytoplasmic antibody in diabetes is represented by decarboxylase. Diabetes 44:1290–1295

Phillips RW, Knox KL, Pierson RE, Tasker JB (1971) Bovine diabetes mellitus. Cornell Vet 61:114–124

Rayfield EJ, Kelly KJ, Yoon JW (1986) Rubella virus-induced diabetes in hamsters. Diabetes 35:1278–1281

Scott FW, Kahrs RF, de Lahunta A, Brown TT, McEntee K, Gillespie JH (1973) Virus-induced congenital anomalies of the bovine fetus. Cornell Vet 63:536–560

See DM, Tilles JG (1998) The pathogenesis of viral-induced diabetes. Clin Diagn Virol 9:85–88

Tajima M, Yazawa T, Hagiwara K, Kurosawa T, Takahashi K (1992) Diabetes mellitus in cattle infected with bovine viral diarrhea mucosal disease virus. J Vet Med A 39:616–620

Taniyama H, Shirakawa T, Furuoka H, Osame S, Kitamura N, Miyazawa K (1993) Spontaneous diabetes mellitus in young cattle: histologic, immunohistochemical, and electron microscopic studies of the islets of Langerhans. Vet Pathol 30:46–54

Taniyama H, Ushiki T, Tajima M, Kurosawa T, Kitamura N, Takahashi K, Matsukawa K, Itakura C (1995) Spontaneous diabetes mellitus associated with persistent bovine viral diarrhea (BVD) virus infection in young cattle. Vet Pathol 32:221–229

Taniyama H, Hirayama K, Kagawa Y, Kurosawa T, Tajima M, Yoshino T, Furuoka H (1999a) Histopathological and immunohistochemical analysis of the endocrine and exocrine pancreas in twelve cattle with insulin-dependent diabetes mellitus (IDDM). J Vet Med Sci 61:803–810

Taniyama H, Hirayama K, Kagawa Y, Ushiki T, Kurosawa T, Furuoka H, Ono T (1999b) Immunohistochemical demonstration of bovine viral diarrhoea virus antigen in the pancreatic islet cells of cattle with insulin-dependent diabetes mellitus. J Comp Pathol 121:149–157

Taniyama H, Oka S, Yokota H, Hirayama K, Kagawa Y, Kurosawa T, Furuoka H, Ono T (1999c) Immunohistochemical detection of the enzyme glutamic acid decarboxylase and hormones of the islets of Langerhans in spontaneous insulin-dependent diabetes mellitus in cattle. Vet Pathol 36:628–631

Tasker JB, Whiteman CE, Martin BR (1966) Diabetes mellitus in the horse. J Am Vet Med Assoc 149:393–399

Toreson WE, Feldman R, Lee JC, Grodsky GM (1964) Pathology of diabetes mellitus produced in rabbits by means of immunization with beef insulin. Am J Clin Pathol 42:531

Tuomilehto J, Zimmet P, Mackey IR, Koskela P, Vidgren G, Toivanen L, Tuomilehto-Wolf E, Kohtamaki K, Stengard J, Rowley MJ (1994) Autoantibodics to glutamic acid decarboxylase as predictors of insulin-dependent diabetes mellitus before clinical onset of disease. Lancet 343:1383–1385

Van Metre DC, Tennant BC, Whitlock RH, Van Metre DC, Tennant BC, Whitlock RH (2008) In: Divers TJ, Peek SF, Divers TJ, Peek SF (eds) Rebhun's diseases of dairy cattle, 2nd edn. Saunders, St. Louis, pp 200–294

von Sandersleben J, von Sandersleben J, Pospischil A, Kraft W (1983) Infection of the pancreas with parvoviruses in cats. Dtsch Tierarztl Wochenschr 90:313–316 (German)

von Tontis A, Wittwer F (1986) Spontaner Diabetes mellitus bei zweieiigen weiblichen Zwillingsrindern. Schweiz Arch Tierheilkd 128:475–482

Yoon JW (1990) Role of viruses and environmental factors in induction of diabetes. Curr Top Microbiol Immunol 164:95–123

Part III
Studies in Humans

Chapter 11
Epidemiology of Viruses in Type 1 Diabetes: Seasonal Incidence, Family Studies, Clustering

Keith W. Taylor

The general epidemiology of diabetes has attracted great interest over very many years, not least because of the light it might throw on the etiology of the disease.

Only certain aspects of the epidemiological pattern of diabetes where it relates to a possible viral etiology are discussed.

Epidemiology and the Nature of Type 1 Diabetes

Classically, type 1 diabetes arises as a sudden event in young people or in children. If untreated, rapid loss of insulin-producing cells leads to gross metabolic disturbance and death in a period of months. These dramatic events, common enough in the preinsulin period, overshadowed the possibility that the disease might have a lengthy induction period and could be episodic in nature, before the final clinical picture emerged.

Epidemiology has therefore been concentrated on the final phase of the disease, when diagnosis would be clear cut. Much of it has been centred on whether external factors might immediately precipitate the disease.

Seasonal Incidence

The first detailed survey of seasonal incidence was carried out in the USA, when Adams (1926) suggested that in acute juvenile diabetes in Minnesota and the surrounding states there were peaks of incidence in the fall, and winter months and a

K.W. Taylor, M.B.B.S., Ph.D., F.R.C.P.
† (Deceased)

K.W. Taylor, H. Hyoty, A. Toniolo, A.J. Zuckerman (eds.), *Diabetes and Viruses*,
DOI 10.1007/978-1-4614-4051-2_11,

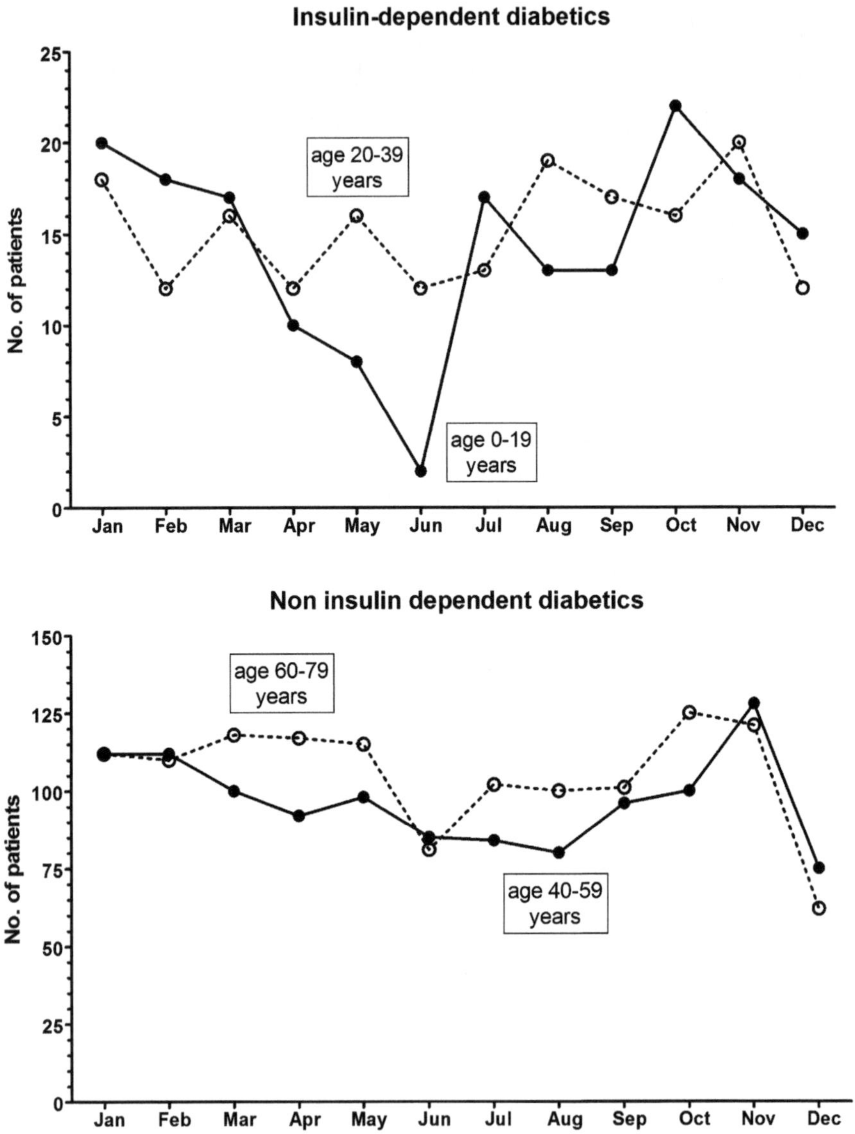

Fig. 11.1 Distribution of new cases of diabetes by month in various age groups (modified from Gamble and Taylor 1969)

trough in the early summer. Thereafter, occasional surveys on seasonal incidence were carried out (e.g., Pannhorst and Rieger 1938) without firm conclusions. The results of a more systematic study of seasonal incidence in two main centres in the UK were published later (Gamble and Taylor 1969). Studies on 293 insulin-requiring diabetics aged 0–19 showed there was a pronounced fall in the numbers of recorded cases in the summer months with a rise in incidence during the winter period. A subsidiary peak in the autumn might have coincided with a virus infection. There was a similar though less pronounced trend in the age group 30–40 (Fig. 11.1).

There have been many further studies on the seasonal incidence of diabetes conducted in many parts of the world since then. Most have been concerned with type 1 patients and children in whom more precise dates for the onset of the disease are more readily obtainable. Most have shown peaks of activity in the winter months, with a pronounced drop in diagnosis in the early summer. This was true for Denmark (Christau et al. 1977), Sweden (Dahlquist et al. 1985), Finland (Padaigha et al. 1999), Greece (Kalliora et al. 2011), Germany (Rosenbauer et al. 1999), and for the Eastern United States (Fishbein et al. 1982).

In the southern hemisphere, a similar seasonal pattern was observed in Australia though with the expected reversal of months (Australian Institute of Health and Welfare Report 2012).

A recent review (Moltchanova et al. 2009) surveyed results from 52 countries and concluded that the seasonality of type 1 diabetes was a worldwide reality, although its extent varied according to geographical location. In the northern hemisphere it appears to be of greater significance in countries which experience colder winters.

Reasons for the seasonality of type 1 diabetes are still controversial. While the pattern could easily reflect infective processes, other explanations have been put forward such as seasonal changes in diet, the start of school sessions, or the timing of vacations.

These studies nevertheless emphasise that environmental factors are highly significant in precipitating the onset of type 1 diabetes.

Regional and National Variations in Incidence of Type 1 Diabetes

National variations in the incidence of type 1 diabetes have also been studied extensively on a worldwide basis. Some very large differences between countries have been recorded. These differences cannot merely reflect racial or genetic characteristics. They must involve environmental factors of a major kind.

Regional differences incidence in smaller geographical areas have attracted much less attention, partly because they are difficult to study and interpret. One such comprehensive study in Yorkshire in the UK has suggested that a lower incidence of diabetes in children is associated with overcrowding in cities (Staines et al. 1997).

This important topic and its relation to infection is discussed later in this book (Chap. 14).

Small-Scale Spatial Clustering

Minor clusters of cases of type 1 diabetes have been reported from time to time in small communities. Thus, Melin and Ursing (1958) reported an outbreak of acute diabetes in a village in Sweden affecting 4 children in a total population of 350.

Two of these children were sibs. This followed an attack of mumps several months earlier.

A more detailed and well-controlled study of spatial clustering was carried out in the English county of Leicestershire by Bodington et al. (1995). In 1986 in Leicestershire, 27 cases of new diabetes were recorded in children of which 9 attended the same school as another new case. Three of the nine cases attended the same primary school which contained a total 297 children. Four new cases of diabetes in children living within an area of 2 sqkm in an outer suburb of London and all diagnosed within 2 months of one another were noted in 2002 (Taylor, unpublished). Such cases, although lacking full statistical and virological analysis, suggest that diabetes was precipitated by a common environmental and probably infective factor.

Family Studies

As has already been suggested, more than one member of a family may be diagnosed with type 1 diabetes within a short space of time following another, or diagnosis may be simultaneous. These cases are uncommon, but several have been studied in great detail.

Enteroviruses seem to be associated with a number of them as illustrated in Table 11.1. It will be seen that no single type of enterovirus predominates. In most instances, titres of enterovirus antibody declined following the initial diagnosis of diabetes, so that infection coincided with diagnosis. It is of interest that most of the cases so far recorded have been in children under the age of 15. All needed insulin.

Table 11.1 Cases of familial type 1 diabetes diagnosed simultaneously or close together in time and associated with enterovirus infection

Reference	Family members involved	Virus type	Spread of time for all affected family to be diagnosed
Nelson et al. (1977)	Four sibs	Coxsackievirus B2[a]	8 months
Phillips and Pauli (1981)	Two sibs	Coxsackievirus B2, B5	Simultaneous
Smith et al. (1998)	Identical twins	Echovirus 6	Simultaneous
Hindersson et al. (2005)	Mother and son	Coxsackievirus B5	Simultaneous

[a] Virus identified only in one child with Bornholm disease

Table 11.2 Single cases of generalised enterovirus infection accompanied by severe type 1 diabetes

Reference	Age of patient	Virus identified	Comments
Wilson et al. (1977)	2	Coxsackievirus B2	
Gladisch et al. (1976)	12	Coxsackievirus B4	Virus in pancreas
Yoon et al. (1979)	11	Coxsackievirus B4	Virus in pancreas Diabetes in mice
Champsaur et al. (1980)	1	Coxsackievirus B5	Virus isolated Diabetes in mice
Nihalani et al. (1982)	20	Coxsackievirus B4	
Vreugdenhil et al. (2000)	6 weeks	Echovirus-9	
Al Hello et al. (2008)	12	Echovirus-11	

Type 1 Diabetes Accompanying Severe Enterovirus Infection

There is now good evidence that type 1 diabetes may occasionally accompany enterovirus infections of some severity. Details of several of these cases have now been published as shown in Table 11.2.

In two cases, virus isolated from the patient produced diabetes in mice. Enterovirus was also isolated from the pancreas in two instances at post-mortem and insulitis was observed.

It is clear that in the face of severe and usually generalised enterovirus infection, the islets of Langerhans may also be involved and insulin-deficient diabetes ensues.

Japanese Fulminant Diabetes

This variant of insulin-requiring diabetes is described in detail in Chap. 22 in this book. In a number of respects it resembles the acute form of diabetes resulting from a severe enterovirus infection as described above. It follows an influenza-like illness, and proceeds to rapid islet destruction leading to extremes of hyperglycaemia and ketosis. Antibodies to enteroviruses have been detected in the blood and enterovirus genome found in the islets of these patients post-mortem. Islet cell antibodies were are not present, however.

General Conclusions

The epidemiological background to type 1 diabetes presented in this chapter is consistent with ideas that the precipitation of the disease is by an environmental agent, whatever the genetic background. In a very few individual cases, it is beyond reasonable doubt that this is an enterovirus. This does not mean that other viruses are not involved (e.g., mumps, rubella, cytomegalovirus). Nevertheless, there is also strong evidence for the involvement of enteroviruses in many other cases, as precipitating factors. This in no way excludes other viruses from participating at an earlier stage in a lengthy process of islet failure.

References

Adams SF (1926) The seasonal variation in the onset of acute diabetes. Arch Int Med 371: 861–864

Al-Hello H, Paananen A, Eskalinen M, Ylipaasto P, Hovi T, Salmela K, Lukaslev AN, Bobegamage S, Roivainen M (2008) An enterovirus strain isolated from a diabetic child belongs to the genetic subcluster of echovirus 11 but is also neutralised with monotypic antisera to coxackievirus A9. J Gen Virol 89:1949–1959

Australian Institute of Health and Welfare Report (2012) Incidence of insulin-treated diabetes in Australia. 2000–2009

Bodington MJ, Muzulu SL, Burden AS (1995) Spatial clustering of childhood diabetes. Evidence for an environmental cause. Diabet Med 12:865–867

Champsaur H, Dussaix E, Samolyk D, Fabre M, Bach C, Assan R (1980) Diabetes and Coxsackie virus B5 infection. Lancet 1:251

Christau B, Kromann H, Andersen OO, Christy M, Buschard K, Arhung K, Kristensen IH, Peitersen B, Steinrud J, Nerup J (1977) Incidence, seasonal and geographical patterns of juvenile insulin dependent diabetes in Denmark. Diabetologia 13:281–284

Dahlquist G, Blom L, Holmgren G, Hagglof R, Larsson Y, Sterky G, Wall S (1985) The epidemiology of diabetes in Swedish children 0–14 years—a 6 year prospective study. Diabetologia 28:802–808

Fishbein HA, Laporte RE, Orchard TJ, Drash AL, Kuller LH, Wagener DK (1982) The Pittsburg insulin dependent diabetes registry; seasonal incidence. Diabetologia 23:83–85

Gamble DR, Taylor KW (1969) Seasonal incidence of diabetes mellitus. BMJ 3:631–633

Gladisch R, Hofmann W, Waldherr R (1976) Myocarditis and insulitis following coxsackie virus infection. Z Kardiol, German 65:837–849

Hindersson M, Elshebani A, Orn A, Tuvemo T, Frisk G (2005) Simultaneous type 1 diabetes onset in mother and son coincident with an enteroviral infection. J Clin Virol 33:158–167

Kalliora MI, Vazeou A, Delis D, Bozas E, Thymelli I, Bartsocas CS (2011) Seasonal variations of type 1 diabetes in Greek Children. Hormones (Athens) 10:67–71

Melin K, Ursing B (1958) Diabetes following mumps. Nord Med 60:1715–1717

Moltchanova EV, Schreier N, Lammi N, Karvonen M (2009) Seasonal variation of diagnosis of type 1 diabetes mellitus in children worldwide. Diabet Med 26:673–678

Nelson PG, Arthur LJH, Gurling KJ, Gamble DR, Taylor KW (1977) Familial juvenile onset diabetes. BMJ 2:1126–1127

Nihalani KD, Pethani RR, Menon PS, Mehtalia SD (1982) Coxsackie B4 virus causing insulin dependent diabetes mellitus, myopericarditis and encephalitis–a case report. J Assoc Physicians India 30:107–109

Padaigha Z, Tuomilehto J, Karvonen M, Dahlquist G, Podar T, Adojaan B, Urbonaite B, Zalinkevicius R, Brigis G, Virtala E, Kohtamaki K, Cepaitis Z, Tuomilehto-Wolf E (1999) Seasonal variation in the incidence of type 1 diabetes during 1983–1992 in countries around the Baltic Sea. Diabet Med 16:736–743

Pannhorst R, Rieger A (1938) Manifestierung des Diabetes und Jahreszeit. Ztschr f klin Med 134:154–161

Phillips WR, Pauli R (1981) Simultaneous onset of type 1 diabetes in siblings. Lancet 2:807

Rosenbauer J, Hertzig P, von Kries R, Neu A, Giani G (1999) Temporal, seasonal and geographical incidence patterns of type 1 diabetes in children under 5 years of age in Germany. Diabetologia 42:1055–1059

Smith CP, Clements GB, Riding MH, Collins P, Bottazzo GF, Taylor KW (1998) Simultaneous onset of type 1 diabetes in identical infant twins with enterovirus infection. Diabet Med 15:515–517

Staines A, Bodansky HJ, McKinney PA, Alexander FE, McNally RJ, Law GR, Lilley HE, Stephenson C, Cartwright RA (1997) Small area variation in the incidence of childhood insulin dependent diabetes in Yorkshire, UK. Links with overcrowding and population density. Int J Epidemiol 26:1307–1313

Vreugdenhil GR, Schloot NC, Hoorens A, Rongen C, Pipeleers DG, Roep DG, Galama JM (2000) Acute onset of type 1 diabetes after severe echovirus 9 infection: putative pathological pathways. Clin Infect Dis 31:1025–1031

Wilson L, Connolly JM, Thomson D (1977) Coxsackie B2 virus infection and acute-onset diabetes in a child. BMJ 1:1008

Yoon JW, Austin M, Onodera T, Notkins AL (1979) Isolation of a virus from the pancreas of a child with diabetic ketoacidosis. N Engl J Med 300:1175–1179

Chapter 12
Molecular Biology and Classification of Enteroviruses

Glyn Stanway

Introduction

Several lines of evidence point to an involvement of virus infection in the onset of type 1 diabetes. The viruses implicated most frequently are those belonging to the genus *Enterovirus*, in particular the species *Human enterovirus B* (HEV-B), of the family *Picornaviridae*. The molecular features of these viruses are discussed.

Importance and Classification of Picornaviruses

At present 12 genera comprise the family *Picornaviridae*: *Enterovirus*, *Cardiovirus*, *Aphthovirus*, *Hepatovirus*, *Parechovirus*, *Erbovirus*, *Kobuvirus*, *Teschovirus*, *Sapelovirus*, *Senecavirus*, *Tremovirus* and *Avihepatovirus* (Stanway et al. 2004; Knowles 2011). The genus *Enterovirus* includes viruses which can cause paralysis, aseptic meningitis, encephalitis and heart disease. Rhinoviruses, the major cause of the common cold, also belong to this genus. Significant human pathogens are found in the genera *Hepatovirus* (hepatitis A virus), *Parechovirus* (human parechoviruses) and *Kobuvirus* (Aichi virus), and recently a human virus of the genus *Cardiovirus* (Saffold virus) has been recognised. However, high-throughput sequencing techniques and studies on an increasing range of organisms have led to the identification

G. Stanway, Ph.D. (✉)
Department of Biological Sciences, University of Essex, Colchester CO4 3SQ, UK
e-mail: stanwg@essex.ac.uk

K.W. Taylor, H. Hyoty, A. Toniolo, A.J. Zuckerman (eds.), *Diabetes and Viruses*,
DOI 10.1007/978-1-4614-4051-2_12, © Springer Science+Business Media New York 2013

of many new picornaviruses in the past 2 or 3 years, including viruses isolated from humans (Klassevirus and Cosavirus), and their diversity makes it likely that the number of genera will soon exceed 20.

The genus *Enterovirus* is the largest among *Picornaviridae* and contains over 250 serotypes (actually types as many new viruses are not characterised serologically and typing is based on sequence identity in one of the capsid proteins, VP1). There are ten species: *Human enterovirus A–D*, *Human rhinovirus A–C*, *Simian enterovirus A*, *Porcine enterovirus B* and *Bovine enterovirus* (Knowles 2011). Originally, human enterovirus serotypes were defined as poliovirus, coxsackie A virus, coxsackie B virus or echovirus depending on pathogenicity and growth properties in mice. More recently, isolated serotypes are designated human enterovirus and a number according to the order of isolation.

Enterovirus Molecular Biology

Like other picornaviruses, enteroviruses have a simple structure consisting of a small (about 28 nm diameter), icosahedral protein coat made up of 60 copies of 4 proteins (VP1–4), which surrounds a single-stranded, positive-sense RNA genome (Stanway et al. 2004). The major capsid proteins (VP1–3) share a common structure, an eight-stranded β-barrel, and presumably originated by gene duplication and divergence (Rossmann et al. 2002). The loops which join the β strands tend to be variable in length and sequence between different virus serotypes and project onto the surface where they make up the antigenic sites, thus giving rise to the diversity of serotypes in this genus. A "canyon" runs around the fivefold axis and is a deep, narrow structure, which is the site of binding to the cellular receptor in a number of cases (Rossmann et al. 2002). There is a hydrophobic pocket underneath the canyon, often filled with a lipid-like molecule termed the pocket factor. Drugs which bind tightly into this pocket prevent uncoating, by interfering with the conformational transitions required and/or receptor binding by distorting the canyon floor (Katpally and Smith 2007).

Enterovirus RNA (Fig. 12.1) is about 7.5 kb in length (7–9 kb in other picornaviruses) and has a small protein (VPg) covalently attached to the 5′terminus (Stanway et al. 2004). The RNA contains a single open reading frame, which is preceded by a long 5′untranslated region (5′UTR) and followed by a much shorter 3′UTR (Whitton et al. 2005). Both these regions contain extensive secondary structures (Witwer et al. 2001). One other secondary structure, the cre, is involved in the generation of new virus genomes (Steil and Barton 2009). It varies in its location in different picornaviruses, but is unusual in that it is often found within the open reading frame, and thus this region can have functions in addition to simply encoding protein. In the four HEV species, the cre lies within the 2C-encoding region (Steil and Barton 2009).

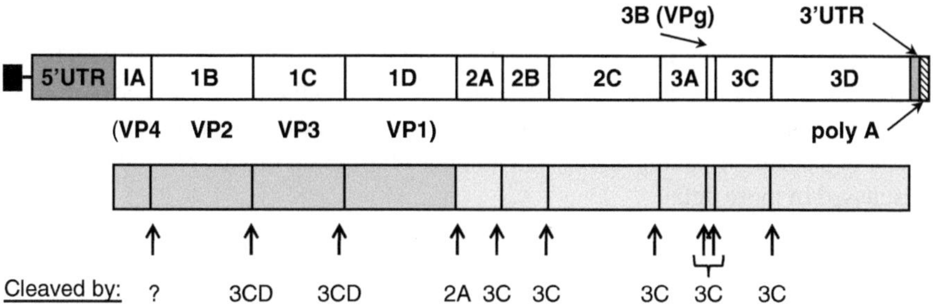

Fig. 12.1 Schematic diagram of the enterovirus genome. The genome is approximately 7.5 kb in length and is made up of a 5′untranslated region (5′UTR), a single open reading frame, a 3′UTR and a poly A tract. The open reading frame encodes a polyprotein of approximately 2,200 amino acids, which is cleaved by virus proteases to give the mature proteins. The protein-encoding regions are numbered according to their location in the open reading frame: 1A–D (encodes the capsid proteins VP4, VP2, VP3 and VP1, respectively), 2A–C and 3A–D. The RNA has a small protein (VPg, the product of the 3B region) covalently attached to its 5′end (indicated by a *filled square*). The polyprotein (*pale grey shading*) is cleaved mainly by the product of the 3C region, 3Cpro, in the case of the capsid proteins while part of the 3CD precursor. 2A cleaves at its own N-terminus. Several other precursors have important roles in the virus life cycle, e.g. 2BC, 3AB and VP0, the precursor of VP4 and VP2 which is assembled into the virus particle and cleaved by an ill-characterised mechanism as the final step of virion assembly

Enterovirus Replication

Following interactions with the receptor at the cell surface, the virus particle is internalised and RNA is released. The RNA functions as an mRNA and the open reading frame is translated to give a single polyprotein, which is cleaved by virus-encoded proteases to give the final virus proteins (Stanway et al. 2004). Proteolytic processing is thus a key feature of replication. Protein-encoding regions are numbered sequentially according to their position in the genome (1A, 1B, 1C, 1D, 2A, 2B, 2C, 3A, 3B, 3C, 3D). 1A, 1B, 1C and 1D encode the capsid proteins VP4, VP2, VP3 and VP1, respectively, while the rest are non-structural and are needed for RNA replication, polyprotein processing and modifying the properties of the host cell to facilitate virus replication and suppress host defences (Whitton et al. 2005; Lin et al. 2009). $3D^{pol}$ is the polymerase protein. This makes negative sense copies of the input RNA and then uses these as templates to give new copies of the virus genome. VP1, VP3 and VP0, the uncleaved precursor of VP4 and VP2, form a protomer and five protomers assemble into a pentamer. Twelve pentamers are then built around the RNA genome to give the virus particle. Cleavage of VP0–4 and VP2 is the final step of assembly (maturation), which stabilises the particle and makes it infectious. Mature virus particles leave the cell by lysis (Lin et al. 2009).

Several of these steps require particular cellular proteins and the distinct repertoire of proteins in different cells may play a role in determining which cell types can be infected or the fate of the infected cell, and therefore the type of disease induced. Some of the stages of the replication cycle known, or suspected to be, the basis of tropism or pathogenicity in a number of different diseases will therefore be discussed in more detail.

Enterovirus Receptors

The interaction between a virus and its cellular receptor is potentially a key determinant of tropism and disease, as only cells possessing a receptor which allows attachment, internalisation of the particle and release of the RNA into an appropriate cellular compartment can be infected (Whitton et al. 2005). However, attachment and entry appear to be somewhat fluid in that closely related viruses may interact with quite different receptors, while diverse viruses may interact with the same receptor. In addition, multiple molecules may be involved in entry, acting as either alternative receptors or coreceptors (Karttunen et al. 2003; Tuthill et al. 2011). Known receptors for enteroviruses are listed in Table 12.1. It is interesting that

Table 12.1 Receptors used by enterovirus species which infect humans (Tuthill et al. 2011)

Species	Example	Receptor	Receptor type
HEV-A	EV-71	PSGL-1	Mucin-like
		SCARB2	Scavenger receptor class B
HEV-B	CBVs	CAR	Immunoglobulin SF
		DAF	Complement control family
	CAV-9	αvβ3, αvβ6	Integrin
		GRP78	HSP70 family
		β2-microglobulin	Part of MHC-1
	E1	α2β1	Integrin
		β2-microglobulin	Part of MHC-1
	Several echovirus types	DAF	Complement control family
		β2-microglobulin	Part of MHC-1
		Heparan sulphate	Carbohydrate
HEV-C	PV	CD155	Immunoglobulin SF
	CAV-21	ICAM-1	Immunoglobulin SF
		DAF	Complement control family
HEV-D	EV-70	Sialic acid	Carbohydrate
HRV-A	HRV-16	ICAM-1	Immunoglobulin SF
	HRV-2	LDLR family	LDLR Family
HRV-B	HRV-14	ICAM-1	Immunoglobulin SF
HRV-C	?	?	

CAV coxsackie A virus, *CBV* coxsackie B virus, *E1* echovirus 1, *PV* poliovirus, *EV* enterovirus, *HRV* human rhinovirus, *HEV* human enterovirus, *Immunoglobulin SF* Immunoglobulin superfamily, *LDLR* low-density lipoprotein receptor several receptors (CD155, ICAM1, CAR) are members of the immunoglobulin superfamily. These are believed to bind within the enterovirus canyon and this may trigger conformational changes needed for uncoating (Rossmann et al. 2002). Integrins, which are heterodimeric proteins made up of one α and one β chain, are also recognised by a few enteroviruses, either by utilising an arginine–glycine–aspartic acid (RGD) motif (coxsackievirus A9 and echovirus 9) to bind to αv-containing integrins, or a non-RGD-dependent interaction with integrin α2β1 (echovirus 1). A diverse range of other receptors is used by enteroviruses, including DAF (decay accelerating factor), exploited by several echoviruses, coxsackie B viruses and some coxsackie A viruses. The carbohydrate molecules heparan sulphate and sialic acid are used by some enteroviruses (Goodfellow et al. 2001; Alexander and Dimock 2002). It is clear that receptors are not entirely species-specific (Table 12.1). Members of *Human rhinovirus A* (HRV-A) contains two subsets of viruses (major and minor receptor groups), which recognise quite different receptors (ICAM-1 and LDLP-family respectively), while HRV-B members recognise ICAM-1. Different HEV-C members also recognise more than one receptor, while HEV-B is particularly diverse in terms of receptor usage. This diversity, together with reported differences in receptor tropism within a serotype, implies some flexibility in receptor use and raises the possibility of changes in tropism within the infected individual which may contribute to the development of disease. Another dimension is potential sequence polymorphism of receptors among individuals, which could lead to an attractive hypothesis to explain differences in susceptibility to enteroviruses or the disease induced, but there seems to be no correlation in the case of type 1 diabetes (Karttunen et al. 2003).

Following interactions with the receptor, enteroviruses exploit one of several host cell endocytotic routes (Mercer et al. 2010). A clathrin-dependent pathway involving endosomes is used frequently and caveolin-dependent routes are also known, as well as several non-clathrin non-caveolin-dependent pathways. The entry route chosen may not be of critical importance as some viruses which use the same receptor seem to enter by different routes (Mercer et al. 2010; Coyne and Bergelson 2006).

Translation

Immediately following genome release into the cytoplasm, the RNA serves as a message for the translation of one long polyprotein (Whitton et al. 2005; Lin et al. 2009). Picornaviruses initiate translation by an internal ribosome entry site (IRES)-mediated mechanism, which differs from cap-dependent initiation of translation of most cellular RNAs, and this provides the opportunity for some picornaviruses to shut-off host cell protein synthesis by interfering with cap-recognition. The picornavirus IRES, located within the 5′UTR, is usually several 100 nucleotides long and falls into one of the four structural types (Lin et al. 2009). The enterovirus IRES belongs to type 1 and has been extensively studied, partly as a key attenuating mutation is found in this region in each of the three Sabin poliovirus vaccine strains.

This mutation leads to destabilisation of one of the several RNA structural domains needed for IRES function and seems to be cell-type-specific (Macadam et al. 2006). In addition, mutations in this region have been implicated in pathogenicity in several other picornaviruses (Whitton et al. 2005).

For full function, the enterovirus IRES is known to require binding of several host proteins in addition to canonical initiation factors (Lin et al. 2009). These include PTB, PCBP2, unr and La. It is likely that differential expression of such proteins plays an important role in defining which cell types can be infected and contributes to pathogenesis (Sarnow 2003).

RNA Replication

Following translation and processing of the polyprotein, the RNA genome must be replicated. RNA replication occurs in peri-nuclear membrane vesicles derived by proliferation of intracellular membranes. 2BC and 3A appear to be involved in this membrane remodelling (Lin et al. 2009). The 5′-most terminal domains of the RNA, which bind to both virus and cellular proteins, are critical for RNA replication, and RNA replication may also involve 3′UTR features, but the role of this region in enteroviruses has not been fully elucidated. Another region of critical importance in RNA replication is the cre, the site of uridylylation of VPg to give the primer required RNA synthesis. This feature, usually a stem-loop with an AAA motif in the loop, is found in different positions in different picornaviruses and is in the 2C-encoding region in enteroviruses (Goodfellow et al. 2000). Both uridylylation of VPg and extension of the primer are brought about by the action of 3Dpol. The involvement of cellular proteins and membranes in enterovirus RNA replication again suggests that this step may potentially be a tropism/pathogenicity determinant. Indeed, mutations in 2B and 3A give rise to host tropism changes in rhinoviruses (Harris and Racaniello 2005).

Host Cell Effects of Enterovirus Infection

In order to protect the infected cell against immune intervention and adapt it for efficient virus replication, assembly and release, in common with other viruses, enteroviruses trigger a multiplicity of changes to host cell structure and function (Whitton et al. 2005). These include the cleavage by virus proteases of a number of proteins involved in host cell translation, transcription and ultrastructure, as well as host cell defence mechanisms such as stress, and interferon responses and NF-κB activation (Lin et al. 2009). In addition to cleavage mechanisms, in enteroviruses, the small hydrophobic proteins 2B and 3A have been shown to interact with cell membranes and interfere with protein trafficking, for instance preventing MHC expression, as well as in the establishment of a novel organelle used in RNA replication (Choe et al. 2005). Again cell type or individual specific differences in these processes could play important roles in defining how enteroviruses cause disease.

Conclusion

The species *Enterovirus* contains many viruses, which although genetically coherent, display a wide variety of properties in terms of receptor tropism and pathogenesis. We need to understand the basis of these properties to facilitate the identification of features and specific virus types which give rise to type 1 diabetes, so that effective drug and vaccine strategies can be devised.

References

Alexander DA, Dimock K (2002) Sialic acid functions in enterovirus 70 binding and infection. J Virol 76:11265–11272

Choe SS, Dodd DA, Kirkegaard K (2005) Inhibition of cellular protein secretion by picornaviral 3A proteins. Virology 337:18–29

Coyne CB, Bergelson JM (2006) Virus-induced Abl and Fyn kinase signals permit coxsackievirus entry through epithelial tight junctions. Cell 124:119–131

Goodfellow I, Chaudhry Y, Richardson A, Meredith J, Almond JW, Barclay W, Evans DJ (2000) Identification of a cis-acting replication element within the poliovirus coding region. J Virol 74:4590–4600

Goodfellow IG, Sioofy AB, Powell RM, Evans DJ (2001) Echoviruses bind heparan sulfate at the cell surface. J Virol 75:4918–4921

Harris JR, Racaniello VR (2005) Amino acid changes in proteins 2B and 3A mediate rhinovirus type 39 growth in mouse cells. J Virol 79:5363–5373

Karttunen A, Pöyry T, Vaarala O, Ilonen J, Hovi T, Roivainen M, Hyypiä T (2003) Variation in enterovirus receptor genes. J Med Virol 70:99–108

Katpally U, Smith TJ (2007) Pocket factors are unlikely to play a major role in the life cycle of human rhinovirus. J Virol 81:6307–6315

Knowles NJ (2011) Picornavirus home page. http://www.picornaviridae.com/. Accessed 2 Jan 2012

Lin J-Y, Chen T-C, Weng K-F, Chang S-C, Chen L-L, Shih S-R (2009) Viral and host proteins involved in picornavirus life cycle. J Biomed Sci 16:103

Macadam AJ, Ferguson G, Stone DM, Meredith J, Knowlson S, Auda G, Almond JW, Minor PD (2006) Rational design of genetically stable, live-attenuated poliovirus vaccines of all three serotypes: relevance to poliomyelitis eradication. J Virol 80:8653–8663

Mercer J, Schelhaas M, Helenius A (2010) Virus entry by endocytosis. Annu Rev Biochem 79:803–833

Rossmann MG, He Y, Kuhn RJ (2002) Picornavirus-receptor interactions. Trends Microbiol 10:324–331

Sarnow P (2003) Viral Internal Ribosome Entry Site elements: novel ribosome-RNA complexes and roles in viral pathogenesis. J Virol 77:2801–2806

Stanway G, Brown F, Christian P, Hovi T, Hyypiä T, King AMQ, Knowles NJ, Lemon SM, Minor PD, Pallansch MA, Palmenberg AC, Skern T (2004) Picornaviridae. In: Fauquet CM, Mayo MA, Maniloff J, Desselberger U, Balleditors LA (eds) Virus taxonomy, VIIIth report of the ICTV. Elsevier/Academic, London, pp p757–778

Steil BP, Barton DJ (2009) Cis-active RNA elements (CREs) and picornavirus RNA replication. Virus Res 139:240–252

Tuthill TJ, Groppelli E, Hogle JM, Rowlands DJ (2011) Picornaviruses. Curr Top Microbiol Immunol 343:43–89

Whitton JL, Cornell CT, Feuer R (2005) Host and virus determinants of picornavirus pathogenesis and tropism. Nat Rev Microbiol 3:765–776

Witwer C, Rauscher S, Hofacker IL, Stadler PF (2001) Conserved RNA secondary structures in *Picornaviridae* genomes. Nucleic Acids Res 29:5079–5089

Chapter 13
Laboratory Diagnosis of Enterovirus Infection: Optimal Methods for Studies of Diabetes

Sami Oikarinen and Maarit Oikarinen

Abstract Laboratory diagnosis of enterovirus infections is more complex than that of many other virus infections. Careful design of study protocols and sample collection procedures is crucial for studies evaluating the role of enteroviruses in type 1 diabetes. Possible viral persistence creates an additional challenge, since the virus may be present in low quantities and in the form of double-stranded RNA. Both direct virus detection and serology have their own advantages and disadvantages, depending on the individual research questions, technologies, and sample types used in the studies. In many cases, their combined use would give the best view on the relationship between enteroviruses and type 1 diabetes. Standardization of enterovirus assays by international collaboration would help identify the optimal diagnostic approaches for type 1 diabetes studies.

Introduction

Enteroviruses have been linked to type 1 diabetes in various stages of beta-cell damaging process, including the time when the autoimmune process begins (the detection of first autoantibodies), during the progression of this process in children with autoantibodies, and at the onset of clinical diabetes (Stene et al. 2010; Oikarinen et al. 2011; Yeung et al. 2011). In addition, some studies have indicated that maternal infections during pregnancy may increase the risk of type 1 diabetes in the child (Hyöty et al. 1995; Elfving et al. 2008). These studies have been based on the "direct" detection of viral RNA or virus proteins in different sample types from prediabetic and type 1 diabetes patients or on the "indirect methods" detecting the markers of human immune response induced by virus infection (Yeung et al. 2011).

S. Oikarinen, M.D. (✉) • M. Oikarinen
School of Medicine, University of Tampere, Tampere, Finland
e-mail: sami.oikarinen@uta.fi

K.W. Taylor, H. Hyoty, A. Toniolo, A.J. Zuckerman (eds.), *Diabetes and Viruses*,
DOI 10.1007/978-1-4614-4051-2_13, © Springer Science+Business Media New York 2013

Reliable detection of enteroviruses in clinical samples is more challenging compared to the detection of many other viruses, mainly because of the high diversity of enteroviruses. They include more than 100 different serotypes with considerable genetic variation. The classification of enteroviruses has been revolutionized by the implementation of molecular methods which can identify genetic relationships by sequencing the viral genome (Oberste et al. 1999a, b). These methods have identified over 30 new enterovirus types and their number is continuously increasing.

Detection of Enteroviruses

Optimal Samples

The majority of the studies addressing the role of enteroviruses in type 1 diabetes have been carried out at the time of diagnosis of the disease. However, it would be essential to study factors that trigger the type 1 diabetes process and which can occur several years before the clinical disease is manifested. Critical infections may be experienced in early infancy or even in utero during pregnancy, and therefore prospective sample series covering the time from birth to the diagnosis of type 1 diabetes would be optimal. In addition, such longitudinal sample series should be taken at relatively short intervals and cover different kinds of sample types where enteroviruses can be detected. Blood and stool samples are usually used when diagnosing enterovirus infections by direct virus detection or antibody assays. In the case of persisting and slowly replicating infection, direct detection of the virus would be particularly challenging, and additional sample types such as tissue samples may be needed. However, it is difficult to obtain biopsies from the pancreas for this type of studies, and such samples are available only in exceptional cases (Imagawa et al. 2001). Prospective studies are very expensive and time consuming, and need a well-organized infrastructure for clinical follow-up. Since the predictive value of the best diabetes risk markers is still far below 100%, a large number of originally non-diabetic subjects have to be followed-up and only few will develop diabetes. So far, extensive prospective studies have been carried out in a few countries including Finland (DIPP and DiMe studies), Germany (BabyDiab study), the USA (DAISY study), and Norway (MIDIA study). Recently, a multicenter TEDDY study has been started in the USA, Finland, Sweden, and Germany. Most of these studies have focused on children who have increased genetic risk for type 1 diabetes and these children have been followed-up from birth.

The selection of the appropriate sample material is crucial for optimal detection of enteroviruses and their possible association with type 1 diabetes. The primary replication of enteroviruses occurs in the intestinal and respiratory mucosa. Infection may be limited to the mucosal surfaces, but in many cases the virus spreads to the blood causing primary viremia. Subsequently, virus may spread to secondary replication sites such as the pancreas and later a secondary episode of viremia may occur. Viremia is short and usually lasts from a few days to no more than 2 weeks. Thus, such a short period reduces the possibility of detecting the virus in the blood, particularly because enterovirus infections are usually asymptomatic and sampling can only be guided occasionally by typical symptoms (Racaniello 2001). This creates a big challenge particularly in prospective studies where blood samples are taken regularly according to a predetermined schedule with relatively long intervals. Viremia can be detected by virus isolation or by PCR (viral RNA in blood). Virus isolation from blood has not been widely used in diabetes studies, but viral RNA has been detected using PCR in whole blood samples as well as in serum or plasma samples taken from type 1 diabetic patients more frequently than in control subjects. In addition, viral RNA has been detected in serum and plasma taken from prediabetic individuals in prospective studies (Tauriainen et al. 2010; Yeung et al. 2011).

In addition to acute infection, viremia may also occur in persistent infection. In such a case, infectious virus is not necessarily present but the virus may be detectable in white blood cells at the RNA or double-stranded RNA level. Some studies have suggested that enteroviruses can be detected in antigen presenting cells in patients with type 1 diabetes (Schulte et al. 2010). These observations suggest that samples enriched for antigen presenting cells might be one of the most optimal targets for virus detection in the peripheral blood of patients with type 1 diabetes.

Enteroviruses are common in young children and are often detected in stool samples collected at random from healthy children. Thus, there is a risk that background infections may mask the possible risk effect of diabetogenic enterovirus types, especially in countries where enterovirus infections are common. In fact, even though enteroviruses have been detected more frequently in the blood of patients with type 1 diabetes than in control subjects, no such difference has been found in stool samples. In a recent study enterovirus was detected equally frequently in stool samples collected from children who developed islet autoantibodies as in control children (Tapia et al. 2011). It is also possible that the frequency of enterovirus infections is about the same in healthy and type 1 diabetes cases, but the susceptibility to the diabetogenic effect of the virus differs between the groups. In such scenario, the risk effect of enteroviruses would be detectable in the subgroup of children who carry this susceptibility (e.g., certain risk genes such as IFIH1).

Detection of enteroviruses in the primary and secondary replication sites, such as biopsies from the intestine and the pancreas, would provide important additional information about type 1 diabetes process. Detection of the virus in the pancreatic islets would be particularly important since it would provide a biological explanation for the islet inflammatory process which is the hallmark of type 1 diabetes. However, due to the anatomic location of the pancreas, it has been difficult to obtain such samples. Currently, large-scale international studies are in progress to collect such samples from prediabetic and diabetic subjects (nPOD study organized in the USA and euroPOD in Europe).

Detection of Viral RNA by RT-PCR

RT-PCR offers certain important advantages for studies evaluating the viral etiology of type 1 diabetes. It is generally more sensitive than other methods used for the detection of viruses directly in clinical specimen. In addition, it makes it possible to study the molecular structure of the viral genome by sequencing PCR amplicons. PCR methods which specifically amplify enterovirus RNA genome have been used widely for detection of the virus in the blood, stool, and tissue samples of patients with type 1 diabetes. These studies have found the virus more frequently in the diabetic patients than in controls (see the recent meta-analysis by Yeung et al. 2011). The sensitivity of virus-specific PCR assays is usually better than that of the new next-generation sequencing methods which are becoming more and more popular, making it possible to detect a wide range of different microbes in a single test. High sensitivity is an important goal because the amount of the virus can be very low due to the nature of the infection such as persistent infection or available sample types, e.g., blood and tissue samples.

Optimization of PCR-based methods is critical for reliable detection of enteroviruses. The assay should be sensitive and specific yet relatively insensitive to PCR inhibitors (Oikarinen et al. 2009). International quality control programs such as Quality Control for Molecular Diagnostics (QCMD) have shown that the sensitivity of enterovirus PCR varies widely between different laboratories, and samples with low virus load are not detected by all methods. These types of quality control panels have indicated that virus laboratories can also have problems with contamination, which may lead to false-positive findings. To achieve high quality in PCR-based diagnosis, the different steps of the PCR assay (RNA extraction, RT and PCR enzymes and reaction conditions, and the sequences of primers) must be optimized and adjusted with the genetic variation of enteroviruses, type of infection to be diagnosed (acute vs. persistent), and sample material available (e.g., amount of PCR inhibitors in the sample). In addition, the technical performance of PCR should be monitored by positive, negative, and internal control samples included in each PCR run. The life cycle of enteroviruses may have relevance in diabetes research, since there is some evidence that the virus may persist in double-stranded RNA form in the pancreas or other tissues in patients with type 1 diabetes and/or the RNA genome may also be in the negative strand form (Klingel et al. 1992; Richardson et al. 2011). In such cases pre-heating of the RNA sample to denaturate double-stranded RNA and use of both sense and antisense primers in the RT-PCR may be important. In addition, deletions may be present in the 5′UTR region of the genome of persisting virus variants (Chapman et al. 2008). Such deletions could have critical consequences, if the primer annealing site is located in that region.

Molecular Typing of Enteroviruses

Enteroviruses cluster into two groups based on their 5′UTR region. On the other hand, the nonstructural gene regions are species-specific, i.e., enteroviruses are clustered into four genogroups A–D. If genotyping is used to identify the serotype of the virus (a correlate of traditional serological typing) the capsid protein coding regions VP1–VP4 should be sequenced. In this region, the intragenotypic divergence in the nucleotide level is up to 25% and in amino acid level up to 12%

(Oberste et al. 1999a, b). This is, of course, an advantage for typing itself, but a disadvantage for primer design, because the use of wobble nucleotides and the long length of resulting PCR amplicons decrease the sensitivity of PCR. The most reliable region for the genotyping of enteroviruses is the VP1 region which contains the major antigenic sites (Oberste et al. 1999a, b; Nix et al. 2006). All known enteroviruses have been sequenced using this region and the sequences can be found in the GenBank. Therefore, VP1 sequencing is becoming more and more important for genotyping of enteroviruses. Genotyping is often performed using blast search, but deeper phylogenetic analysis provides more reliable results. For successful phylogenetic analysis expertise and understanding of different methods as well as the influence of different assay parameters are needed. For example, (Kroneman et al. 2011) have published Web-based genotyping tools using optimized algorithms and parameters for phylogenetic analyses.

Virus Isolation

Enteroviruses can be isolated from several types of samples, including tissue and blood samples as well as respiratory secretions and stools. The concentration of the virus is the highest in stool and in respiratory secretions (e.g., throat swabs). Stool samples are widely used as primary samples for enterovirus isolation. In some cases, enteroviruses can be detected in stool samples for prolonged time periods after infection, ranging up to several weeks or even months. However, this is not true for all infections as certain enteroviruses are preferentially excreted via respiratory route. The main advantage of virus isolation is the possibility of using the isolated virus strains in further experimental studies in different model systems and to characterize them molecularly in detail (complete sequence). For example, virus strains isolated at the start of the beta cell damaging process could be studied in human pancreatic islet-cell cultures to see possible specific interactions with these cells. The main disadvantage of virus isolation is its relatively low sensitivity compared to PCR. In addition, many enterovirus serotypes do not grow well in cell lines, a fact that may cause false negative results. Virus isolation is also labor-intensive requiring sterile cell culture work. Therefore, PCR has largely replaced virus isolation in diagnostic laboratories. In diabetes research, PCR can be done first, and virus isolation can be attempted from PCR positive samples. However, it should be noted that in samples which contain large concentrations of PCR inhibitors, the sensitivity of virus isolation can actually be better than that of PCR. Such inhibitors are frequent, for instance, in stool samples (Oikarinen et al. 2009).

Tissue Tests

Immunohistochemistry (IHC) and in situ hybridization (ISH) are the methods most commonly used for detection of enteroviruses in tissue samples (Fig. 13.1). In addition, RT-PCR can be used, even though its sensitivity may not be optimal in formalin-fixed

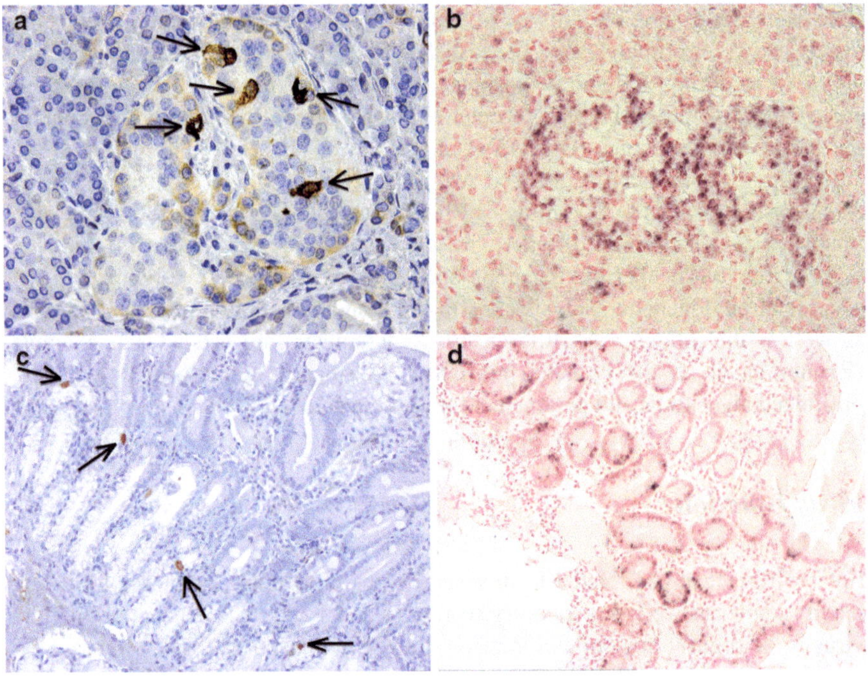

Fig. 13.1 Detection of enterovirus in the pancreas (**a**, **b**) and small intestinal mucosa (**c**, **d**). In panels **a** and **c** the *brown color* indicates the presence of enterovirus VP1 protein (immunohistochemistry) and in panels **b** and **d** the *dark purple* precipitate indicates the presence of enteroviral genome (in situ hybridization). Pancreas samples were provided by nPOD and small intestinal mucosa samples by Professor Markku Mäki

samples and usually needs frozen or fresh tissue or samples which have been treated with special RNA preservation buffers. IHC is based on specific antibodies against enterovirus proteins, whereas ISH uses probes designed to hybridize with the enteroviral genome. Pancreatic tissue samples are usually obtained from autopsy or organ donation, although pancreas biopsies have also been performed occasionally (Imagawa et al. 2001).

IHC and ISH have been developed and optimized to be used for both formalin-fixed paraffin-embedded and frozen samples. Unlike other enterovirus screening methods, IHC and ISH enable the localization of virus in different regions and anatomical sites of the target tissue and, using double-staining with specific antibodies against enterovirus and pancreatic islet hormones such as insulin, glucagon, somatostatin, they allow localization of the virus in different cell types.

The antibody used most frequently for detection of enterovirus by IHC is a commercial monoclonal antibody clone 5-D8/1 (DakoCytomation). This antibody was developed in 1987 (Yousef et al. 1987) and it recognizes a conserved group-specific epitope in enteroviral VP1 capsid protein (Samuelson et al. 1995). It reacts with a wide range of different enterovirus serotypes in infected cell culture samples (Trabelsi et al. 1995; Oikarinen et al. 2010). On the other hand, it has been shown in some studies (Terletskaia-Ladwig et al. 2008; Miao et al. 2009) that this antibody fails to detect several CAV and echovirus serotypes, as well as EV68-71 serotypes. One important concern is that this clone has been reported to cross-react with certain host proteins such as HSP60/65 and IA-2 which are expressed in the pancreas (Harkonen et al. 2000, 2002). It may also react with uninfected human cardiomyocytes (Klingel et al. 2004), vascular smooth muscle cells, and centroacinar cells in the exocrine pancreas (Richardson et al. 2009). The Enterovirus Screening Set (Chemicon) includes four species-specific antibody blends (Coxsackievirus B Blend, Echovirus Blend, Enterovirus Blend and Poliovirus Blend, and Pan-Enterovirus Blend). The Pan-Enterovirus Blend is a mixture of two monoclonal antibodies 9D5 and 2E11, presumably designed against a virus-encoded, non-virion determinant (Yagi et al. 1992). These antibody mixtures have also been reported to react widely with different enterovirus serotypes and also with viruses other than enteroviruses (Miao et al. 2009). Thus, IHC has been used widely for detection of enterovirus proteins in tissue samples. However, caution is needed in the interpretation of positive findings since cross-reactivity with host antigens may occur. Optimally stringent assay conditions are critical for avoiding this cross-reactivity, and confirmation of positive staining with other methods is recommended.

Non-isotopic ISH applications have became more and more popular and have replaced isotopic methods. For example, digoxigenin-labeled probes are used commonly in enterovirus detection (Hohenadl et al. 1991; Oikarinen et al. 2010). The probes are generally designed to hybridize with a conserved sequence which is common to all known enterovirus serotypes, and also species-specific probes can be used (Foulis et al. 1997; Ylipaasto et al. 2004). ISH is technically quite challenging due to multiple assay steps, and it is always a compromise to obtain a sufficiently strong enough positive signal without gaining high levels of background staining. As in PCR, positive and negative controls should be included in each test run to monitor the quality and reproducibility of ISH assays.

It has been suggested that persistent enterovirus infection in the small intestine and/or in the pancreas may be an important factor in the pathogenesis of type 1 diabetes. Detection of this kind of slowly replicating persisting virus is much more challenging than that of actively replicating virus. During an acute enterovirus infection, the amount of positive-stranded RNA is 50–100 times higher than that of negative-stranded RNA (Chehadeh et al. 2000). In contrast, in persistent infection the amount of positive- and negative-stranded RNA is about equal and the synthesis of capsid proteins can be decreased. In this case virus replication occurs mainly at the RNA level (Klingel et al. 1992) where double-stranded viral RNA complexes are formed (Tam and Messner 1999). Antibodies against double-stranded RNA and PKR (dsRNA-dependent protein kinase) have been used to detect these molecules in cell and tissue samples (Richardson et al. 2009, 2010), which offers one option for detection of viral persistence. It is also possible to use strand-specific ISH or RT-PCR to study the balance between positive- and negative-stranded viral RNA (Klingel et al. 1992; Foulis et al. 1997; Ylipaasto et al. 2004).

The limited availability of samples from the pancreas of patients with type 1 diabetes has hindered the progress in this research field. The process leading to type 1 diabetes progress is usually slow and it would be vital to show the presence of the virus at the beginning of the process. This creates a huge challenge for studies with pancreatic tissue, since such samples should be taken long before type 1 diabetes is diagnosed. Another important aspect relates to the processing of tissue samples. Pancreatic enzymes start to degrade the tissue very rapidly; therefore, the sample should be either fixed or frozen immediately, and in case of autopsy samples, the post-mortem time should be as short as possible. Formalin-fixation is also known to degrade some of the RNA.

The diabetic process in the pancreas is often "patchy" (Foulis 1996), i.e., morphological changes can vary in different parts of the pancreas. Thus, a single sample from one part of the pancreas does not necessarily provide a representative view of the disease process. Possible enterovirus infection might be restricted to certain part of the organ and several sections of the pancreas should be examined to detect the virus using ISH or IHC.

In summary, the detection of enteroviruses in tissue samples is demanding and the results have been varied. Because of the reported cross-reactivity with host tissue and possible low sensitivity of some enterovirus antibodies, it is highly recommended to confirm the result of IHC using other methods. It is also important to take into consideration the possible viral persistence when interpreting the results. In persistent infections, the viral genome should be detectable using ISH, while synthesis of viral proteins may be at a very low level leading to negative result by IHC.

Antibody Assays

Virus antibodies are usually measured in serum or plasma, but other sample types such as whole blood, dried blood spots, and stool samples can also be used. The sensitivity of these assays for detection of enterovirus infections depends on several factors including the technical set-up of the assay and the sample type. Acute infections are typically diagnosed by the presence of virus-specific IgM in a single sample or by detection of increases in antibody levels (IgG, IgM, or IgA) between two serial samples. Usually, enterovirus antibodies are measured using enzyme immunoassay (EIA) and previously also by using radioimmunoassay. Infection history can be studied by measuring IgG class antibodies, but it should be noted that IgG responses which are detected by EIA can be transient lasting only for a few months. In longitudinal sample series, such as those collected in prospective follow-up studies, the length of sampling interval has a critical impact on assay sensitivity. Long sample intervals can lead easily to false-negative findings since transient antibody responses remain undetected. In contrast to antibodies measured by EIA, the antibodies measured by a neutralization assay last longer and can be used as a marker of past infection ("serological scar").

Serological diagnosis of enterovirus infections is more complex than that of many other virus infections. The reason for this is the large number of different enterovirus serotypes which makes it difficult to cover them all by antibody assays. EIA is the method used most widely. Its ability to detect enterovirus antibodies depends on the antigen which is used in the assay, immunoglobulin isotype measured, set-up of the assay, and the nature of sample material. It is important to realize that the antibodies which are detected by EIA are not usually specific for any particular enterovirus serotype, since enteroviruses contain antigenic structures which are common to several different enterovirus serotypes. These cross-reactive epitopes become exposed on the virus surface when it becomes attached to the plastic surface of the EIA plate (Torfason et al. 1988). In spite of this broadly reactive nature of enterovirus antigens, the EIA antibody assays do not cover all enterovirus serotypes. Thus, using a single serotype as an EIA antigen, it is not possible to detect antibodies against all enteroviruses, but rather a subgroup of antigens. Synthetic peptides carrying epitopes common to several enteroviruses have been successfully used as broadly reactive antigens in EIAs (Hovi and Roivainen 1993; Samuelson et al. 1994; Hyöty et al. 1995).

The standard indirect EIA where the virus antigen is bound to the plastic is usually suitable for the measurement of IgG and IgA class antibodies against enteroviruses. However, reliable measurement of IgM usually requires an antibody-capture format where IgM class immunoglobulins are first captured by anti-human IgM antibodies on EIA plate, followed by incubation of serum, virus, and detection layer, respectively. This type of capture assay can be also used for IgG and IgA measurements. The advantage of such antibody-capture assays is that they eliminate the competition between different antibody classes in the binding to the virus antigen.

The most sensitive and specific serological method is the measurement of neutralizing antibodies. By definition, these antibodies can neutralize the infectivity of the virus in vitro. Neutralizing antibodies do not usually cross-react between different enterovirus serotypes and can therefore identify the type of the virus causing the infection. These antibodies also remain elevated for several years (even decades) after the infection making it possible to study the past infection history. However, transient antibody responses also occur, and serological responses may even be absent, particularly if the titer of virus exposure has been small (Saliba et al. 1968). In addition, immune protection against enteroviruses is mainly based on neutralizing antibodies giving an important biological correlate for the antibody results. In spite of these important advantages, neutralizing antibody assays have not been widely used in studies evaluating enterovirus-diabetes association. This is due to the expensive and labor-intensive techniques (sterile cell culture work) as well as the required special knowledge of enterovirus biology which have hindered the extension of these methods outside specialized virus laboratories. In addition, the serotype-specific nature of these antibodies means that several assays must be carried out in parallel to measure antibodies against different enterovirus serotypes. Thus, large-scale screening of neutralizing antibodies against several enterovirus types is extremely more expensive compared to standard EIA techniques.

In conclusion, antibody assays have important advantages over direct virus detection methods. They make it possible to study past enterovirus exposures (serological scar) and diagnose acute infections from samples which have been taken after the infection has been cleared and the virus is no longer detectable. The combined use of direct virus detection and virus serology can considerably increase the sensitivity of enterovirus diagnosis in studies evaluating the viral etiology of type 1 diabetes.

Acknowledgments We wish to thank the Network for Pancreatic Organ Donors with Diabetes (nPOD), a collaborative type 1 diabetes research project sponsored by the Juvenile Diabetes Research Foundation International (JDRF) for the pancreas samples. Organ Procurement Organizations (OPO) partnering with nPOD to provide research resources are listed at www.jdrfnpod.org/our-partners.php. We thank Professor Markku Mäki and Dosent Katri Kaukinen for small intestine biopsy samples. Also we are grateful to The Juvenile Diabetes Research Foundation International, EU commission (PEVET Research Program, Grant Agreement Number 261441), and The Academy of Finland for funding our research.

References

Chapman NM, Kim KS, Drescher KM, Oka K, Tracy S (2008) 5′ terminal deletions in the genome of a coxsackievirus B2 strain occurred naturally in human heart. Virology 2:480–491

Chehadeh W, Kerr-Conte J, Pattou F, Alm G, Lefebvre J, Wattre P, Hober D (2000) Persistent infection of human pancreatic islets by coxsackievirus B is associated with alpha interferon synthesis in beta cells. J Virol 21:10153–10164

Elfving M, Svensson J, Oikarinen S, Jonsson B, Olofsson P, Sundkvist G, Lindberg B, Lernmark A, Hyoty H, Ivarsson SA (2008) Maternal enterovirus infection during pregnancy as a risk factor in offspring diagnosed with type 1 diabetes between 15 and 30 years of age. Exp Diabetes Res 2008:271958

Foulis AK (1996) The pathology of the endocrine pancreas in type 1 (insulin-dependent) diabetes mellitus. Acta Pathol Microbiol Immunol Scand 3:161–167

Foulis AK, McGill M, Farquharson MA, Hilton DA (1997) A search for evidence of viral infection in pancreases of newly diagnosed patients with IDDM. Diabetologia 1:53–61

Harkonen T, Puolakkainen M, Sarvas M, Airaksinen U, Hovi T, Roivainen M (2000) Picornavirus proteins share antigenic determinants with heat shock proteins 60/65. J Med Virol 3:383–391

Harkonen T, Lankinen H, Davydova B, Hovi T, Roivainen M (2002) Enterovirus infection can induce immune responses that cross-react with beta-cell autoantigen tyrosine phosphatase IA-2/IAR. J Med Virol 3:340–350

Hohenadl C, Klingel K, Mertsching J, Hofschneider PH, Kandolf R (1991) Strand-specific detection of enteroviral RNA in myocardial tissue by in situ hybridization. Mol Cell Probes 1:11–20

Hovi T, Roivainen M (1993) Peptide antisera targeted to a conserved sequence in poliovirus capsid VP1 cross-react widely with members of the genus Enterovirus. J Clin Microbiol 5:1083–1087

Hyöty H, Hiltunen M, Knip M, Laakkonen M, Vähäsalo P, Karjalainen J, Koskela P, Roivainen M, Leinikki P, Hovi T, Åkerblom HK, DiMe-study group (1995) A prospective study of the role of Coxsackie B and other enteroviruses in the pathogenesis of IDDM. Diabetes 44: 652–657

Imagawa A, Hanafusa T, Tamura S, Moriwaki M, Itoh N, Yamamoto K, Iwahashi H, Yamagata K, Waguri M, Nanmo T, Uno S, Nakajima H, Namba M, Kawata S, Miyagawa JI, Matsuzawa Y (2001) Pancreatic biopsy as a procedure for detecting in situ autoimmune phenomena in type 1 diabetes: close correlation between serological markers and histological evidence of cellular autoimmunity. Diabetes 6:1269–1273

Klingel K, Hohenadl C, Canu A, Albrecht M, Seemann M, Mall G, Kandolf R (1992) Ongoing enterovirus-induced myocarditis is associated with persistent heart muscle infection: quantitative analysis of virus replication, tissue damage, and inflammation. Proc Natl Acad Sci U S A 1:314–318

Klingel K, Sauter M, Bock CT, Szalay G, Schnorr JJ, Kandolf R (2004) Molecular pathology of inflammatory cardiomyopathy. Med Microbiol Immunol 2–3:101–107

Kroneman A, Vennema H, Deforche K, v d Avoort H, Peñaranda S, Oberste MS, Vinjé J, Koopmans M (2011) An automated genotyping tool for enteroviruses and noroviruses. J Clin Virol 51(2):121–125

Miao LY, Pierce C, Gray-Johnson J, DeLotell J, Shaw C, Chapman N, Yeh E, Schnurr D, Huang YT (2009) Monoclonal antibodies to VP1 recognize a broad range of enteroviruses. J Clin Microbiol 10:3108–3113

Nix WA, Oberste MS, Pallansch MA (2006) Sensitive, seminested PCR amplification of VP1 sequences for direct identification of all enterovirus serotypes from original clinical specimens. J Clin Microbiol 8:2698–2704

Oberste MS, Maher K, Kilpatrick DR, Flemister MR, Brown BA, Pallansch MA (1999a) Typing of human enteroviruses by partial sequencing of VP1. J Clin Microbiol 5:1288–1293

Oberste MS, Maher K, Kilpatrick DR, Pallansch MA (1999b) Molecular evolution of the human enteroviruses: correlation of serotype with VP1 sequence and application to picornavirus classification. J Virol 3:1941–1948

Oikarinen S, Tauriainen S, Viskari H, Simell O, Knip M, Virtanen S, Hyoty H (2009) PCR inhibition in stool samples in relation to age of infants. J Clin Virol 3:211–214

Oikarinen M, Tauriainen S, Penttila P, Keim J, Rantala I, Honkanen T, Hyoty H (2010) Evaluation of immunohistochemistry and in situ hybridization methods for the detection of enteroviruses using infected cell culture samples. J Clin Virol 3:224–228

Oikarinen S, Martiskainen M, Tauriainen S, Huhtala H, Ilonen J, Veijola R, Simell O, Knip M, Hyoty H (2011) Enterovirus RNA in blood is linked to the development of type 1 diabetes. Diabetes 1:276–279

Racaniello VR (2001) Picornaviridae: the viruses and their replication. In: Knipe DM, Howley PM (eds) Fields virology. Lippincott Williams & Wilkins, Philadelphia, pp 685–722

Richardson SJ, Willcox A, Bone AJ, Foulis AK, Morgan NG (2009) The prevalence of enteroviral capsid protein vp1 immunostaining in pancreatic islets in human type 1 diabetes. Diabetologia 6:1143–1151

Richardson SJ, Willcox A, Hilton DA, Tauriainen S, Hyoty H, Bone AJ, Foulis AK, Morgan NG (2010) Use of antisera directed against dsRNA to detect viral infections in formalin-fixed paraffin-embedded tissue. J Clin Virol 3:180–185

Richardson SJ, Willcox A, Bone AJ, Morgan NG, Foulis AK (2011) Immunopathology of the human pancreas in type-I diabetes. Semin Immunopathol 1:9–21

Saliba GS, Franklin SL, Jackson GG (1968) ECHO-11 as a respiratory virus: quantitation of infection in man. J Clin Invest 6:1303–1313

Samuelson A, Forsgren M, Johansson B, Wahren B, Sallberg M (1994) Molecular basis for serological cross-reactivity between enteroviruses. Clin Diagn Lab Immunol 3:336–341

Samuelson A, Forsgren M, Sallberg M (1995) Characterization of the recognition site and diagnostic potential of an enterovirus group-reactive monoclonal antibody. Clin Diagn Lab Immunol 3:385–386

Schulte BM, Bakkers J, Lanke KH, Melchers WJ, Westerlaken C, Allebes W, Aanstoot HJ, Bruining GJ, Adema GJ, Van Kuppeveld FJ, Galama JM (2010) Detection of enterovirus RNA in peripheral blood mononuclear cells of type 1 diabetic patients beyond the stage of acute infection. Viral Immunol 1:99–104

Stene LC, Oikarinen S, Hyoty H, Barriga KJ, Norris JM, Klingensmith G, Hutton JC, Erlich HA, Eisenbarth GS, Rewers M (2010) Enterovirus infection and progression from islet autoimmunity to type 1 diabetes: the Diabetes and Autoimmunity Study in the Young (DAISY). Diabetes 12:3174–3180

Tam PE, Messner RP (1999) Molecular mechanisms of coxsackievirus persistence in chronic inflammatory myopathy: viral RNA persists through formation of a double-stranded complex without associated genomic mutations or evolution. J Virol 12:10113–10121

Tapia G, Cinek O, Rasmussen T, Witso E, Grinde B, Stene LC, Ronningen KS (2011) Human enterovirus RNA in monthly fecal samples and islet autoimmunity in Norwegian children with high genetic risk for type 1 diabetes: the MIDIA study. Diabetes Care 1:151–155

Tauriainen S, Oikarinen S, Oikarinen M, Hyoty H (2010) Enteroviruses in the pathogenesis of type 1 diabetes. Semin Immunopathol 1:45–55

Terletskaia-Ladwig E, Meier S, Hahn R, Leinmuller M, Schneider F, Enders M (2008) A convenient rapid culture assay for the detection of enteroviruses in clinical samples: comparison with conventional cell culture and RT-PCR. J Med Microbiol 8:1000–1006

Torfason EG, Galindo R, Keyserling HL (1988) Comparison of five ELISA assays for IgG antibody against coxsackievirus B1. J Med Virol 1:53–60

Trabelsi A, Grattard F, Nejmeddine M, Aouni M, Bourlet T, Pozzetto B (1995) Evaluation of an enterovirus group-specific anti-VP1 monoclonal antibody, 5-D8/1, in comparison with neutralization and PCR for rapid identification of enteroviruses in cell culture. J Clin Microbiol 9:2454–2457

Yagi S, Schnurr D, Lin J (1992) Spectrum of monoclonal antibodies to coxsackievirus B-3 includes type- and group-specific antibodies. J Clin Microbiol 9:2498–2501

Yeung WC, Rawlinson WD, Craig ME (2011) Enterovirus infection and type 1 diabetes mellitus: systematic review and meta-analysis of observational molecular studies. Br Med J 342:d35

Ylipaasto P, Klingel K, Lindberg AM, Otonkoski T, Kandolf R, Hovi T, Roivainen M (2004) Enterovirus infection in human pancreatic islet cells, islet tropism in vivo and receptor involvement in cultured islet beta cells. Diabetologia 2:225–239

Yousef GE, Brown IN, Mowbray JF (1987) Derivation and biochemical characterization of an enterovirus group-specific monoclonal antibody. Intervirology 3:163–170

Chapter 14
Enterovirus Immunity and the "Hygiene Hypothesis"

Heikki Hyöty

Abstract The current knowledge fits with a scenario where one or more commonly circulating enteroviruses initiate the beta-cell damaging process in early childhood. These viruses may represent certain enterovirus types and strains which have special properties explaining their ability to cause beta-cell damage (e.g., tropism to islet cells). Particular susceptibility for the virus is probably needed and only a small fraction of infected individuals may develop beta-cell damage. In this sense type 1 diabetes has many similarities with polio, the well-known enterovirus disease. Based on experience from polio, both the increasing incidence of type 1 diabetes and the remarkable geographical variation in the incidence rates can be related to varying circulation of diabetogenic enteroviruses in these populations. On the other hand, animal experiments have suggested that under certain conditions enteroviruses may also have a protective effect, which seem to be mediated by their ability to activate immunoregulatory mechanisms. Both these aspects (risk vs. protective effect) should be taken into account when possible viral effects on the epidemiology of type 1 diabetes are investigated. Large-scale birth cohort studies, such as the TEDDY study, will play a key role in the identification of these effects and virus–host interactions which determine the outcome of the infection.

Introduction

To understand possible viral effects on the epidemiological patterns of type 1 diabetes, one needs to combine epidemiological observations with the information on the mechanisms of enterovirus-induced beta-cell damage. Currently, there is no generally

H. Hyöty, M.D., Ph.D. (✉)
Department of Virology, University of Tampere and Tampere University Hospital, Biokatu 10, 33520 Tampere, Finland
e-mail: heikki.hyoty@uta.fi

K.W. Taylor, H. Hyoty, A. Toniolo, A.J. Zuckerman (eds.), *Diabetes and Viruses*,
DOI 10.1007/978-1-4614-4051-2_14,

accepted single mechanism whereby enteroviruses could cause diabetes, but there are to be two main scenarios, which are relevant to these considerations. The traditional "virus disease" scenario is based on the idea that certain enterovirus destroys pancreatic beta cells, either directly or by immune-mediated mechanisms. This fits well with previous experience from many enterovirus diseases, such as polio. Another scenario is based on the hypothesis that under certain conditions enteroviruses may also have a protective effect: they can induce immunoregulatory responses which can downregulate autoreactive immune responses in analogy with hygiene hypothesis in allergy. This chapter summarizes epidemiological observations in light of these two scenarios.

Epidemic Features of the Beta-Cell Damaging Process

Beta cells are destroyed by an immune-mediated process, which usually progress slowly and may continue for years before clinical type 1 diabetes is diagnosed. The process starts early, and the majority of children who develop type 1 diabetes have turned positive for islet autoantibodies before the age of 3 years, in many cases already before the age of 1 year (Siljander et al. 2009). Therefore, to understand the dynamic relationships between the epidemiology of enterovirus infections type 1 diabetes, the initiation phase which occurs long before the appearance of symptoms of diabetes should be studied.

In fact, several publications support the idea that enterovirus infections play a role in this initiation phase. One of the most interesting observations is the clear seasonal pattern of the onset of this autoimmune process; children turn positive for autoantibodies usually during the late summer and fall, which parallels the seasonality of enterovirus infections (Kimpimaki et al. 2001). This pattern is seen in almost every year. Altogether, these seasonal peaks support the "epidemic" nature of the autoimmune process. Accordingly, assuming that a virus initiates the process, this should happen quite soon after the infection with a relatively constant time-lag. These observations indicate that the search for diabetogenic enteroviruses should focus on viruses occurring at this early age and circulating almost every year. This conclusion is supported by prospective studies showing a "peak" in enterovirus infections a few months before autoantibodies are first detected (Hiltunen et al. 1997; Oikarinen et al. 2011; Salminen et al. 2003) as well as animal experiments showing that enterovirus infection induces islet autoantibodies and diabetes with a few weeks' time-lag (Chatterjee et al. 1992; Gerling et al. 1991). The diagnosis of clinical type 1 diabetes follows a similar seasonal pattern, even though it may not be as clear as that observed in autoantibody seroconversions (Gamble and Taylor 1969; Moltchanova et al. 2009). Interestingly, the seasonality of type 1 diabetes is the strongest in countries with high incidence of type 1 diabetes and clear seasonality in enterovirus infections (this correlates with the latitude). It is also stronger in boys than in girls which parallels the gender bias in severe enterovirus diseases (Moltchanova et al. 2009).

It is possible that the same infectious agent which initiates the process can also accelerate it and precipitate the symptoms of type 1 diabetes. In fact, enteroviruses have been detected both at the initiation phase (autoantibody seroconversion) and at the diagnosis of clinical type 1 diabetes possibly reflecting this kind of "multiple hit" effect (Yeung et al. 2011).

Type 1 Diabetes as a Viral Disease

In this scenario enteroviruses are considered as a necessary causal factor for the development of type 1 diabetes. In other words, diabetes is considered as a complication of enterovirus infection. This has been the dominating hypothesis in studies evaluating the association between enteroviruses and diabetes and it has been supported by a number of epidemiological and experimental observations.

One of the key observations is the strong tropism of enteroviruses to human pancreatic islets. Several studies have suggested that human islet cells are highly permissive for a number of different enterovirus types in vitro (Chehadeh et al. 2000; Roivainen et al. 2002; Yin et al. 2002) and that enterovirus proteins and genome are present in the pancreatic islets of both diabetic patients and children who have died of enterovirus infections (Dotta et al. 2007; Foulis et al. 1990; Oikarinen et al. 2008; Richardson et al. 2009; Tauriainen et al. 2011). The virus has been found predominantly in beta cells while the exocrine pancreas has been mostly negative. This tropism correlates also with the expression of one major enterovirus receptor, coxsackie and adenovirus receptor, by the islet cells (Oikarinen et al. 2008). Thus, one key mechanism of enterovirus-induced beta-cell damage may be related to the capability of the virus to reach the pancreatic islets. It is possible that such tropism is characteristic to only certain specific enterovirus types and/or strains, analogously with the tropism of the three poliovirus serotypes to motoneurons (polioviruses belong to enteroviruses). In mouse models enteroviruses infect mostly exocrine pancreas but some strains can also infect the islets (Jaidane et al. 2009). In addition, studies carried out with a closely related picornavirus, encephalomyocarditis virus, support this scenario: this virus has two variants (strains), one being highly tropic to insulin-producing beta cells and causing diabetes in infected animals (Jun and Yoon 2001).

If the diabetogenic enterovirus types are common, as indirectly suggested by the epidemiological observations described above, the virus needs to hit a particularly susceptible host to be able to cause diabetes. Susceptibility to these viruses may be linked to the genes which modulate the risk of both type 1 diabetes and immune response to enteroviruses (such as IFIH1, IRF7 network, HLA-DR) as well as other individual factors such as age and gender (young age and male gender increase the risk of complications of enterovirus infections). Again, this scenario resembles closely that previously described in polio (polioviruses used to be very common causing paralytic disease in less than 1% of infected individuals). In addition, the virus may have interactions with other risk factors, such as cow's milk proteins (Makela et al. 2006).

Lessons from Other Enterovirus Diseases

Type 1 diabetes and the well-known enterovirus disease, polio, share common features which may help to understand the role of enteroviruses in type 1 diabetes (Table 14.1). These similarities include seasonality, time-trends in disease incidence, high incidence in "high hygiene" areas, inflammation in the target organ, and highly selective cell damage. In fact, poliomyelitis was once suggested to be an autoallergic disease, where poliovirus infection induces immune-mediated paralysis in genetically predisposed individuals (Wyatt 1976). One major difference is the long prodromal subclinical phase of disease in type 1 diabetes compared to the rapid appearance of polio paralysis after the infection. This is quite logical since even a limited damage in motoneurons leads immediately to clinical symptoms while up to 90% of beta cells can be destroyed without causing any symptoms leading to the diagnosis of type 1 diabetes.

During the polio era, polioviruses were very common and almost every individual became infected by adulthood and the majority by the age of 5 years. Only less than 1% of the infected individuals developed the paralytic disease while the great majority (about 90%) had subclinical infection. Certain host factors increased the risk of the severe disease including male gender, older age, physical exercise, and muscle damage during infection, but the mechanisms regulating the risk are still largely unknown. Some reports have indicated association with certain HLA genes.

Interestingly, a change in the dynamics of poliovirus circulation turned out to have a major effect on the risk of paralytic disease. The increasing hygiene gradually decreased the transmission of polioviruses during the nineteenth century, which paradoxically led to the first clear epidemics of paralytic disease at the end of the century (Monto 1999). These epidemics started in young infants (the disease was first called infant paralysis) in countries with high standard of living (e.g., Nordic countries and USA) and continued thereafter until the vaccine was developed in the 1950s. These epidemics and the rapid increase in paralytic polio are believed to be

Table 14.1 Epidemiological characteristics of enterovirus infections and type 1 diabetes

	Enterovirus	Type 1 diabetes
Seasonality	Seasonal pattern in temperate climate, peaking in late summer and autumn	Seasonal pattern in the initiation of the process peaking in late summer and autumn (similar pattern in the diagnosis of clinical diabetes)
Time-trends	Decreasing during the last decades	Increasing during the last decades
Geography	Common in low hygiene and low standard of living	Rare in low hygiene and low standard of living
Age	Most frequent in young children, especially severe illness	Beta-cell damaging process starts early, often <2 year of age
Gender	Male gender is associated with severe illness	Males are affected more frequently than females

caused by a decrease in herd immunity in young children. The basis of this phenomenon was a delay in the age of the first infections; some children were not exposed to polioviruses until later in childhood when they were no more protected by maternal antibodies (Nathanson and Kew 2010). As hygiene, sanitation, and housing improved the proportion of children escaping infection in infancy rose and the number of paralytic diseases rose in parallel.

The Polio Hypothesis

Viskari et al. (2000) launched a new hypothesis to explain the conspicuous increase seen in the incidence of type 1 diabetes in developed countries after World War II as well as the marked geographical variation in the incidence rates. They claimed that these phenomena are connected to the circulation of enteroviruses. They named this hypothesis as the "polio hypothesis" due to the analogy with the epidemiological pattern previously seen in polio. The same idea has also been discussed elsewhere (Zinkernagel 2001).

According to the polio hypothesis, a low frequency of enterovirus infections in the background population increases the risk of type 1 diabetes by making children more susceptible to enterovirus-induced beta-cell damage. In a population with a high prevalence of enterovirus infections and a low risk of type 1 diabetes, the first infections are experienced soon after birth when maternal antibodies protect the child (Fig. 14.1). Infections which occur at the presence of maternal antibodies induce an immune response but remain superficial. This has been called a natural vaccination of the child (Zinkernagel 2001) and is illustrated in Fig. 14.2. The child's own enterovirus immunity develops gradually as the child experiences serial infections inducing a progressively expanding repertoire of memory T-cells which are known to cross-react between different enterovirus types (Cello et al. 1996; Juhela et al. 1998a). These memory T-cells can boost enterovirus-specific immune responses when the child become infected by the same or different enterovirus type (anamnestic response) speeding up the elimination of the virus and limiting its systemic spread.

In contrast, in a population with a low prevalence of enteroviruses and a high risk of type 1 diabetes, infants lack maternal antibodies because increasing proportion of the mothers has not experienced the virus which is infecting the child. Immune protection against enteroviruses depends largely on neutralizing antibodies. These antibodies are serotype specific making the infant completely unprotected against those virus types which the mother has never experienced. In addition, the infections occur later when maternal antibodies have already disappeared and breast-feeding has been discontinued. This creates a susceptibility period in early childhood when both passive and acquired immunity are weak and maternal antibodies would be needed to compensate this defect. Thus, the polio hypothesis could be particularly relevant for such enteroviruses which circulate in very young children infecting

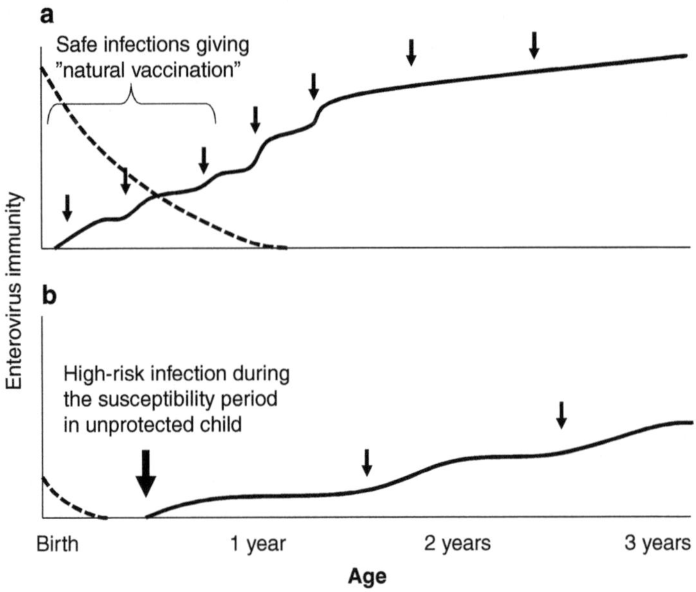

Fig. 14.1 Immunological basis of the polio hypothesis. In a population with high prevalence of enterovirus infections and low risk of type 1 diabetes (**a**), the child experiences the first infections (*black arrows*) soon after birth when maternal antibodies (*dotted curve*) still protect the child. The *solid curve* indicates the development of child's own enterovirus immunity as a consequence of serial infections and maturation of immune system. In a population with low prevalence of enterovirus infections and high risk of type 1 diabetes (**b**), the infant lacks maternal antibodies and/or experiences the infections later when maternal antibodies have already disappeared. This creates a susceptibility period in early infancy. If the child becomes infected by a diabetogenic enterovirus (*large arrow*) during this period, the risk of diabetes is high. In addition, child's long-term enterovirus immunity develops more slowly and remains at relatively low level due to the lack of booster infections

them during this susceptibility period. Figure 14.3 summarized the changes in the population dynamics of enterovirus infections which led to the past polio epidemics and which may now contribute to the ongoing "epidemic" of type 1 diabetes.

Observations Supporting Polio Hypothesis

The polio hypothesis has been supported by studies comparing the frequency and time-trends of enterovirus infections and type 1 diabetes different populations. Viskari et al. (2000) carried out the first studies in Finland where the incidence of type 1 diabetes is the highest in the world and has increased fivefold during the past 50 years (currently about 60 per 100,000 children). In samples taken from pregnant women during 1983–1995 as a part of the national infectious disease screening

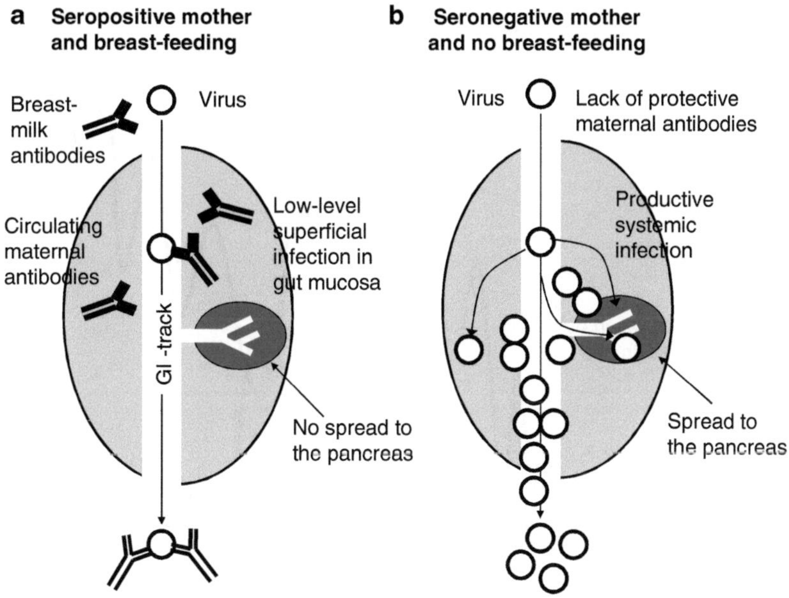

Fig. 14.2 Biological basis of the polio hypothesis: the role of maternal antibodies in the protection of young infants against the diabetogenic effect of enteroviruses. Panel (**a**) illustrates a child whose mother has high titers of neutralizing antibodies against the infecting virus serotype. Protective maternal antibodies are transferred to the child through placenta (systemic effect) and via breast-milk (local mucosal effect), both protecting the child against the spread of the virus to the pancreas. Maternal antibodies limit the infection to mucosal surfaces, but the child becomes immunized against the virus. The virus is illustrated with an *open circle*, the child with a *large oval*, and the target organ (pancreas) with a *gray circle*. Panel (**b**) illustrates a child whose mother does not have antibodies against the infecting serotype and cannot provide protection for the child. The virus replicates effectively in intestinal mucosa and gut-associated immune system, spreading to the pancreas and causing inflammation in the infected tissues. Infection before the age of 1 year is a particular risk factor as child's own immune system is still immature and the protection depends largely on maternal antibodies

program they found a significant decrease in enterovirus antibody levels over time. In a further study they confirmed this finding in larger series in both Finland and Sweden (Viskari et al. 2005). These analyses were done using assays which detect antibodies against several different enterovirus types suggesting that the overall exposure to enteroviruses has decreased. This implies that the proportion of newborns who lack maternal antibodies has increased, making new born children now more susceptible than before. The same group has also compared the prevalence of enterovirus infections in seven countries with either exceptionally high or low/intermediate incidence of type 1 diabetes. They found that enterovirus antibodies were less frequent and at lower levels in countries with high diabetes incidence compared to countries with low diabetes incidence (Viskari et al. 2004, 2005). Finland, in particular, had lower antibody levels compared to other counties. For example, altogether 42% of Finnish pregnant women lacked neutralizing antibodies

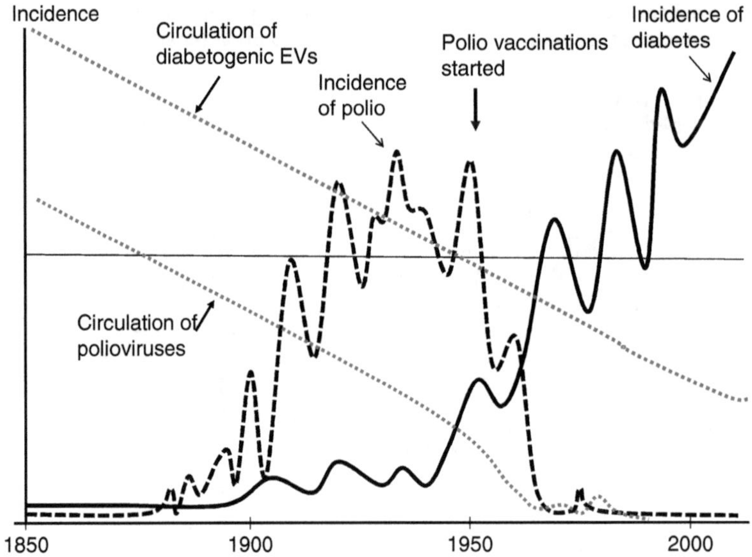

Fig. 14.3 Epidemiological basis of the polio hypothesis: inverse correlation between the circulation of enteroviruses and the risk of severe infection. The incidence of paralytic polio (*dotted black line*) is shown in relation to the circulation of polioviruses in background population (*gray dotted line*). Similarly, the incidence of type 1 diabetes (*solid black line*) is shown in relation to the circulation of diabetogenic enteroviruses in the population (*dotted gray line*). The threshold of virus circulation which can maintain population immunity and low risk of complications is shown as a *horizontal line*. In the case of polio, the circulation of the virus dropped below this threshold at the end of nineteenth century and, according to polio hypothesis, in type 1 diabetes 50 years later. The main reason in both cases is an increase in infections which occur in the absence of protective maternal antibodies in young infants

to coxsackievirus B4, which has been linked to type 1 diabetes in previous studies. The corresponding figure in the neighboring Estonia and Karelian Republic of Russia was only 14%. Such a big difference between these countries has probably biologically relevance and, assuming that this is a diabetogenic virus, could contribute to the high incidence of type 1 diabetes in Finland. Further, Viskari et al. found that while the overall incidence of enterovirus meningitis has decreased during the past decades, the proportion of less than 6 months old cases has increased, supporting also the role of maternal antibodies in the protection of young infants against severe enterovirus infections (Viskari et al. 2000). Same kind of secular increase in severe neonatal enterovirus disease has also been reported in other countries (Shattuck and Chonmaitree 1992). Recently, a rapid increase in severe coxsackievirus B1 infections has documented among young infants in the USA (Wikswo et al. 2009). Altogether, there seems to be an inverse relationship between the frequency of enterovirus infections and type 1 diabetes in the background population, a pattern which is in line with the polio hypothesis.

Even though enterovirus infections seem to have become less frequent during the last decades, they are still common and occur in young children during this susceptibility period. A study carried out in Finland during 1990s indicated that altogether 30% of healthy children had experienced enterovirus infection by the age of 6 months and 60% of children by 12 months (Juhela et al. 1998b). Another study followed newborns during an enterovirus season in New York in 1981, indicating an incidence of 13% during the first month of life (Jenista et al. 1984). Thus, considerable proportion of children becomes infected during this susceptibility period. However, only one study has been carried out to find out how often these infections actually occur in the absence of maternal antibodies. In that study children were followed from birth in Estonia where the incidence of type 1 diabetes is about 17 per 100,000 children. Infections were diagnosed by detecting viral RNA in regularly collected stool samples and the serotype was identified by sequencing the viral genome. Neutralizing antibodies were measured against that particular serotype from cord-blood. The results indicated that 38% of infections, which were experienced during the first months of age, occurred in the absence of maternal antibodies against the causative virus type (Salur et al. 2011).

Breast-feeding is known to protect children against many virus infections including enterovirus infections (Jenista et al. 1984; Sadeharju et al. 2007). In fact, maternal antibodies in breast-milk may provide even stronger protection against enteroviruses than transplacentally acquired antibodies in circulation (Sadeharju et al. 2007). This is logical since the transmission of enteroviruses happens via mucosal route and ingested breast-milk antibodies can neutralize the virus before it can infect the host. Thus, one can argue that breast-feeding should also protect against enterovirus-induced diabetes. Breast-feeding may indeed have a protective effect against type 1 diabetes, even though conflicting observations also exist (Knip and Akerblom 2005).

In summary, epidemiological observations support polio hypothesis and more studies are indicated to test it further. It would be important to find out if diabetogenic enteroviruses belong to certain serotypes and to study if these observations hold true for these particular virus types. Large-scale prospective studies in different populations will play a key role in this effort.

Type 1 Diabetes and the "Hygiene Hypothesis"

In this scenario type 1 diabetes is considered as an autoimmune disease which develops spontaneously or is initiated by other factors than enteroviruses. The idea of a spontaneous autoimmune process as an underlying mechanism is largely based on studies in NOD mice, the most widely used animal model for type 1 diabetes. These mice develop an autoimmune process which starts "spontaneously" and damages beta cells leading to type 1 diabetes-like disease. Several microbes can prevent or delay this process, suggesting that under certain conditions microbes can have a protective effect resembling that previously suggested in allergies (hygiene hypothesis). In fact, recent studies have widened the scope of the hygiene hypothesis from allergies to type 1 diabetes and other immune-mediated diseases, and the concept that microbes can be important regulators of immune system is under vigorous investigation (Bach 2005). This idea has also been supported by a recent observation showing that of type 1 diabetes and IgE-mediated allergic sensitization co-occur in such subjects who are seronegative for hepatitis A virus (Seiskari et al. 2010). Thus, living in very hygienic conditions may lead to a defect in immune regulations and predispose to immune-mediated diseases.

Among other microbes enteroviruses prevent the development of diabetes in NOD mice (Drescher et al. 2004). This effect seems to be mediated by the induction of immunoregulatory mechanisms such as regulatory T-cells which can suppress the autoimmune phenotype in these animals (Filippi et al. 2009; Tracy and Drescher 2007). The timing of the infection is critical, since this protection can be seen only when the mice are infected before the autoimmune process has started. In older mice, which are already affected by an inflammation process in the pancreas, the virus can even accelerate the process, particularly when given in high doses (Serreze et al. 2000). Thus far, two enterovirus serotypes have been studied in NOD mouse model (coxsackieviruses B3 and B4) and it is not known if other serotypes could have a similar protective effect.

These observations have raised the question if certain enteroviruses could also have a protective effect in man. However, the number of human studies addressing this question is very limited. One study showed that neutralizing antibodies against coxsackieviruses B3 and B4, the same serotypes which have had a protective effect in NOD mice, are decreased in type 1 diabetes patients compared to controls (Palmer et al. 1982). Theoretically, this difference could reflect a protective effect. Coxsackievirus B3 can also selectively inhibit major histocompatibility complex class I presentation pathway which reduces cytotoxic T-cell responses to infected cells (Cornell et al. 2007). Interestingly, recent studies have suggested that enterovirus infections may protect from IgE-mediated allergic sensitization (Seiskari et al. 2007) also supporting possible immunoregulatory effects. Such an effect could be mediated, e.g., by induction of immunoregulatory cytokines such as IL-10 by the virus (bystander suppression mechanism). However, further studies are still needed to find out if enteroviruses have immunoregulatory effects which play a role in the pathogenesis of human type 1 diabetes.

Acknowledgments I wish to thank Hanna Viskari, M.D., Ph.D., Maria Lönnrot, M.D., Ph.D., Anita Kondrashova, M.D., Ph.D., Sisko Tauriainen, Ph.D., Sami Oikarinen, M.Sc. and all other collaborators contributing to the findings summarized in this chapter. The studies summarized in this chapter have been financially supported by the Juvenile Diabetes Research Foundation, European Commission, Sohlberg's Foundation, and the Academy of Finland.

References

Bach JF (2005) Infections and autoimmune diseases. J Autoimmun 25(Suppl):74–80

Cello J, Strannegard O, Svennerholm B (1996) A study of the cellular immune response to enteroviruses in humans: identification of cross-reactive T cell epitopes on the structural proteins of enteroviruses. J Gen Virol 77(Pt 9):2097–2108

Chatterjee NK, Hou J, Dockstader P, Charbonneau T (1992) Coxsackievirus B4 infection alters thymic, splenic, and peripheral lymphocyte repertoire preceding onset of hyperglycemia in mice. J Med Virol 38:124–131

Chehadeh W, Kerr-Conte J, Pattou F, Alm G, Lefebvre J, Wattre P, Hober D (2000) Persistent infection of human pancreatic islets by coxsackievirus B is associated with alpha interferon synthesis in beta cells. J Virol 74:10153–10164

Cornell CT, Kiosses WB, Harkins S, Whitton JL (2007) Coxsackievirus B3 proteins directionally complement each other to downregulate surface major histocompatibility complex class I. J Virol 81:6785–6797

Dotta F, Censini S, van Halteren AG, Marselli L, Masini M, Dionisi S, Mosca F, Boggi U, Muda AO, Prato SD, Elliott JF, Covacci A, Rappuoli R, Roep BO, Marchetti P (2007) Coxsackie B4 virus infection of beta cells and natural killer cell insulitis in recent-onset type 1 diabetic patients. Proc Natl Acad Sci U S A 104:5115–5120

Drescher KM, Kono K, Bopegamage S, Carson SD, Tracy S (2004) Coxsackievirus B3 infection and type 1 diabetes development in NOD mice: insulitis determines susceptibility of pancreatic islets to virus infection. Virology 329:381–394

Filippi CM, Estes EA, Oldham JE, von Herrath MG (2009) Immunoregulatory mechanisms triggered by viral infections protect from type 1 diabetes in mice. J Clin Invest 119:1515–1523

Foulis AK, Farquharson MA, Cameron SO, McGill M, Schonke H, Kandolf R (1990) A search for the presence of the enteroviral capsid protein VP1 in pancreases of patients with type 1 (insulin-dependent) diabetes and pancreases and hearts of infants who died of coxsackieviral myocarditis. Diabetologia 33:290–298

Gamble DR, Taylor KW (1969) Seasonal incidence of diabetes mellitus. Br Med J 3:631–633

Gerling I, Chatterjee NK, Nejman C (1991) Coxsackievirus B4-induced development of antibodies to 64,000-Mr islet autoantigen and hyperglycemia in mice. Autoimmunity 10:49–56

Hiltunen M, Hyoty H, Knip M, Ilonen J, Reijonen H, Vahasalo P, Roivainen M, Lonnrot M, Leinikki P, Hovi T, Akerblom HK (1997) Islet cell antibody seroconversion in children is temporally associated with enterovirus infections. Childhood Diabetes in Finland (DiMe) Study Group. J Infect Dis 175:554–560

Jaidane H, Sane F, Gharbi J, Aouni M, Romond MB, Hober D (2009) Coxsackievirus B4 and type 1 diabetes pathogenesis: contribution of animal models. Diabetes Metab Res Rev 25:591–603

Jenista JA, Powell KR, Menegus MA (1984) Epidemiology of neonatal enterovirus infection. J Pediatr 104:685–690

Juhela S, Hyoty H, Lonnrot M, Roivainen M, Simell O, Ilonen J (1998a) Enterovirus infections and enterovirus specific T-cell responses in infancy. J Med Virol 54:226–232

Juhela S, Hyöty H, Lonnrot M, Roivainen M, Simell O, Ilonen J (1998b) Enterovirus infections and enterovirus specific T-cell responses in infancy. J Med Virol 54:226–232

Jun HS, Yoon JW (2001) The role of viruses in type 1 diabetes: two distinct cellular and molecular pathogenic mechanisms of virus-induced diabetes in animals. Diabetologia 44:271–285

Kimpimaki T, Kupila A, Hamalainen AM, Kukko M, Kulmala P, Savola K, Simell T, Keskinen P, Ilonen J, Simell O, Knip M (2001) The first signs of beta-cell autoimmunity appear in infancy in genetically susceptible children from the general population: the Finnish Type 1 Diabetes Prediction and Prevention Study. J Clin Endocrinol Metab 86:4782–4788

Knip M, Akerblom HK (2005) Early nutrition and later diabetes risk. Adv Exp Med Biol 569:142–150

Makela M, Vaarala O, Hermann R, Salminen K, Vahlberg T, Veijola R, Hyoty H, Knip M, Simell O, Ilonen J (2006) Enteral virus infections in early childhood and an enhanced type 1 diabetes-associated antibody response to dietary insulin. J Autoimmun 27:54–61

Moltchanova EV, Schreier N, Lammi N, Karvonen M (2009) Seasonal variation of diagnosis of type 1 diabetes mellitus in children worldwide. Diabet Med 26:673–678

Monto AS (1999) Francis field trial of inactivated poliomyelitis vaccine: background and lessons for today. Epidemiol Rev 21:7–23

Nathanson N, Kew OM (2010) From emergence to eradication: the epidemiology of poliomyelitis deconstructed. Am J Epidemiol 172:1213–1229

Oikarinen M, Tauriainen S, Honkanen T, Vuori K, Karhunen P, Vasama-Nolvi C, Oikarinen S, Verbeke C, Blair GE, Rantala I, Ilonen J, Simell O, Knip M, Hyoty H (2008) Analysis of pancreas tissue in a child positive for islet cell antibodies. Diabetologia 51:1796–1802

Oikarinen S, Martiskainen M, Tauriainen S, Huhtala H, Ilonen J, Veijola R, Simell O, Knip M, Hyoty H (2011) Enterovirus RNA in blood is linked to the development of type 1 diabetes. Diabetes 60:276–279

Palmer JP, Cooney MK, Ward RH, Hansen JA, Brodsky JB, Ray CG, Crossley JR, Asplin CM, Williams RH (1982) Reduced Coxsackie antibody titres in type 1 (insulin-dependent) diabetic patients presenting during an outbreak of Coxsackie B3 and B4 infection. Diabetologia 22:426–429

Richardson SJ, Willcox A, Bone AJ, Foulis AK, Morgan NG (2009) The prevalence of enteroviral capsid protein vp1 immunostaining in pancreatic islets in human type 1 diabetes. Diabetologia 52:1143–1151

Roivainen M, Ylipaasto P, Savolainen C, Galama J, Hovi T, Otonkoski T (2002) Functional impairment and killing of human beta cells by enteroviruses: the capacity is shared by a wide range of serotypes, but the extent is a characteristic of individual virus strains. Diabetologia 45:693–702

Sadeharju K, Knip M, Virtanen SM, Savilahti E, Tauriainen S, Koskela P, Akerblom HK, Hyoty H (2007) Maternal antibodies in breast milk protect the child from enterovirus infections. Pediatrics 119:941–946

Salminen K, Sadeharju K, Lonnrot M, Vahasalo P, Kupila A, Korhonen S, Ilonen J, Simell O, Knip M, Hyoty H (2003) Enterovirus infections are associated with the induction of beta-cell autoimmunity in a prospective birth cohort study. J Med Virol 69:91–98

Salur L, Oikarinen S, Tauriainen S, Mandel M, Hyöty H, Uibo R (2011) Enterovirus infections in young infants – are children still protected by maternal antibodies? Hum Vaccin 7(9):966–971

Seiskari T, Kondrashova A, Viskari H, Kaila M, Haapala AM, Aittoniemi J, Virta M, Hurme M, Uibo R, Knip M, Hyoty H (2007) Allergic sensitization and microbial load – a comparison between Finland and Russian Karelia. Clin Exp Immunol 148:47–52

Seiskari T, Viskari H, Kondrashova A, Haapala AM, Ilonen J, Knip M, Hyoty H (2010) Co-occurrence of allergic sensitization and type 1 diabetes. Ann Med 42:352–359

Serreze DV, Ottendorfer EW, Ellis TM, Gauntt CJ, Atkinson MA (2000) Acceleration of type 1 diabetes by a coxsackievirus infection requires a preexisting critical mass of autoreactive T-cells in pancreatic islets. Diabetes 49:708–711

Shattuck KE, Chonmaitree T (1992) The changing spectrum of neonatal meningitis over a fifteen-year period. Clin Pediatr (Phila) 31:130–136

Siljander HT, Simell S, Hekkala A, Lahde J, Simell T, Vahasalo P, Veijola R, Ilonen J, Simell O, Knip M (2009) Predictive characteristics of diabetes-associated autoantibodies among children with HLA-conferred disease susceptibility in the general population. Diabetes 58:2835–2842

Tauriainen S, Oikarinen S, Oikarinen M, Hyoty H (2011) Enteroviruses in the pathogenesis of type 1 diabetes. Semin Immunopathol 33:45–55

Tracy S, Drescher KM (2007) Coxsackievirus infections and NOD mice: relevant models of protection from, and induction of, type 1 diabetes. Ann N Y Acad Sci 1103:143–151

Viskari HR, Koskela P, Lonnrot M, Luonuansuu S, Reunanen A, Baer M, Hyoty H (2000) Can enterovirus infections explain the increasing incidence of type 1 diabetes? Diabetes Care 23:414–416

Viskari H, Ludvigsson J, Uibo R, Salur L, Marciulionyte D, Hermann R, Soltesz G, Fuchtenbusch M, Ziegler AG, Kondrashova A, Romanov A, Knip M, Hyoty H (2004) Relationship between the incidence of type 1 diabetes and enterovirus infections in different European populations: results from the EPIVIR project. J Med Virol 72:610–617

Viskari H, Ludvigsson J, Uibo R, Salur L, Marciulionyte D, Hermann R, Soltesz G, Fuchtenbusch M, Ziegler AG, Kondrashova A, Romanov A, Kaplan B, Laron Z, Koskela P, Vesikari T, Huhtala H, Knip M, Hyoty H (2005) Relationship between the incidence of type 1 diabetes and maternal enterovirus antibodies: time trends and geographical variation. Diabetologia 48:1280–1287

Wikswo ME, Khetsuriani N, Fowlkes AL, Zheng X, Penaranda S, Verma N, Shulman ST, Sircar K, Robinson CC, Schmidt T, Schnurr D, Oberste MS (2009) Increased activity of Coxsackievirus B1 strains associated with severe disease among young infants in the United States, 2007–2008. Clin Infect Dis 49:e44–e51

Wyatt HV (1976) Is poliomyelitis an auto-allergic disease triggered by virus? Med Hypotheses 2:262–268

Yeung WC, Rawlinson WD, Craig ME (2011) Enterovirus infection and type 1 diabetes mellitus: systematic review and meta-analysis of observational molecular studies. Br Med J 342:d35

Yin H, Berg AK, Westman J, Hellerstrom C, Frisk G (2002) Complete nucleotide sequence of a Coxsackievirus B-4 strain capable of establishing persistent infection in human pancreatic islet cells: effects on insulin release, proinsulin synthesis, and cell morphology. J Med Virol 68:544–557

Zinkernagel RM (2001) Maternal antibodies, childhood infections, and autoimmune diseases. N Engl J Med 345:1331–1335

Chapter 15
Enteroviruses in Blood

Antonio Toniolo, Alessandro Salvatoni, Giovanni Federico, Giuseppe Maccari, Oscar Díaz-Horta, and Andreina Baj

Abstract At the time of clinical diagnosis, the majority of children with type 1 diabetes carry EVs of different species in their blood. Controls rarely carry EVs in blood. In the blood, these viruses are present at very low titers and are minimally able to replicate in cell culture. At the time of clinical diagnosis, the presence of asymptomatic enterovirus infections is common among family members. The enterovirus types involved remain to be defined, but enteroviruses belonging to the B species appear particularly prevalent. Geographic and temporal clusters of enterovirus infection and type 1 diabetes have been documented in Northern Italy. It will be important to determine the length of persistence of enteroviruses in the blood of diabetic children. The results do not provide direct evidence for a causal relationship between enterovirus infection and diabetes, but strongly suggest that the association is not fortuitous.

A. Toniolo, M.D. (✉) • G. Maccari • O. Diaz-Horta
Laboratory of Microbiology, University of Insubria and Ospedale di Circolo, Varese, Italy
e-mail: antonio.toniolo@uninsubria.it

A. Salvatoni
Pediatric Endocrinology Unit, Department of Experimental Medicine,
University of Insubria Medical School, Varese, Italy

G. Federico
Unit of Pediatric Endocrinology and Diabetes, Department of Reproductive Medicine
and Child Development, University of Pisa Medical School, Pisa, Italy

A. Baj
Laboratory of Microbiology, Department of Experimental Medicine,
University of Insubria Medical School, Varese, Italy

K.W. Taylor, H. Hyoty, A. Toniolo, A.J. Zuckerman (eds.), *Diabetes and Viruses*,
DOI 10.1007/978-1-4614-4051-2_15,

Introduction

In just a few cases scientists have been able to isolate viruses from patients with type 1 diabetes (reviewed in: Notkins 1979; Toniolo et al. 1988; Hober and Sauter 2010; Roivainen and Klingel 2010; Tauriainen et al. 2011). Overt hyperglycemia, in fact, is the final consequence of a prolonged pathologic process that, if linked to a viral infection, may involve slowly replicating agents that challenge current methods of virus isolation in vitro.

A recent meta-analysis of molecular studies aimed at virus detection in newly diagnosed cases of type 1 diabetes lends support to the idea that RNA agents (enteroviruses) are present at varying but remarkable frequencies in the early phase of the disease (Yeung et al. 2011). It remains obscure whether these correlations were merely coincidental with the "failing" of an altered immune system or had more profound implications.

However, pathologic studies initiated over 30 years ago (Jenson et al. 1980), have demonstrated enterovirus proteins and/or genome in pancreatic islets, particularly in beta cells (Richardson et al. 2009; Willcox et al. 2011). Indirect signs of an ongoing viral infection in the pancreas of diabetic individuals have been also found. Among these, the local expression of interferon and the detection of double-stranded RNA (indicating replicating viruses) in islet cells (Richardson et al. 2010, 2011).

EVs have been considered agents capable of causing acute clinical conditions, either asymptomatic or symptomatic. However, evidence has been accumulating for these viruses being able of both producing slow pathology in vivo and persisting in the host in spite of a perceptible immune response. Examples include evolution of myocarditis into dilated myocardiopathy (Fujioka et al. 2004; Chapman and Kim 2008; Gorbea et al. 2010) as well as progressive disorders of the central nervous system (Baj et al. 2007; Feuer et al. 2009; Cavalcante et al. 2010; Rhoades et al. 2011). These clinical studies were stimulated by the results of experiments showing that—as in the Theiler virus model of demyelinating disease—a continuous low-level enterovirus replication can be established in a variety of cultured cell types (Conaldi et al. 1997; Kelly et al. 2010) and in animal models (Destombes et al. 1997; Rahnefeld et al. 2011).

It has also been recognized that viruses belonging to the enterovirus genus undergo remarkable genetic variation and evolution (McWilliam Leitch et al. 2010; Harvala et al. 2011). These events are linked to the high mutation rates proper of RNA viruses, recombination among genomes of different enterovirus types (based on little understood mechanisms), and "codon deoptimization," i.e., the accumulation of "silent mutations" that—beyond a certain threshold—slow down virus replication (Coleman et al. 2008). The end result of these multiple events is that the genetic structure of current enterovirus isolates is often different from that of enterovirus prototype strains that were collected 30–60 years ago (Tracy et al. 2010; Hu et al. 2011). In addition to that, inclusion of over 100 virus types in the enterovirus genus (Yozwiak et al. 2010) makes identification and detection of all these agents extremely complex. Difficulties not only apply to tissue culture and serology. Gene amplification methods and genome sequencing have revolutionized our ability to detect viruses both in infected hosts and in the environment (Foxman and Iwasaki

2011), but these procedures require the targeting of stable sequences (Nix et al. 2006; Pallansch and Oberste 2010). The conserved 5′ untranslated region (5′UTR) has been chosen most frequently for enterovirus investigations, although, unfortunately, it carries little or no information for identifying the type of the infecting virus.

Seeking Viruses in Pediatric Patients at the Clinical Onset of Type 1 Diabetes

There is now a very large body of evidence on this subject. It is not our intention to review this in any detail. The complexities of this area of investigation may be best illustrated by referring to our own experience of researching the problem locally, in Italy. Two studies (DAISY and MIDIA that have been carried out in the USA and Norway, respectively) have shown that detection of enterovirus in feces does not predict the development of type 1 diabetes, whereas enterovirus detection in blood is associated with an increased risk of developing diabetes (Stene et al. 2010; Tapia et al. 2011).

Based on our previous experience, a search for enteroviruses in diabetes was initiated by setting up methods capable of detecting viruses in blood under conditions in which the agents had extremely low titers and little or no ability of replicating in cultured cells. Basically, a double approach was followed: direct detection in plasma by gene amplification (reverse transcriptase polymerase chain reaction) and co-culture of peripheral blood leukocytes (PBLs) with a panel of EV-susceptible cell lines. Enterovirus genomes were then detected in RNA extracted from culture medium and/or cultured cells using RT-PCR. Viral antigens were detected by antibody assays to conserved capsid epitopes of different enterovirus types.

As shown in Fig. 15.1a, non-degenerated primers were produced to cover different genomic regions of approximately 100 different enterovirus types. The variability of genome sequences of different enterovirus types is reported in the lower panel (Fig. 15.1b). To enhance the sensitivity of tests, non-degenerated primers have been used throughout. Samples were considered EV-positive when a signal could

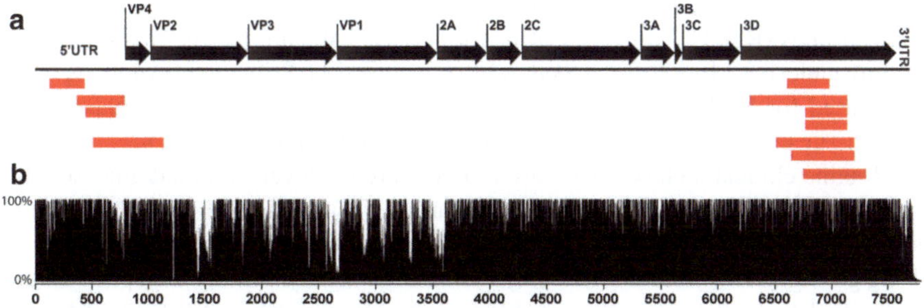

Fig. 15.1 Diagram of the enterovirus genome that is approximately 7.5 kb in length. Panel (**a**): the genome regions that have been amplified by molecular tests are represented in *red*. To this end, primers were designed to target sequences that are conserved among 96 enterovirus types. Panel (**b**): full-genome similarity plot showing the percent identity of nucleotide sequences among 96 enterovirus types (gap open cost = 10; gap extension cost = 1)

be obtained from at least two different genome regions. Results presented here refer to blood samples analyzed with primers directed to three genomic regions: 5′UTR, 5′UTR-VP2 (a region comprising the VP4 capsid gene), and 3D (RNA polymerase gene). With regard to the 3D gene, seven primer pairs had to be used, since this region is not markedly conserved. This extra effort, however, provided partial information with regard to the identification of the A–D species (and of subgroups within the B and C species). However, due to possible recombination events, identification of enterovirus types based on 3D sequences alone is not always possible. Reliable identification of the exact type of enterovirus, in fact, is based on their typing by hyperimmune sera which can neutralize their infectivity in cell culture in a serotype specific manner or on the measurement of increases in neutralizing antibodies in serum [for those agents able to attain sufficient titers and produce detectable cytopathic effects (CPE) in vitro]. Alternatively, identification can be obtained using gene sequence analysis of the highly variable capsid proteins (Nix et al. 2006; Yozwiak et al. 2010).

Subjects Investigated: Children at the Clinical Onset of Type 1 Diabetes, Consenting Family Members, Non-diabetic Controls

Two Pediatric Endocrine Units participated in this study which involved 112 consecutive patients diagnosed from January 2007 to January 2010 (61 boys and 51 girls). The median age of diabetic children was 9.0 years (range 2–16 years). Forty-one consenting family members of 16 diabetic probands were also investigated. Non-diabetic controls ($n=69$) were composed of adult blood donors ($n=34$) and children ($n=35$; matched by age, time, and location) who had been diagnosed with either short stature or overweight/obesity. At the time of writing, the subjects participating to the study have been followed for at least 16 months. On the day of diagnosis, blood samples (K_2-EDTA and serum) were obtained from each patient and from his/her consenting family members. Plasma and serum were stored at −80°C for further analysis. EDTA blood samples were processed immediately for separating PBLs. After washing, PBLs were co-cultured with enterovirus-susceptible cell lines (RD, HeLa, AV3, CaCo) for at least 1 month. During this time, each culture underwent four to six passages by trypsinization. Primers shown in Fig. 15.1 were used for RT-PCR assays that were run on RNA extracted from plasma and from tissue culture medium of cell lines exposed to the patients' PBLs.

Routine clinical methods were used to measure the levels of blood glucose, glycosylated hemoglobin (HbA1c), and, 1 year after diagnosis, the individual insulin requirement (IU/kg/day). C-peptide levels (time 0 and 6 min after stimulation with 1 mg glucagon i.v.) were measured by radioimmunoassay. Fasting C-peptide values in the undetectable range (<0.2 ng/ml) were assigned a value of 0.1 ng/ml for the analyses. The Class-II human leukocyte antigen (HLA) genotype was determined by molecular methods. Titers of autoantibodies to glutamic acid decarboxylase (GAD), tyrosine phosphatase-like molecule (IA-2), zinc transporter-8 (ZnT8), and insulin (IA) were measured by radiobinding assays (Bonifacio et al. 1995a, b; Lampasona et al. 2010; Naserke et al. 1998). Levels of 12 cytokines/chemokines were measured by ELISA in the supernatant of HeLa cell monolayers co-cultured for at least 1 month with PBL of patients or controls. In the same cultures, expression of the viral VP1 capsid protein was evaluated by immunofluorescence using monoclonal antibodies directed to conserved epitopes of this antigen (Miao et al. 2009).

Clinical Data and Virology

At the time of diagnosis, high glucose levels were present in the group of children investigated (median 386 mg/dl; SD 175). As compared to normal ranges (Fig. 15.2a), fasting serum levels of C-peptide were reduced significantly (median 0.30 ng/ml; SD 0.61) and were increasing modestly 6 min after glucagon stimulation (median 0.75 ng/ml; SD 1.34). The substantial reduction of C-peptide levels as compared to expected values [range 0.7–2.1 ng/ml, basal; 1.7–6.0 ng/ml, after glucagon stimulation (Sosenko et al. 2008)] indicates that a sizeable loss of the beta cell reserve had already occurred at the time of clinical onset. As shown in Fig. 15.2b, levels of HbA1c were increased markedly at the time of diagnosis [median 11.20%; SD 2.70 (normal upper level=6.1%)], indicating that, before diagnosis, elevated glucose levels had been present in newly diabetic children for considerable periods of time (i.e., ≥2–3 months). As shown in Fig. 15.2b, 1 year after diagnosis insulin therapy brought HbA1c levels to near-normal levels (median 6.95%; SD 1.02).

All diabetic children were carrying high-risk HLA alleles and were positive for at least one diabetes-related autoantibody. Significant titers of GAD, IA-2, ZnT8, IA autoantibodies were present in 53%, 66%, 51%, and 88% of cases, respectively. These autoantibodies could not be detected in blood donors, non-diabetic control children, and non-diabetic family members of children with type 1 diabetes. Thus, the children investigated could not be classified as cases of "fulminant diabetes" (a form of insulin-dependent diabetes characterized by shorter duration from onset of hyperglycemic symptoms to first hospital visit, near-normal levels of HbA1c at onset, negativity for islet autoantibodies and ketosis; Shibasaki et al. 2010; see Chap. 22).

In our cohort, the diagnosis of slowly progressing autoimmune type 1 diabetes was made on the basis of increased levels of HbA1c at the time of diagnosis, low C-peptide levels, and positivity for diabetes-related autoantibodies.

Highly sensitive molecular methods and attempts at isolating viruses in cell cultures demonstrated that enterovirus detection in blood is a rare event in non-diabetic subjects. In our patients, only 2/69 (3%) non-diabetic controls carried the enterovirus genome in blood. In contrast, at the time of diagnosis, 89/112 (79%) children with type 1 diabetes were positive for enterovirus genomes and had low-level infectivity in leukocytes (Toniolo et al. 2010). The data are summarized in Fig. 15.3 together with the detected enterovirus species. It should be borne in mind that species identification has been achieved through characterization of the 3D genome region. Thus, as stated previously, identification cannot be held as conclusive, mainly due to recombination events possibly occurring among different enterovirus types.

The results indicate that a single enterovirus species is not involved in all the cases investigated. However, most of the enteroviruses were associated with the B

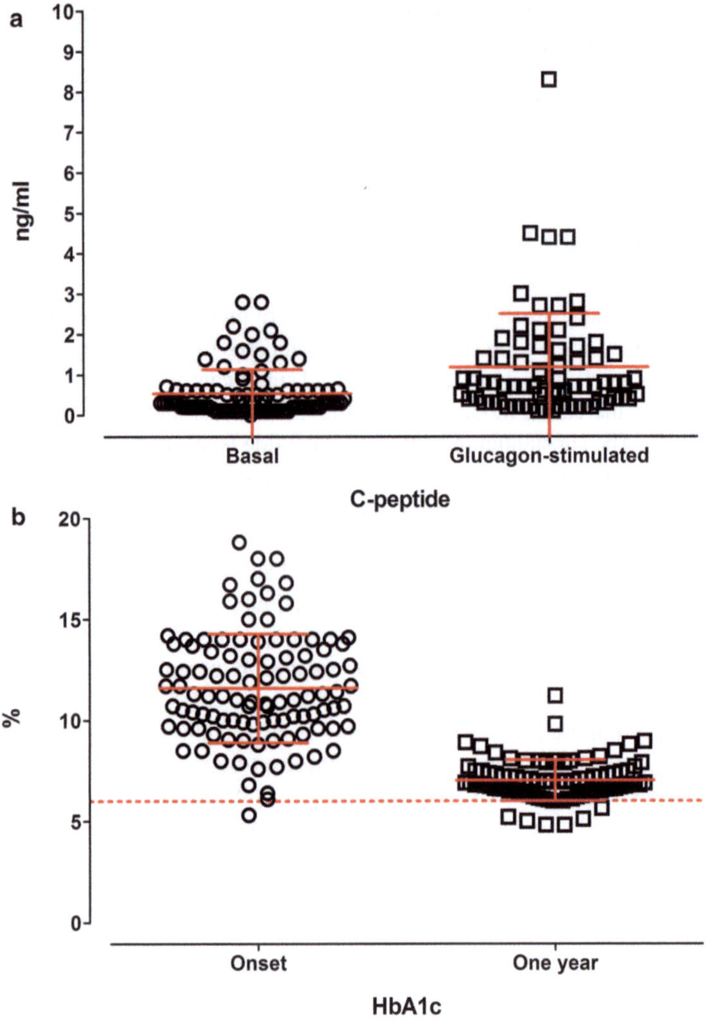

Fig. 15.2 Levels of C-peptide and glycosylated hemoglobin (HbA1c) in newly diagnosed children with type 1 diabetes ($n=112$). Each *dot* represents a value of a single patient. Panel (**a**): C-peptide levels were measured at the time of diagnosis (basal and 6 min after stimulation with 1 mg glucagon i.v.). The median ± SD is shown. Most values are well below the expected ranges (see text). Panel (**b**): the percentage of HbA1c was measured both at the time of diagnosis (strongly elevated levels) and after 1 year of insulin therapy (levels slightly above upper reference values). The median ± SD is also shown. The upper level in healthy controls is represented by the *red dotted line*

species. It should be noted that the B species is the largest one of the enterovirus genus, containing at least 58 serotypes. Molecular tests based on the 3D region allow identifying three subgroups within this species (B1–B3).

Early studies of viruses in type 1 diabetes and the few cases in which virus isolation/typing has been achieved (e.g., Yoon et al. 1979; Champsaur et al. 1980; Dotta et al. 2007) indicate members of the B species as those most probably implicated in

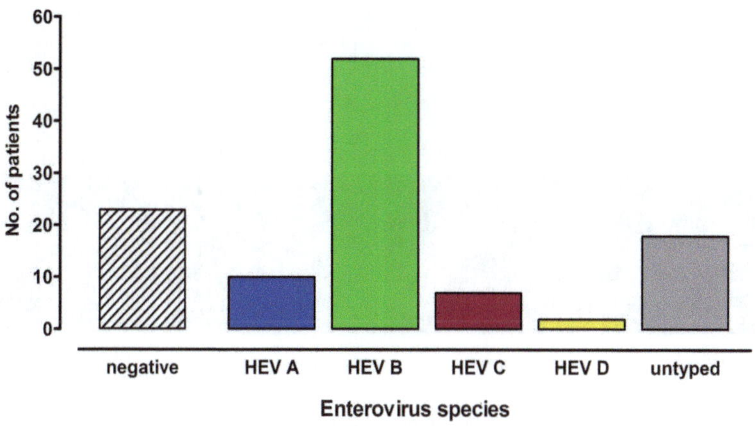

Fig. 15.3 Detection and identification of enterovirus species in the blood of children with newly diagnosed type 1 diabetes. Among 112 probands, 23 children were enterovirus negative (21%), 89 were enterovirus positive (79%). Among enterovirus-positive patients, analysis of the 3D genome region allowed to assign the majority of cases to enteroviruses of the B species. Approximately 20% of cases could not be typed

diabetes (Coxsackievirus B4 in particular). It should be considered, however, that involvement of the "B1 subgroup" of the B species has been particularly frequent in our patients. Based on the 3D amplification method developed by us, the B1 subgroup contains 20 echovirus types together with coxsackieviruses B1–B3. This is in agreement with reports from Finland and the tropics that have shown an association of echoviruses with type 1 diabetes (Vreugdenhil et al. 2000; Otonkoski et al. 2000; Díaz-Horta et al. 2001; Paananen et al. 2003; Cabrera-Rode et al. 2003, 2005; Williams et al. 2006; Al-Hello et al. 2008).

When cultured for prolonged periods with human cell lines, the infectious agents derived from PBLs of type 1 diabetes patients may produce weak but perceptible CPE. Figure 15.4 shows phase-contrast microscopy of HeLa cells exposed to PBLs obtained from four members of a single family (panel a) and the expression of enterovirus antigens (immunofluorescence) in cultures of corresponding samples (panel b). As shown in panel a, co-culture of cell lines with samples of the mother and father of the type 1 diabetes proband failed to produce CPE, whereas samples from the diabetic proband and his brother produced faint CPE. CPE production was in agreement with the expression of the enteroviral VP1 capsid protein: fluorescent dots are seen in the cytoplasm of HeLa cells co-cultured with PBLs of the diabetic proband and his brother, but not in cells co-cultured with PBLs of their parents. RT-PCR confirmed the findings: the type 1 diabetes proband and his brother were carrying an enterovirus of the B1 subgroup of the B species. This, and extensive observations made in additional families, indicate that, at the time of clinical diagnosis, enteroviruses are present often not only in type 1 diabetes probands but also in their family members.

Notably, in the families investigated, the virus-carrier status of non-diabetic family members was not associated with noticeable clinical symptoms. This observation speaks openly against dramatic events being produced by low-level enterovirus infection. Analogous observations were made in the course of polio epidemics

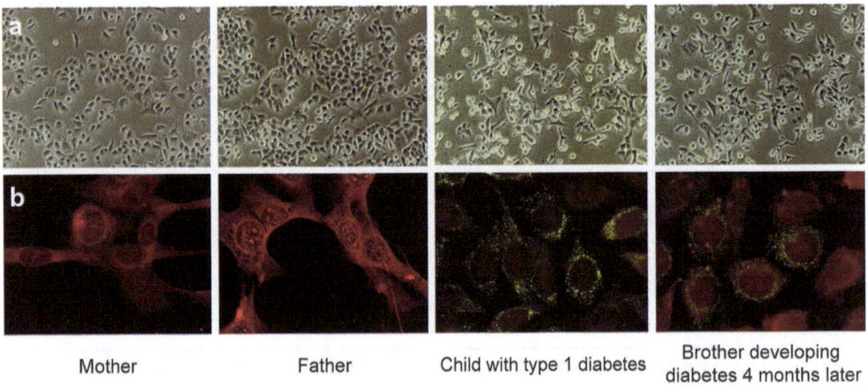

Fig. 15.4 Cytopathic effects (CPE) and expression of enterovirus capsid antigen in HeLa cell cultures exposed for 1 month to peripheral blood leukocytes (PBLs) of four members of the same family. Samples were obtained at the time of clinical diagnosis in the type 1 diabetes proband. Panel (**a**) (200× magnification): phase-contrast microscopy of subconfluent monolayers of HeLa cells co-cultured with PBL (*left* to *right*) of the mother, father, diabetic proband, and his brother. No CPE is seen in the first two images. Faint CPE is observable in the third and fourth images (diabetic proband and his brother). The virus-positive brother developed overt diabetes 5 months later. Panel (**b**) (1,000× magnification): same cultures as in panel (**a**). Indirect immunofluorescence with monoclonal antibodies to conserved capsid antigens of enteroviruses. Expression of the enterovirus capsid VP1 protein is seen as fluorescent dots in the cytoplasm of HeLa cells. *Left* to *right*: virus-negative cultures (father and mother) and virus-positive cultures (diabetic proband and his brother). The two virus-positive children were carrying an enterovirus of the B species (B1 subgroup)

(in which, generally, fewer than 1:100–1:500 infected people manifested neurologic symptoms) and seem to fit with the original definition of the echovirus genus (i.e., viruses not associated frequently with detectable symptoms or disease).

With regard to the potential pathogenetic mechanisms, examination of cell culture media derived from HeLa cells co-cultured with PBLs (either from healthy blood donors or EV-carrying diabetic children) showed that the production of the monocyte chemoattractant protein 1 (MCP1) was significantly enhanced in cultures exposed to PBLs of diabetic children. Levels of other cytokines/chemokines remained comparable. This observation suggests that the slowly replicating agent(s) detected in children at the clinical onset of type 1 diabetes can promote inflammation by attracting monocyte/macrophages. Recent data indicate that MCP1 also induces amylin expression in pancreatic beta cells (Cai et al. 2011), the main constituent of amyloid deposits in islets of patients with type 2 diabetes.

Geographic and Temporal Clusters of EV-Positive Cases of Type 1 Diabetes

In the course of these studies, 76 EV-positive cases of type 1 diabetes have been detected in the Northern part of the Province of Varese, Italy (Fig. 15.5a). This area is bordered by high mountains to the North and set apart by lakes and rivers to the

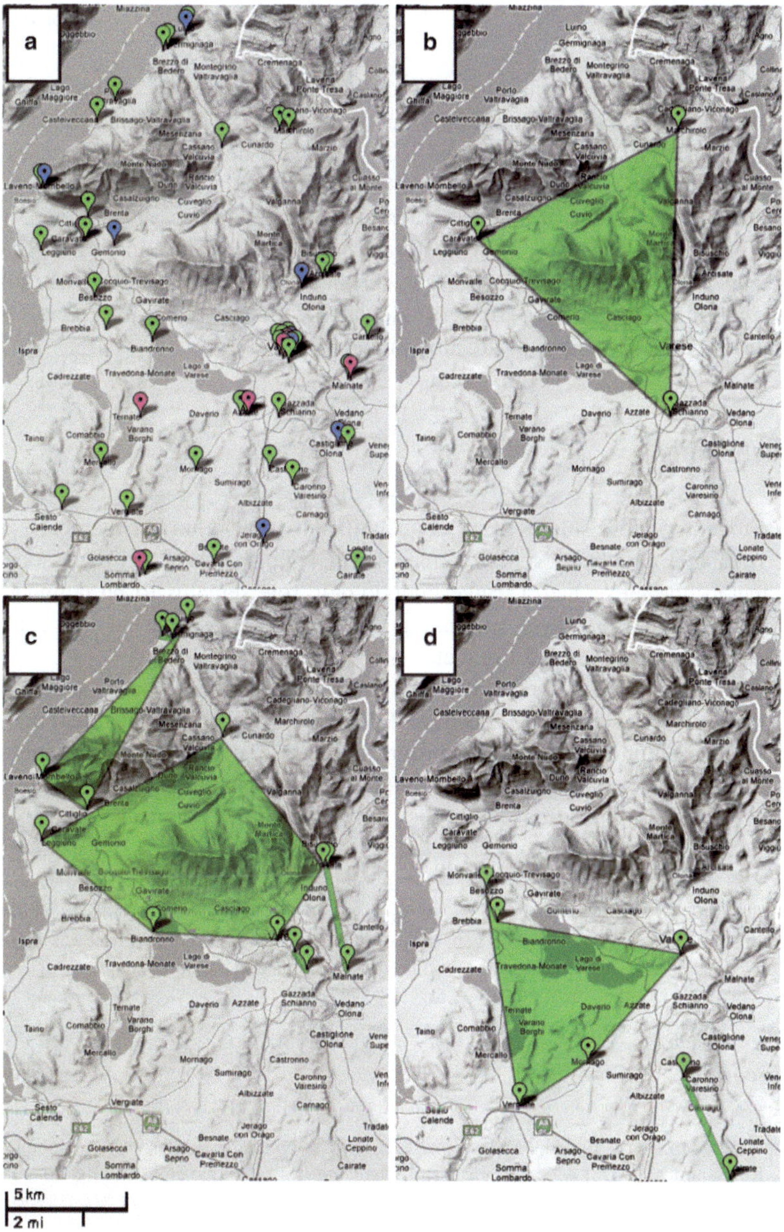

Fig. 15.5 Google® map of the Northern part of the Province of Varese, Italy, with the location of the enterovirus-positive cases of type 1 diabetes (**a**) and examples of geographic and temporal clusters of cases (**b–d**). (**a**) Location of 76 enterovirus-positive cases of type 1 diabetes that have been detected in the course of the study. Cases associated with different species of enteroviruses are represented with markers of different colors: A species (*blue*); B species (*green*); C species (*fuchsia*). (**b**) One cluster of 3 cases associated with an enterovirus of the B1 subgroup (the B1 subgroup comprises 20 echovirus types and Coxsackieviruses B1–B3). (**c**) Five clusters (total 14 cases) associated with an enterovirus of the B1 subgroup (as above). (**d**) Two clusters (total 7 cases) associated with an enterovirus of the B2 subgroup (the B2 subgroup comprises selected echoviruses, selected numbered enteroviruses, and Coxsackievirus B5)

West and the East. As a consequence, most children developing symptoms of type 1 diabetes are seen at a single hospital. Over a 4-year period, geographic and temporal clustering of cases have been noted. Clusters were defined as those cases that occurred with a maximum lag of 9 months within 20 km from each other. It was found that each cluster was associated with a single enterovirus species, and, in the case of the B and C species, each cluster was associated with a single B or C virus subgroup. The sites of the new cases that have been investigated during the study are shown in Fig. 15.5 (panel a). Clusters associated with enterovirus B1 (panels b and c) and B2 (panel d) subgroups are also shown.

Conclusions

Taken together, these studies point to a significant association between early stage type 1 diabetes and the presence of enteroviruses in blood. Whether the association was merely casual or had a triggering/etiologic role remains to be defined. The above observations, however, substantiate further the long-suspected implication of these agents in juvenile type 1 diabetes.

It is well known that C-peptide levels decline rapidly in the perionset period of type 1 diabetes (Sosenko et al. 2008) and that preservation of the residual beta cell reserve is essential for a favorable prognosis. Thus, early postdiagnosis interventions need to be developed for possible therapy as close to the diagnosis as possible. Among these, antiviral drugs/antibodies may be taken into consideration (Norder et al. 2011; Chen et al. 2011).

Acknowledgments This work was financed by the CARIPLO Foundation, Milano, Italy (grant Ricerca Biomedica 2009-2577). The generous support of Gianni Valcavi, Attorney is gratefully acknowledged. Thanks to Abner Louis Notkins for inspiration and encouragement.

References

Al-Hello H, Paananen A, Eskelinen M, Ylipaasto P, Hovi T, Salmela K, Lukashev AN, Bobegamage S, Roivainen M (2008) An enterovirus strain isolated from diabetic child belongs to a genetic subcluster of echovirus 11, but is also neutralised with monotypic antisera to coxsackievirus A9. J Gen Virol 89:1949–1959

Baj A, Monaco S, Zanusso G, Dall'Ora E, Bertolasi A, Toniolo A (2007) Virology of the post-polio syndrome. Fut Virol 2:183–192

Bonifacio E, Genovese S, Braghi S, Bazzigaluppi E, Lampasona V, Bingley PJ, Rogge L, Pastore MR, Bognetti E, Bottazzo GF, Gale EAM, Bosi E (1995a) Islet autoantibody markers in IDDM: risk assessment strategies yielding high sensitivity. Diabetologia 38:816–822

Bonifacio E, Lampasona V, Genovese S, Ferrari M, Bosi E (1995b) Identification of protein tyrosine phosphatase-like IA2 (islet cell antigen 512) as the insulin-dependent diabetes-related 37/40K autoantigen and a target of islet-cell antibodies. J Immunol 155:5419–5426

Cabrera-Rode E, Sarmiento L, Tiberti C, Molina G, Barrios J, Hernández D, Díaz-Horta O, Di Mario U (2003) Type 1 diabetes islet associated antibodies in subjects infected by echovirus 16. Diabetologia 46:1348–1353

Cabrera-Rode E, Sarmiento L, Molina G, Pérez C, Arranz C, Galvan JA, Prieto M, Barrios J, Palomera R, Fonseca M, Mas P, Díaz-Díaz O, Díaz-Horta O (2005) Islet cell related antibodies and type 1 diabetes associated with echovirus 30 epidemic: a case report. J Med Virol 76:373–377

Cai K, Qi D, Hou X, Wang O, Chen J, Deng B, Qian L, Liu X, Le Y (2011) MCP-1 upregulates amylin expression in murine pancreatic β cells through ERK/JNK-AP1 and NF-κB related signaling pathways independent of CCR2. PLoS One 6:e19559

Cavalcante P, Barberis M, Cannone M, Baggi F, Antozzi C, Maggi L, Cornelio F, Barbi M, Didò P, Berrih-Aknin S, Mantegazza R, Bernasconi P (2010) Detection of poliovirus-infected macrophages in thymus of patients with myasthenia gravis. Neurology 74:1118–1126

Champsaur H, Dussaix E, Samolyk D, Fabre M, Bach C, Assan R (1980) Diabetes and Coxsackie virus B5 infection. Lancet 1:251

Chapman NM, Kim KS (2008) Persistent coxsackievirus infection: enterovirus persistence in chronic myocarditis and dilated cardiomyopathy. Curr Top Microbiol Immunol 323:275–292

Chen Z, Chumakov K, Dragunsky E, Kouiavskaia D, Makiya M, Neverov A, Rezapkin G, Sebrell A, Purcell R (2011) Chimpanzee-human monoclonal antibodies for treatment of chronic poliovirus excretors and emergency postexposure prophylaxis. J Virol 85:4354–4362

Coleman JR, Papamichail D, Skiena S, Futcher B, Wimmer E, Mueller S (2008) Virus attenuation by genome-scale changes in codon pair bias. Science 320:1784–1787

Conaldi PG, Biancone L, Bottelli A, De Martino A, Camussi G, Toniolo A (1997) Distinct pathogenic effects of group B coxsackieviruses on human glomerular and tubular kidney cells. J Virol 71:9180–9187

Destombes J, Couderc T, Thiesson D, Girard S, Wilt SG, Blondel B (1997) Persistent poliovirus infection in mouse motoneurons. J Virol 71:1621–1628

Díaz-Horta O, Bello M, Cabrera-Rode E, Suárez J, Más P, García I, Abalos I, Jofra R, Molina G, Díaz-Díaz O, Di Mario U (2001) Echovirus 4 and type 1 diabetes mellitus. Autoimmunity 34:275–281

Dotta F, Censini S, van Halteren AG, Marselli L, Masini M, Dionisi S, Mosca F, Boggi U, Muda AO, Prato SD, Elliott JF, Covacci A, Rappuoli R, Roep BO, Marchetti P (2007) Coxsackie B4 virus infection of beta cells and natural killer cell insulitis in recent-onset type 1 diabetic patients. Proc Natl Acad Sci USA 104:5115–5120

Feuer R, Ruller CM, An N, Tabor-Godwin JM, Rhoades RE, Maciejewski S, Pagarigan RR, Cornell CT, Crocker SJ, Kiosses WB, Pham-Mitchell N, Campbell IL, Whitton JL (2009) Viral persistence and chronic immunopathology in the adult central nervous system following Coxsackievirus infection during the neonatal period. J Virol 83:9356–9369

Foxman EF, Iwasaki A (2011) Genome-virome interactions: examining the role of common viral infections in complex disease. Nat Rev Microbiol 9:254–264

Fujioka S, Kitaura Y, Deguchi H, Shimizu A, Isomura T, Suma H, Sabbah HN (2004) Evidence of viral infection in the myocardium of American and Japanese patients with idiopathic dilated cardiomyopathy. Am J Cardiol 94:602–605

Gorbea C, Makar KA, Pauschinger M, Pratt G, Bersola JL, Varela J, David RM, Banks L, Huang CH, Li H, Schultheiss HP, Towbin JA, Vallejo JG, Bowles NE (2010) A role for Toll-like receptor 3 variants in host susceptibility to enteroviral myocarditis and dilated cardiomyopathy. J Biol Chem 285:23208–23223

Harvala H, Sharp CP, Ngole EM, Delaporte E, Peeters M, Simmonds P (2011) Detection and genetic characterization of enteroviruses circulating among wild populations of chimpanzees in Cameroon: relationship with human and simian enteroviruses. J Virol 85:4480–4486

Hober D, Sauter P (2010) Pathogenesis of type 1 diabetes mellitus: interplay between enterovirus and host. Nat Rev Endocrinol 6:279–289

Hu YF, Yang F, Du J, Dong J, Zhang T, Wu ZQ, Xue Y, Jin Q (2011) Complete genome analysis of coxsackievirus A2, A4, A5, and A10 strains isolated from hand-foot-and-mouth disease patients in China revealing frequent recombination of human enterovirus A. J Clin Microbiol 49:2426–2434

Jenson AB, Rosenberg HS, Notkins AL (1980) Pancreatic islet-cell damage in children with fatal viral infections. Lancet 2:354–358

Kelly EJ, Hadac EM, Cullen BR, Russell SJ (2010) MicroRNA antagonism of the picornaviral life cycle: alternative mechanisms of interference. PLoS Pathog 6:e1000820

Lampasona V, Petrone A, Tiberti C, Capizzi M, Spoletini M, di Pietro S, Songini M, Bonicchio S, Giorgino F, Bonifacio E, Bosi E, Buzzetti R (2010) Zinc transporter 8 antibodies complement GAD and IA-2 antibodies in the identification and characterization of adult-onset autoimmune diabetes. Diabetes Care 33:104–108

McWilliam Leitch EC, Cabrerizo M, Cardosa J, Harvala H, Ivanova OE, Kroes AC, Lukashev A, Muir P, Odoom J, Roivainen M, Susi P, Trallero G, Evans DJ, Simmonds P (2010) Evolutionary dynamics and temporal/geographical correlates of recombination in the human enterovirus echovirus types 9, 11, and 30. J Virol 84:9292–9300

Miao LY, Pierce C, Gray-Johnson J, DeLotell J, Shaw C, Chapman N, Yeh E, Schnurr D, Huang YT (2009) Monoclonal antibodies to VP1 recognize a broad range of enteroviruses. J Clin Microbiol 47:3108–3113

Naserke HE, Dozio N, Ziegler AG, Bonifacio E (1998) Comparison of a novel micro-assay for insulin autoantibodies with the conventional radiobinding assay. Diabetologia 41:681–683

Nix WA, Oberste MS, Pallansch MA (2006) Sensitive, seminested PCR amplification of VP1 sequences for direct identification of all enterovirus serotypes from original clinical specimens. J Clin Microbiol 44:2698–2704

Norder H, De Palma AM, Selisko B, Costenaro L, Papageorgiou N, Arnan C, Coutard B, Lantez V, De Lamballerie X, Baronti C, Solà M, Tan J, Neyts J, Canard B, Coll M, Gorbalenya AE, Hilgenfeld R (2011) Picornavirus non-structural proteins as targets for new anti-virals with broad activity. Antiviral Res 89:204–218

Notkins AL (1979) The causes of diabetes. Sci Am 241:62–73

Otonkoski T, Roivainen M, Vaarala O, Dinesen B, Leipälä JA, Hovi T, Knip M (2000) Neonatal type I diabetes associated with maternal echovirus 6 infection: a case report. Diabetologia 43:1235–1238

Paananen A, Ylipaasto P, Rieder E, Hovi T, Galama J, Roivainen M (2003) Molecular and biological analysis of echovirus 9 strain isolated from a diabetic child. J Med Virol 69:529–537

Pallansch MA, Oberste MS (2010) High degree of genetic diversity of non-polio enteroviruses identified in Georgia by environmental and clinical surveillance, 2002-2005. J Med Microbiol 59:1340–1347

Rahnefeld A, Ebstein F, Albrecht N, Opitz E, Kuckelkorn U, Stangl K, Rehm A, Kloetzel PM, Voigt A (2011) Antigen presentation capacity of dendritic cells is impaired in ongoing enterovirus-myocarditis. Eur J Immunol 41:2774–2781

Rhoades RE, Tabor-Godwin JM, Tsueng G, Feuer R (2011) Enterovirus infections of the central nervous system. Virology 411:288–305

Richardson SJ, Willcox A, Bone AJ, Foulis AK, Morgan NG (2009) The prevalence of enteroviral capsid protein vp1 immunostaining in pancreatic islets in human type 1 diabetes. Diabetologia 52:1143–1151

Richardson SJ, Willcox A, Hilton DA, Tauriainen S, Hyoty H, Bone AJ, Foulis AK, Morgan NG (2010) Use of antisera directed against dsRNA to detect viral infections in formalin-fixed paraffin-embedded tissue. J Clin Virol 49:180–185

Richardson SJ, Willcox A, Bone AJ, Morgan NG, Foulis AK (2011) Immunopathology of the human pancreas in type-I diabetes. Semin Immunopathol 33:9–21

Roivainen M, Klingel K (2010) Virus infections and type 1 diabetes risk. Curr Diab Rep 10:350–356

Shibasaki S, Imagawa A, Tauriainen S, Iino M, Oikarinen M, Abiru H, Tamaki K, Seino H, Nishi K, Takase I, Okada Y, Uno S, Murase-Mishiba Y, Terasaki J, Makino H, Shimomura I, Hyöty H, Hanafusa T (2010) Expression of toll-like receptors in the pancreas of recent-onset fulminant type 1 diabetes. Endocr J 57:211–219

Sosenko JM, Palmer JP, Rafkin-Mervis L, Krischer JP, Cuthbertson D, Matheson D, Skyler JS (2008) Glucose and C-peptide changes in the perionset period of type 1 diabetes in the Diabetes Prevention Trial-Type 1. Diabetes Care 31:2188–2192

Stene LC, Oikarinen S, Hyöty H, Barriga KJ, Norris JM, Klingensmith G, Hutton JC, Erlich HA, Eisenbarth GS, Rewers M (2010) Enterovirus infection and progression from islet autoimmunity to type 1 diabetes: the Diabetes and Autoimmunity Study in the Young (DAISY). Diabetes 59:3174–3180

Tapia G, Cinek O, Rasmussen T, Witsø E, Grinde B, Stene LC, Rønningen KS (2011) Human enterovirus RNA in monthly fecal samples and islet autoimmunity in Norwegian children with high genetic risk for type 1 diabetes: the MIDIA study. Diabetes Care 34:151–155

Tauriainen S, Oikarinen S, Oikarinen M, Hyöty H (2011) Enteroviruses in the pathogenesis of type 1 diabetes. Semin Immunopathol 33:45–55

Toniolo A, Federico G, Manocchio I, Onodera T (1988) Aetiology and pathogenesis of type-i diabetes: viruses. In: Besser GM, Bodanski JM, Cudworth AG (eds) Clinical diabetes: an illustrated text. Gower Medical Publ, London, pp 10.1–20

Toniolo A, Maccari G, Federico G, Salvatoni A, Bianchi G, Baj A (2010) Are enterovirus infections linked to the early stages of type 1 diabetes? In: American Society for Microbiology 110th General Meeting, San Diego, CA, ST-902

Tracy S, Drescher KM, Jackson JD, Kim K, Kono K (2010) Enteroviruses, type 1 diabetes and hygiene: a complex relationship. Rev Med Virol 20:106–116

Vreugdenhil GR, Schloot NC, Hoorens A, Rongen C, Pipeleers DG, Melchers WJ, Roep BO, Galama JM (2000) Acute onset of type I diabetes mellitus after severe echovirus 9 infection: putative pathogenic pathways. Clin Infect Dis 31:1025–1031

Willcox A, Richardson SJ, Bone AJ, Foulis AK, Morgan NG (2011) Immunohistochemical analysis of the relationship between islet cell proliferation and the production of the enteroviral capsid protein, VP1, in the islets of patients with recent-onset type 1 diabetes. Diabetologia 54:2417–2420

Williams CH, Oikarinen S, Tauriainen S, Salminen K, Hyoty H, Stanway G (2006) Molecular analysis of an echovirus 3 strain isolated from an individual concurrently with appearance of islet cell and IA-2 autoantibodies. J Clin Microbiol 44:441–448

Yeung WC, Rawlinson WD, Craig ME (2011) Enterovirus infection and type 1 diabetes mellitus: systematic review and meta-analysis of observational molecular studies. BMJ 342:d35

Yoon JW, Austin M, Onodera T, Notkins AL (1979) Isolation of a virus from the pancreas of a child with diabetic ketoacidosis. N Engl J Med 300:1173–1179

Yozwiak NL, Skewes-Cox P, Gordon A, Saborio S, Kuan G, Balmaseda A, Ganem D, Harris E, DeRisi JL (2010) Human enterovirus 109: a novel interspecies recombinant enterovirus isolated from a case of acute pediatric respiratory illness in Nicaragua. J Virol 84:9047–9058

Chapter 16
Coxsackieviruses and Insulitis

Letizia Galleri, Fabio Arturo Grieco, Guido Sebastiani, Isabella Spagnuolo, Francesco Vendrame, and Francesco Dotta

Abstract Coxsackievirus infections are believed to be a relevant risk factor in the induction of pancreatic beta cell damage and autoimmune response in type 1 diabetes mellitus. Genomic RNA and proteins of coxsackieviruses have been detected in tissues of type 1 diabetes patients, supporting the involvement of enteroviruses in the pathogenesis of type 1 diabetes. Coxsackieviruses may infect beta cells, trigger the activation of innate immune systems, or accelerate the autoimmune process leading to the disease. Local inflammatory changes generated in pancreatic islets and the mechanisms leading to its generation and progression have been studied in animal models of type 1 diabetes and in humans. The role of coxsackieviruses in the insulitic process is discussed in this chapter, together with the ability of selected coxsackievirus serotypes to protect against type 1 diabetes.

Introduction

Type 1 diabetes mellitus is a multistep autoimmune disease characterized by the specific immune-mediated destruction of pancreatic beta cells, causing a progressive loss of insulin secretion.

L. Galleri • F.A. Grieco • G. Sebastiani • I. Spagnuolo • F. Dotta, M.D. (✉)
Diabetes Unit, Department of Internal Medicine, Endocrine and Metabolic Sciences and Biochemistry, University of Siena, Siena, Italy

Umberto Di Mario ONLUS Research Foundation, c/o Toscana Life Science Park, Siena, Italy

F. Vendrame
Umberto Di Mario ONLUS Research Foundation, c/o Toscana Life Science Park, Siena, Italy

Diabetes Research Institute, University of Miami, Miami, FL, USA

F. Dotta (✉)
University of Siena Medical School, Policlinico "Le Scoette", Siena, Italy
e-mail: francesco.dotta@alice.it

K.W. Taylor, H. Hyoty, A. Toniolo, A.J. Zuckerman (eds.), *Diabetes and Viruses*, DOI 10.1007/978-1-4614-4051-2_16,

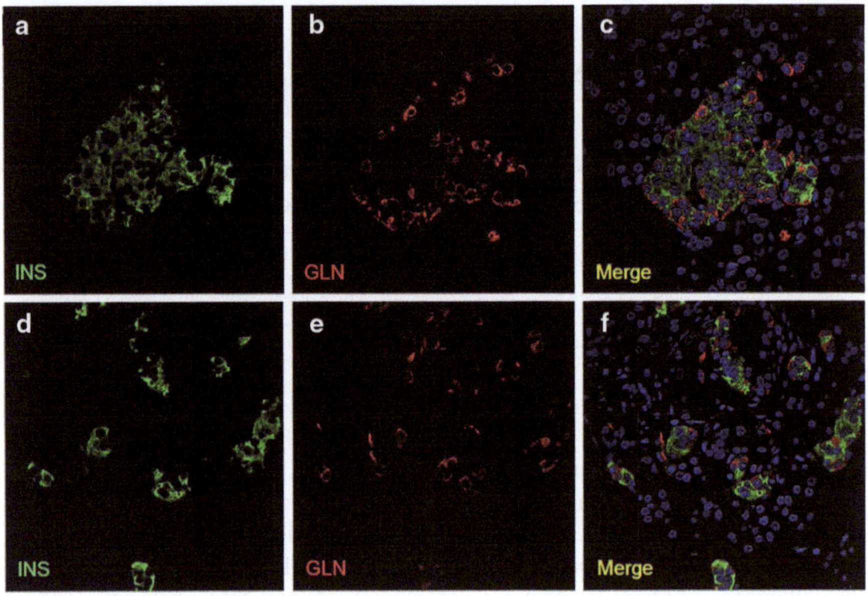

Fig. 16.1 Variable content of beta cells in an enterovirus-infected human pancreas. Confocal microscopy images for insulin (**a** and **d** in *green*), glucagon (**b** and **e** in *red*), and their *merge* (**c** and **f**) show an apparently intact islet with a preserved insulin content (*upper panel*) and an islet with a small amount of residual beta cells (*lower panel*). Nuclear staining has been performed with DAPI (in *blue*)

In both humans and animal models, such as non-obese diabetic (NOD) mice that develop spontaneously autoimmune diabetes similar to the human disease, the progressive beta cell destruction is a T-lymphocyte-mediated event: diabetic patients develop autoaggressive T lymphocytes subsets directed against their own pancreatic beta cells, generating an islet inflammation process defined insulitis.

Insulitis is a chronic inflammatory infiltrate of the islets of Langerhans and is observed commonly in the majority of type 1 diabetic cases, at the time when most beta cells have been either destroyed or are inactive functionally (Eizirik et al. 2009). Insulitis may show different degrees of severity (Fig. 16.1), ranging from a very mild inflammatory process with almost intact beta cells to an intense lymphocytic infiltrate with massive beta cell destruction. However, a minimum of five mononuclear cells per islet section is considered a threshold for defining human insulitis (Richardson et al. 2011). The inflammatory infiltrate shows a predominance of $CD8^+$ cytotoxic T cells which reach the peak when the beta cell number drops. $CD4^+$ T cells are also present in lower numbers, together with different subsets of B lymphocytes, macrophages, and dendritic cell (DCs). Natural Killer (NK) cells and regulatory T cells (T-reg) are rarely observed (Richardson et al. 2011). The presence of most cell subsets in infiltrated islets from recent-onset type 1 diabetes patients confirms the involvement of both innate and adaptive immune systems in the pro-inflammatory process leading to the beta cell impairment, damage, or death.

Causal Factors for Type 1 Diabetes

It is well known that genetic susceptibility plays a key role in the pathogenesis and that it is linked to HLA class I and class II genes. However, since identical twins lack concordance in the disease rate, and type 1 diabetes incidence across the world has been increasing constantly by 2–5% over the last decades (Dotta et al. 2007), different factors are thought to be involved. Indeed, genetic factors together with environmental triggers may open the route for type 1 diabetes. In this scenario, viral infections are thought to play a key role. Among the wide range of viruses suspected to play such a role [e.g., retroviruses, reoviruses, and rotaviruses (Jun and Yoon 2003)], human enteroviruses are held as major players (Yeung et al. 2011). Infections by these agents may increase the incidence of the disease, inducing and/or amplifying the immune assault to pancreatic beta cells. However, depending on the virus strain and/or the timing of infection, coxsackieviruses may have opposite effect on insulitis and type 1 diabetes pathogenesis, protecting beta cells from damage and T cell-mediated destruction (Hober and Sauter 2010).

Coxsackieviruses in Type 1 Diabetes

Coxsackieviruses are single-stranded RNA enteroviruses belonging to the *Picornaviridae* family. They are small non-enveloped virus particles with icosahedral symmetry of the capsid surrounding the RNA genome. The capsid consists of 60 capsomers each containing four structural proteins, VP1 to VP4. Twenty-four serotypes are known in group A coxsackieviruses and six serotypes in the group B coxsackieviruses. Coxsackievirus infections usually start from the respiratory or gastrointestinal tract, spread to the bloodstream, then localize to secondary sites of infection such as pancreas (Hober and Sauter 2010), heart, or central nervous system. Beside some evidence that coxsackieviruses A may be associated with type 1 diabetes (Jun and Yoon 2003), group B coxsackieviruses—possibly in concert with unknown environmental factors—have the strongest correlation with type 1 diabetes both in humans and in animal models (Table 16.1). Strains of coxsackievirus B4 and coxsackievirus B5 isolated from the pancreas of individuals with newly diagnosed type 1 diabetes have been shown to induce islet cell damage and diabetes in animal models (Jun and Yoon 2003). Several authors have detected coxsackievirus B genome and proteins in the pancreas (Dotta et al. 2007; Oikarinen et al. 2008a; Richardson et al. 2009; Ylipaasto et al. 2004), gut (Oikarinen et al. 2008b), and peripheral blood mononuclear cells (PBMCs) (Jaïdane and Hober 2008) of type 1 diabetes patients. These observations reinforced the thesis of the link between coxsackieviruses and the autoimmune process leading to type 1 diabetes. Upon infection, these agents may interact with the innate and adaptive immune systems, thus inducing an aberrant immune response against pancreatic self-antigens, a necessary step in the pathogenetic process.

Table 16.1 Association of groups A and B coxsackieviruses with the development of type 1 diabetes and/or beta cell damage in cell cultures, animal models, and humans

Virus group	Serotype	Host	Remarks	Reference
Coxsackie A viruses		Humans	Presence of IgM antibodies against coxsackievirus A in recent-onset type 1 diabetes patients	Jun and Yoon (2003)
	A7, A9	In vitro models	Signs of viral infection of in vitro mouse pancreatic islet cells	Dotta et al. (2010)
Coxsackie B viruses	B1	In vitro models	Infection and damage of human beta cells	Hober and Sauter (2010)
	B2	Humans	Induction of positivity to (islet) autoantibodies; RNA detected in serum	Yeung et al. (2011)
	B3	Cultured human islet cells	Infection and damage of human beta cells	Hober and Sauter (2010)
		Mouse	Induction or prevention of T1D depends on the timing of infection	Filippi et al. (2009)
		Humans	Induction of positivity to (islet) autoantibodies; RNA detected in stool samples and/or serum	Hober and Sauter (2010), Jaïdane et al. (2009), Yeung et al. (2011)
	B4	Rat cell line	Persistent infection of rat insulinoma cell line RIN without induction of metabolism alteration	Dotta et al. (2010)
		Mouse	Induction or prevention of type 1 diabetes depends on the timing of infection. In vitro studies show induction of beta cell functional damage and/or destruction	Jaïdane et al. (2009)
		Nonhuman primates	Type 1 diabetes induction is influenced by genetic background	Jun and Yoon (2003)
		Humans	Signs of viral infections in beta cells of type 1 diabetes patients. Evidence of damaged and impaired beta cell function after in vitro infection of human pancreatic islets	Dotta et al. (2007), Hober and Sauter (2010)
	B4 (Tuscany strain)	Humans	Detected in 3 of 6 recent-onset type 1 diabetes donors. Induction of impaired beta cell function in human islets infected in vitro	Dotta et al. (2007)
	B4 (JVB strain)	Mouse	Nondiabetogenic	Jun and Yoon (2003)
	B4 (E2 strain)		Diabetogenic	Jun and Yoon (2003)
	B5	Mouse cell line	infection of murine insulinoma cell line MIN-6	Roivainen and Klingel (2010)
		Mouse	Damage to islet cells and glucose intolerance	Jun and Yoon (2003)
		Humans	Induction of positivity to (islet) autoantibodies. The viral isolate obtained from a patient was able to induce glucose intolerance in mice	Jun and Yoon (2003), Yeung et al. (2011)
	B6	Humans	Induction of islet autoantibodies. Detection of viral RNA in serum	Yeung et al. (2011)

Coxsackieviruses and the Immune System in Insulitis

In vertebrates, both the innate and the adaptive immune systems evolved for detecting microbes and clearing the infection. While the innate immune system represents the first line of defense against invading microorganisms, providing an immediate first line defense, the adaptive immune system acts more specifically against a pathogen, thus conferring protection against re-exposure to the same pathogen. Both immune systems have a role when a virus infects the pancreas of a susceptible host. As stated previously, coxsackieviruses B show cytolytic activity and are capable of inducing beta cells destruction. No T cell infiltration is observed, and insulitis does not take place. In this case, an acute form of diabetes may occur, both in humans and in animal models (Elshebani et al. 2007). However, cytolytic activity is not sufficient to explain the extensive loss of beta cells upon coxsackievirus infection and does not explain the autoimmune process typical of type 1 diabetes. The mechanisms causing viral-induced beta cell damage and the attraction of immunocompetent cells to the islets have to be defined. In this regard, mechanisms other than direct beta cells lysis have been suggested for the induction of virus-mediated type 1 diabetes:

1. The innate immune system is activated by the binding of microbial components to pattern recognition receptors (PRRs), such as Toll-like receptor (TLR)-3, RIG-1, and IFIH1 (also designed as MDA5). These receptors bind to double-stranded RNA generated during virus replication (Grieco et al. 2011; Ylipaasto et al. 2005; Yoneyama et al. 2004). Once intracellular dsRNA and extracellular dsRNA (released by damaged cells) have been detected, antiviral activities are triggered, contributing to the subsequent adaptive immune response (Grieco et al. 2011; Kato et al. 2006). Indeed, when dsRNA binds to PPRs, transcription factors like NFkB and IRF-3 are activated. NFkB thus initiates the transcription of inflammatory cytokine genes such as interleukin (IL)-1β, interferon (IFN)-γ IL-15, and chemoattractant chemokines such as chemokine (C–C motif) ligand 2 (CCL2) (Fig. 16.2) and C–X–C motif chemokine 10 (CXCL10). Activated

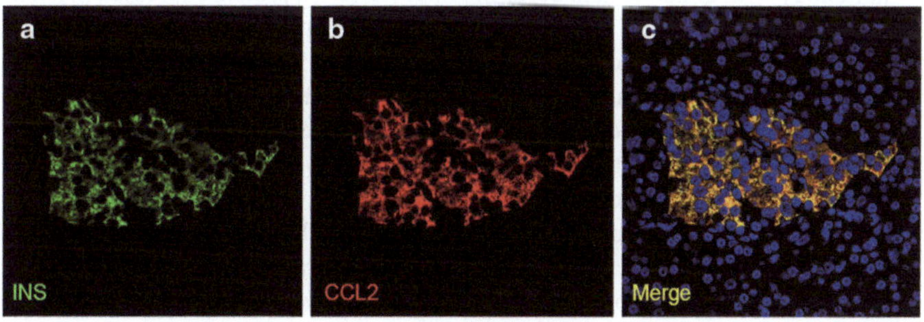

Fig. 16.2 Beta cell-specific expression of pro-inflammatory chemokine CCL2 in an enterovirus-infected human pancreas. Representative images of human pancreas infected by enterovirus showing the colocalization of insulin and the pro-inflammatory chemokine CCL2 in beta cells. Insulin (**a**) is in *green*, CCL2 (**b**) in *red*, and their *merge* (**c**) shows double-positive cells for insulin and CCL2 in *yellow*. Nuclear staining has been performed with DAPI (in *blue*)

IRFs induce transcription of IFN-α and IFN-β. The above cytokines activate resident antigen presenting cells (APCs), and recruit T lymphocytes, NK cells, monocytes, and DCs. Local inflammation, coupled with antiviral defenses, is supposed to eradicate infection. However, when this sequel of events happens in a diabetes-prone individual, these mechanisms of defense may trigger an abnormal response. In patients with type 1 diabetes, pancreas may show insulitic areas with increased islet inflammation and beta cell dysfunction and damage, which in turn may contribute to the release of beta cell antigens (Dotta et al. 2007; Eizirik et al. 2009; Grieco et al. 2011; Hober and Sauter 2010). Activation of PRRs also induces apoptosis and upregulation of major histocompatibility complex class I (MHC-I) in beta cells (Hober and Sauter 2010). These events may promote IFN-induced autoimmunity in both recent-onset type 1 diabetes and fulminant diabetes (Aida et al. 2011; Richardson et al. 2011; Roep et al. 2010).

Recently, the intervention of endoplasmic reticulum (ER) stress in the induction and amplification of inflammatory processes has been proposed. dsRNA and/or inflammatory cytokines are associated with ER stress. Inflamed cells try to protect themselves and respond to ER stress by increasing the degradation of altered proteins. However, if homeostasis is not restored, apoptosis is triggered and modified antigens can be presented to APCs. These events contribute to the autoimmune process in association with the upregulation of MHC class I molecules (Eizirik et al. 2009). So far, however, the involvement of ER stress in the induction of type 1 diabetes induction has still to be proven.

2. The adaptive immune response stimulated by these events can be enhanced by different and not mutually exclusive phenomena. For instance, subsequent infections by different enterovirus types may induce *epitope spreading*, creating novel autoantigens for autoreactive T cells (Hober and Sauter 2010).
3. *Molecular mimicry* is an additional mechanism capable of enhancing the autoimmune process in type 1 diabetes (Hober and Sauter 2010). Viral components may share antigenic properties with specific beta cell antigens, thus promoting cross-reactivity between antiviral T cells and autoreactive T cells. One example is the p2C protein, a non-capsid component of coxsackievirus B4 that shares sequence homology with glutamic acid decarboxylase (GAD), a protein expressed in islets. T cell receptors may fail to properly differentiate between these molecules; thus the similarity of sequences is supposed to generate antigen-specific T effector cells and/or antibodies that recognize beta cells bearing the GAD epitope. Events of this type may also occur due to similarities of the enterovirus capsid protein VP1 (and procapsid protein VP0) with tyrosine phosphatase IA2/IAR and heat shock protein 60. Molecular mimicry seems more relevant for amplifying an ongoing autoimmune process than to initiate one (Filippi and von Herrath 2008; Hober and Sauter 2010). In order to make molecular mimicry effective, viral and beta cell antigens have to share 100% of homology which appears not to occur in human type 1 diabetes (Christen et al. 2004).
4. Rather than molecular mimicry, a *bystander effect* is thought to cause activation of autoreactive T effector cells (Hober and Sauter 2010). Several data from coxsackievirus B infection of human and animal models support this hypothesis.

A local infection of beta cells can cause tissue damage with the consequent release of self-antigens from the islets. The bystander effect is mediated by APCs, which present self-antigens to preexisting autoreactive T cells, but not against viral antigens (Horwitz et al. 2004). In viral infections, an enhancement of autoantigen presentation through MHC class I may also lead to the bystander activation of autoreactive T cells.

Finally, coxsackievirus infection has also been shown as beneficial to the host, preventing type 1 diabetes. This opposite effect depends on the virus serotype, on the cell tropism, and on the timing of infection. Studies in animal models have shown that selected coxsackievirus B3 and B4 strains may prevent type 1 diabetes, conferring a long-term protection from the disease. Indeed, NOD mice develop infection-mediated type 1 diabetes through a bystander effect just if the infection occurs when a sufficient number of preexisting autoreactive T cells are in place (Filippi et al. 2009). These events are mediated by IL-4 and IFN-γ, both required for triggering the onset of type 1 diabetes (Serreze et al. 2005). When mice are infected at an early stage, before sufficient numbers of infiltrating cells accumulate into the islets, it may protect from type 1 diabetes (Filippi et al. 2009). When beta cell destruction is not induced, as in the case of coxsackievirus B3 infection of prediabetic NOD mouse, viruses may be tolerogenic. In the case of coxsackievirus B3, diabetes may be delayed and disease incidence may be decreased by induction of T-reg in the proximity of islets, and upregulation of the programmed death-1 ligand (PD-L1) in lymphocytes. The latter seem to avoid the expansion of diabetogenic cytotoxic $CD8^+$ T cells bearing programmed death-1 (PD-1) immunoinhibitory receptor (Filippi et al. 2009).

Different viruses may be responsible for induction or abrogation of type 1 diabetes. The same agent may either induce or prevent type 1 diabetes depending on the autoimmune status of the infected host. This complex scenario helps to explain how multiple events can be involved in the etiology of type 1 diabetes, and makes more difficult to understand it considering that different degrees of insulitis may be produced. For instance, the "Tuscany strain" of coxsackievirus B4 leads to a nondestructive islet inflammation mediated mainly by NK cells, and human islets infected in vitro by this strain show impaired beta cell function, as demonstrated by lower levels of insulin release in response to glucose compared to not infected islets (Dotta et al. 2007).

Conclusion

Mounting evidence supports the hypothesis that enteroviruses (and particularly coxsackieviruses) can participate, at least in some individuals, in beta cell damage and islet inflammation through a series of mechanisms, not mutually exclusive (Fig. 16.3). These may involve the capacity of a given virus to infect directly beta cells with the consequent impairment of insulin secretion associated or not with beta cell destruction. In addition, coxsackieviruses through interaction with PRR

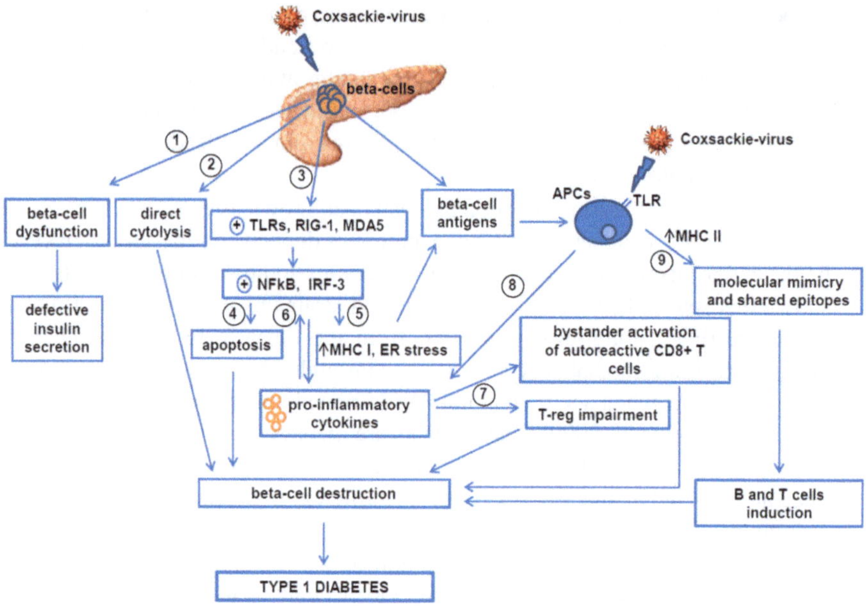

Fig. 16.3 Potential coxsackievirus-mediated mechanisms involved in type 1 diabetes mellitus induction. Beta cell infection by a coxsackievirus may trigger or aggravate type 1 diabetes mellitus through a number of mechanisms not mutually exclusive. (*1*) The virus may impair beta cell function altering glucose sensing mechanism and/or insulin secretion, without beta cell lysis. (*2*) Coxsackievirus may directly destroy insulin-producing-beta cells. (*3*) PRRs (such as TLR-3, RIG-1, and MDA 5) detect viral dsRNA and induce transcriptional factors (e.g., NFkB and IRF-3) activation, which in turn (*4*) lead to beta cell apoptosis, and/or (*5*) increase MHC class I expression on beta cells and/or induce ER stress, with consequent increased beta cell antigen(s) presentation. (*6*) The activation of transcriptional factors induces the secretion of pro-inflammatory cytokines such as IL-1β, TNF, IFN-αβγ which (*7*) may recruit preexisting autoreactive effector CD8⁺ T cells or impair T-reg cell function. (*8*) The enhanced presentation of beta cell antigens, and/or the activation of TLRs which sense dsRNA, induce APCs activation, which increases the secretion of pro-inflammatory mediators, thus amplifying the already ongoing process (*step 7*). (*9*) Molecular mimicry, triggered by enhancement of MHC class II expression on APCs presenting viral peptides, might participate to recruitment of virus-specific B and T cells as well to the events leading to beta cell destruction

(e.g., TLR-3, RIG-1, MDA) may activate transcription factors (e.g., NFkB and IRF-3) with the consequent increase of MHC class I expression on beta cells and/or ER stress and expression of pro-inflammatory cytokines that may contribute to the recruitment of preexisting autoreactive effector $CD8^+$ T cells or impair T-reg cell functions. Finally, molecular mimicry, triggered by the enhancement of MHC class II expression on APCs, may contribute to recruiting virus-specific B and T cells, thus favoring events causing beta cell destruction.

Ongoing collaborative studies involving international research networks, such as nPOD (Network for Pancreatic Organ donors with Diabetes) supported by the Juvenile Diabetes Research Foundation or VIDIS (Viruses in Diabetes International

Study Group), performed on serum samples and on tissues specimens obtained from type 1 diabetic patients and from individuals with islet autoimmunity, will certainly contribute to elucidate the complex interplay among viruses, the immune system, and pancreatic beta cells.

References

Aida K, Nishida Y, Tanaka S, Maruyama T, Shimada A, Awata T, Suzuki M, Shimura H, Takizawa S, Ichijo M, Akiyama D, Furuya F, Kawaguchi A, Kaneshige M, Itakura J, Fujii H, Endo T, Kobayashi T (2011) RIG-I- and MDA5-initiated innate immunity linked with adaptive immunity accelerates beta cell death in fulminant type 1 diabetes. Diabetes 60:884–889

Christen U, Edelmann KH, McGavern DB, Wolfe T, Coon B, Teague MK, Miller SD, Oldstone MB, von Herrath MG (2004) A viral epitope that mimics a self antigen can accelerate but not initiate autoimmune diabetes. J Clin Invest 114:1290–1298

Dotta F, Censini S, van Halteren AG, Marselli L, Masini M, Dionisi S, Mosca F, Boggi U, Muda AO, Prato SD, Elliott JF, Covacci A, Rappuoli R, Roep BO, Marchetti P (2007) Coxsackie B4 virus infection of beta cells and natural killer cell insulitis in recent-onset type 1 diabetic patients. Proc Natl Acad Sci USA 104:5115–5120

Dotta F, Galleri L, Sebastiani G, Vendrame F (2010) Virus infections: lessons from pancreas histology. Curr Diab Rep 10:357–361

Eizirik DL, Colli ML, Ortis F (2009) The role of inflammation in insulitis and beta cell loss in type 1 diabetes. Nat Rev Endocrinol 5:219–226

Elshebani A, Olsson A, Westman J, Tuvemo T, Korsgren O, Frisk G (2007) Effects on isolated human pancreatic islet cells after infection with strains of enterovirus isolated at clinical presentation of type 1 diabetes. Virus Res 124:193–203

Filippi CM, von Herrath MG (2008) Viral trigger for type 1 diabetes: pros and cons. Diabetes 57:2863–2871

Filippi CM, Estes EA, Oldham JE, von Herrath MG (2009) Immunoregulatory mechanisms triggered by viral infections protect from type 1 diabetes in mice. J Clin Invest 119:1515–1523

Grieco FA, Vendrame F, Spagnuolo I, Dotta F (2011) Innate immunity and the pathogenesis of type 1 diabetes. Semin Immunopathol 33:57–66

Hober D, Sauter P (2010) Pathogenesis of type 1 diabetes mellitus: interplay between enterovirus and host. Nat Rev Endocrinol 6:279–289

Horwitz MS, Ilic A, Fine C, Balasa B, Sarvetnick N (2004) Coxsackieviral-mediated diabetes: induction requires antigen-presenting cells and is accompanied by phagocytosis of beta cells. Clin Immunol 110:134–144

Jaïdane H, Hober D (2008) Role of coxsackievirus B4 in the pathogenesis of type 1 diabetes. Diabetes Metab 34:537–548

Jaïdane H, Sané F, Gharbi J, Aouni M, Romond MB, Hober D (2009) Coxsackievirus B4 and type 1 diabetes pathogenesis: contribution of animal models. Diabetes Metab Res Rev 25:591–603

Jun HS, Yoon JW (2003) A new look at viruses in type 1 diabetes. Diabetes Metab Res Rev 19:8–31

Kato H, Takeuchi O, Sato S, Yoneyama M, Yamamoto M, Matsui K, Uematsu S, Jung A, Kawai T, Ishii KJ, Yamaguchi O, Otsu K, Tsujimura T, Koh CS, Reise Sousa C, Matsuura Y, Fujita T, Akira S (2006) Differential roles of MDA5 and RIG-I helicases in the recognition of RNA viruses. Nature 441:101–105

Oikarinen M, Tauriainen S, Honkanen T, Vuori K, Karhunen P, Vasama-Nolvi C, Oikarinen S, Verbeke C, Blair GE, Rantala I, Ilonen J, Simell O, Knip M, Hyöty H (2008a) Analysis of pancreas tissue in a child positive for islet cell antibodies. Diabetologia 51:1796–1802

Oikarinen M, Tauriainen S, Honkanen T, Oikarinen S, Vuori K, Kaukinen K, Rantala I, Mäki M, Hyöty H (2008b) Detection of enteroviruses in the intestine of type 1 diabetic patients. Clin Exp Immunol 151:71–75

Richardson SJ, Willcox A, Bone AJ, Foulis AK, Morgan NG (2009) The prevalence of enteroviral capsid protein vp1 immunostaining in pancreatic islets in human type 1 diabetes. Diabetologia 52:1143–1151

Richardson SJ, Willcox A, Bone AJ, Morgan NG, Foulis AK (2011) Immunopathology of the human pancreas in type-I diabetes. Semin Immunopathol 33:9–21

Roep BO, Kleijwegt FS, van Halteren AG, Bonato V, Boggi U, Vendrame F, Marchetti P, Dotta F (2010) Islet inflammation and CXCL10 in recent-onset type 1 diabetes. Clin Exp Immunol 159:338–343

Roivainen M, Klingel K (2010) Virus infections and type 1 diabetes risk. Curr Diab Rep 10:350–356

Serreze DV, Wasserfall C, Ottendorfer EW, Stalvey M, Pierce MA, Gauntt C, O'Donnell B, Flanagan JB, Campbell-Thompson M, Ellis TM, Atkinson MA (2005) Diabetes acceleration or prevention by a coxsackievirus B4 infection: critical requirements for both interleukin-4 and gamma interferon. J Virol 79:1045–1052

Yeung WC, Rawlinson WD, Craig ME (2011) Enterovirus infection and type 1 diabetes mellitus: systematic review and meta-analysis of observational molecular studies. BMJ 342:d35. doi:10.1136/bmj.d35

Ylipaasto P, Klingel K, Lindberg AM, Otonkoski T, Kandolf R, Hovi T, Roivainen M (2004) Enterovirus infection in human pancreatic islet cells, islet tropism in vivo and receptor involvement in cultured islet beta cells. Diabetologia 47:225–239

Ylipaasto P, Kutlu B, Rasilainen S, Rasschaert J, Salmela K, Teerijoki H, Korsgren O, Lahesmaa R, Hovi T, Eizirik DL, Otonkoski T, Roivainen M (2005) Global profiling of coxsackievirus- and cytokine-induced gene expression in human pancreatic islets. Diabetologia 48:1510–1522

Yoneyama M, Kikuchi M, Natsukawa T, Shinobu N, Imaizumi T, Miyagishi M, Taira K, Akira S, Fujita T (2004) The RNA helicase RIG-I has an essential function in double-stranded RNA-induced innate antiviral responses. Nat Immunol 5:730–737

Chapter 17
Viruses in the Human Pancreas

S.J. Richardson, A. Willcox, A.J. Bone, N.G. Morgan, and A.K. Foulis

Abstract There is extensive epidemiological evidence highlighting a possible aetiological role for enteroviral infection in type 1 diabetes. Direct evidence of the presence of enterovirus in the pancreatic islets of type 1 diabetics is, however, limited, due mainly to the paucity of samples from patients diagnosed recently with the disease. This chapter summarises the evidence implicating enteroviral infection in the human pancreas in type 1 diabetes and considers both factors indicating the presence of virus (viral capsid protein, electron microscopic visualisation of viral particles or expression of viral RNA) and the host response to a viral infection (a "viral footprint"). The relationship of these two indicators of viral infection with two apparently different forms of type 1 diabetes (autoimmune versus fulminant) is discussed. It is hypothesised that differing host responses, and perhaps different genetic variation among the viruses involved, may determine whether an enteroviral infection of beta cells causes (1) A rapid lytic cell death—characteristic of fulminant type 1 diabetes and neonatalcoxsackievirus infection, (2) A persistent infection which evokes an autoimmune reaction to beta cells, eventually resulting in their destruction—autoimmune type 1 diabetes, (3) A persistent infection with little host response and little damage.

Overall, it is concluded that the evidence for viral involvement is persuasive but that there is still a long way to go in order to elucidate the precise role of enteroviral infections in the aetiology of type 1 diabetes.

S.J. Richardson, Ph.D. (✉) • A. Willcox, Ph.D. • N.G. Morgan, Ph.D.
Peninsula Medical School, University of Exeter, Plymouth PL6 8BU, UK
e-mail: sarah.richardson@pms.ac.uk

A.J. Bone, Ph.D.
School of Pharmacy and Biomolecular Sciences, University of Brighton, Moulescoomb, Brighton BN2 4GJ, UK

A.K. Foulis, M.D.
Department of Pathology, Royal Infirmary, Glasgow G4 OSF, UK
e-mail: Alan.Foulis@glasgow.ac.uk

K.W. Taylor, H. Hyoty, A. Toniolo, A.J. Zuckerman (eds.), *Diabetes and Viruses*, 167
DOI 10.1007/978-1-4614-4051-2_17, © Springer Science+Business Media New York 2013

Mounting epidemiological evidence has implicated viruses, particularly (but not exclusively) enteroviruses, in the pathogenesis of type 1 diabetes (Andreoletti et al. 1997; Clements et al. 1995; Coutant et al. 2002; Lonnrot et al. 2000; Moya-Suri et al. 2005; Nairn et al. 1999; Richardson et al. 2011; Sarmiento et al. 2007; Schulte et al. 2010; Tauriainen et al. 2011; Yin et al. 2002). However, it is unclear whether viral infection plays a role in the initial development of beta cell autoimmunity or if it accelerates the progression of a pre-existing autoimmune reaction. Indeed, both possibilities may be true depending on the circumstances. In this chapter, we review the evidence implicating viral infection within the pancreas as a causative factor in human type 1 diabetes.

Type 1 diabetes results from destruction of insulin-secreting beta cells in the islets of Langerhans and can occur in two forms. Most commonly, an autoimmune reaction proceeds over a protracted time course leading to the onset of clinical symptoms when ~75% of beta cells have been destroyed. By contrast, a second form of the disease has been described, principally in Japanese populations, in which the process of beta cell destruction occurs much more quickly, leading to "fulminant" type 1 diabetes. In both cases there is increasing evidence that enteroviral infection may underlie the process of beta cell destruction, but the role of the virus appears to be quite different in these two forms of the disease (Dotta et al. 2007; Richardson et al. 2009; Shibasaki et al. 2010; Tanaka et al. 2009).

In order to examine the relationships between enteroviral infection and the pathogenesis of type 1 diabetes, it is important to conduct studies of the target organ (i.e. the pancreas). Such studies have revealed that an early event within the pancreatic islets in autoimmune diabetes is the expression of interferon-alpha by insulin-secreting beta cells (Foulis et al. 1987b). Interferon-alpha expression is closely associated with Class I MHC hyperexpression by the endocrine cells of the affected islet, a hallmark feature of type 1 diabetes (Bottazzo et al. 1985; Foulis et al. 1987a; Pujol-Borrell et al. 1986). Later there is an infiltrate into the islet of immune cells (insulitis), dominated by cytotoxic $CD8^+$ T lymphocytes, with smaller numbers of B lymphocytes, $CD4^+$ cells and macrophages (Willcox et al. 2009). Beta cells are destroyed within the inflamed islet and, following their loss, the inflammatory cell infiltrate disappears, leaving an insulin-deficient islet containing glucagon-secreting alpha cells, somatostatin-secreting delta cells and pancreatic polypeptide-secreting PP cells, which have not been harmed (Foulis and Stewart 1984). The whole process in the pancreas, from the first to last beta cell being destroyed, is thought to take many months or even years in autoimmune diabetes.

By contrast, in fulminant diabetes, there is a diffuse infiltrate of inflammatory cells within the pancreas resulting in exocrine inflammation (and clinical acute pancreatitis) as well as islet inflammation. The predominant infiltrating cell type in the islets in fulminant type 1 diabetes is the macrophage, closely followed by $CD8^+$ T cells (Shibasaki et al. 2010; Tanaka et al. 2009), and the inflammatory process occurs rapidly with most beta cells (and many alpha cells) being destroyed in a matter of days or weeks (Shibasaki et al. 2010; Tanaka et al. 2009). Within 1 month, the insulitis declines and, after this time, evidence of immune cell infiltration is rarely detected in pancreas biopsy samples.

The presence of interferon-alpha in beta cells in autoimmune type 1 diabetes (Foulis et al. 1987b) and in fulminant type 1 diabetes (Aida et al. 2011) is consistent with the hypothesis that the beta cells might be infected by a virus, as viral infection stimulates beta cells to secrete this cytokine (Chehadeh et al. 2000). Equally, the appearance of massive rapid destruction of the beta cells in fulminant diabetes led some to hypothesise that this was due to a widespread lytic infection by a virus (Foulis et al. 1988). Examination of the literature reveals, however, only limited evidence of direct infection of pancreatic beta cells in patients with type 1 diabetes. At first sight, this appears problematic to the viral hypothesis but, in reality, the paucity of data relates to a still more fundamental problem; namely that pancreatic tissue obtained from patients diagnosed recently with type 1 diabetes is available in only very limited quantity. By contrast, tissue from patients with longer disease duration can be obtained more readily, but this is less profitable to study since, as described above, the pancreas becomes largely depleted of beta cells over the course of time and many of the clues which might hint at the cause of their destruction are lost. Despite these constraints, evidence has been marshalled which supports a viral aetiology for type 1 diabetes.

In the late 1970s, two case studies were published which documented the development of acute-onset diabetes in each of two children who had been infected with a coxsackievirus of the B4 serotype (CVB4) (Gladisch et al. 1976; Yoon et al. 1979). Gladisch et al. (1976) confirmed the presence of virus in the pancreas using FITC-labelled CVB antibody and they also noted extensive insulitis and lysis of islet cells. In the second study, Yoon et al. prepared pancreas homogenates from a diabetic patient and inoculated mouse, monkey and human cell cultures with these extracts. They were then able to isolate and type an enterovirus and found that, again, this was classified as the CVB4 serotype. As in the case studied by Gladisch et al. (1976), analysis of the pancreas revealed evidence of islet inflammation and lysis of beta cells. When used in subsequent mouse studies, the virus isolated by Yoon et al. (1979) was capable of infecting beta cells and causing islet inflammation. This was accompanied by beta cell necrosis and the development of hyperglycaemia in infected mice (Yoon et al. 1979).

Such studies prompted others to search for evidence of the presence of enteroviruses in the pancreases of more patients with type 1 diabetes. In a very comprehensive immunohistochemical analysis of formalin-fixed paraffin-embedded pancreases by Foulis et al. in 1990, 88 patients with recent-onset type 1 diabetes were studied. The investigators sought evidence for the presence of the highly conserved enteroviral capsid protein, VP1, but failed to detect any positive staining among the samples (Foulis et al. 1990). This outcome was especially unexpected given that the technique used was able to detect enteroviral VP1$^+$ cells in the heart and pancreas of neonates who had died of culture proven coxsackie viral myocarditis, which suggested a sensitivity for the immunohistochemical technique similar to that of viral culture from autopsy tissue. This group went on to extract DNA from the samples of diabetic pancreases and performed PCR looking for viral-specific sequences. They also looked for evidence of enteroviral RNA by in situ hybridisation (Foulis et al. 1997). Amplification of extracted DNA from 47 of the patients with primers designed to detect Epstein–Barr virus or cytomegalovirus did not reveal any positive signals. Non-radioactive in situ hybridisation with enteroviral probes revealed infection in coxsackie-infected neonatal pancreas, but no positive signal was observed in the 29 pancreas samples of diabetic patients in which good RNA preservation was established by the detection of insulin mRNA. Subsequently, however, the use of a radioactive enteroviral-specific in situ hybridisation probe did reveal the presence of enteroviral RNA in 4 out of 65 diabetic pancreas samples (Ylipaasto et al. 2004), suggesting that detection of virus in autopsy diabetic pancreases may be possible with more sensitive techniques.

Recently, new antibodies with improved detection efficiency for enteroviral VP1 (notably the 5D8/1 clone marketed by Dako), coupled with the development of heat-induced antigen retrieval (HIER) techniques for immunohistochemistry (which enables antigens that were previously masked to be revealed), have improved the sensitivity of virus detection in fixed archival tissues. Dotta et al. (2007) reported evidence of enteroviral VP1 in the pancreases of two recent-onset type 1 diabetes patients and in a third patient who had undergone whole pancreas graft. The immunohistochemical evidence of VP1 staining was supported by the isolation of a virus (once again belonging to the CVB4 serotype) from the pancreas of one of the patients studied. The VP1 staining was largely confined to beta cells and viral particles were observed in the cytoplasm of these cells when examined under the electron microscope. More importantly, it was revealed that the viral isolate was capable of infecting human islets cultured in vitro and that this resulted in an impairment of insulin secretion (without any change in islet insulin content). This implies that enteroviral infection can lead to functional impairment of beta cells under conditions where the virus does not induce specific beta cell cytotoxicity. In summarising these important findings, it is also worth noting that the virus isolated by Dotta et al. had a high degree of homology with a laboratory reference strain dating from a much earlier period, which has led some to question its provenance (Tracy et al. 2010).

The advent of improved techniques for enteroviral detection in formalin-fixed paraffin-embedded tissue allowed our group to revisit the original collection of recent-onset type 1 diabetic pancreases studied by Foulis et al. (Richardson et al. 2009). Whereas these had previously been thought to be negative for viral protein, we found evidence of enteroviral VP1 immunopositivity in the pancreas of 44 of 72 (61%) of patients. This compared with a total of only 4 positive endocrine cells detected in the pancreas of 3 of 39 (7.7%) non-diabetic paediatric cases. Serial sections showed that the VP1 staining was restricted to insulin-containing islets (ICIs) (insulin-deficient islets being negative), and dual immunofluorescence demonstrated that VP1 was confined to beta cells within these islets (Richardson et al. 2009; Willcox et al. 2011).

Virally infected cells can be expected to mount a specific pattern of molecular responses, the function of which is to minimise the impact of the infection on the cell by, for example, shutting down host protein synthesis. Alteration of the expression of certain markers in a cell may therefore act as a "footprint", indicating the likely presence of a virus within the cell. One such marker, the inducible pathogen recognition receptor, protein kinase R (PKR) was found to be frequently present in $VP1^+$ islets (Richardson et al. 2009), and more recent work has confirmed that VP1 and PKR co-localise in the beta cells of patients with type 1 diabetes (Richardson, unpublished observations, Fig. 17.1). Interestingly, the level of VP1 expression is

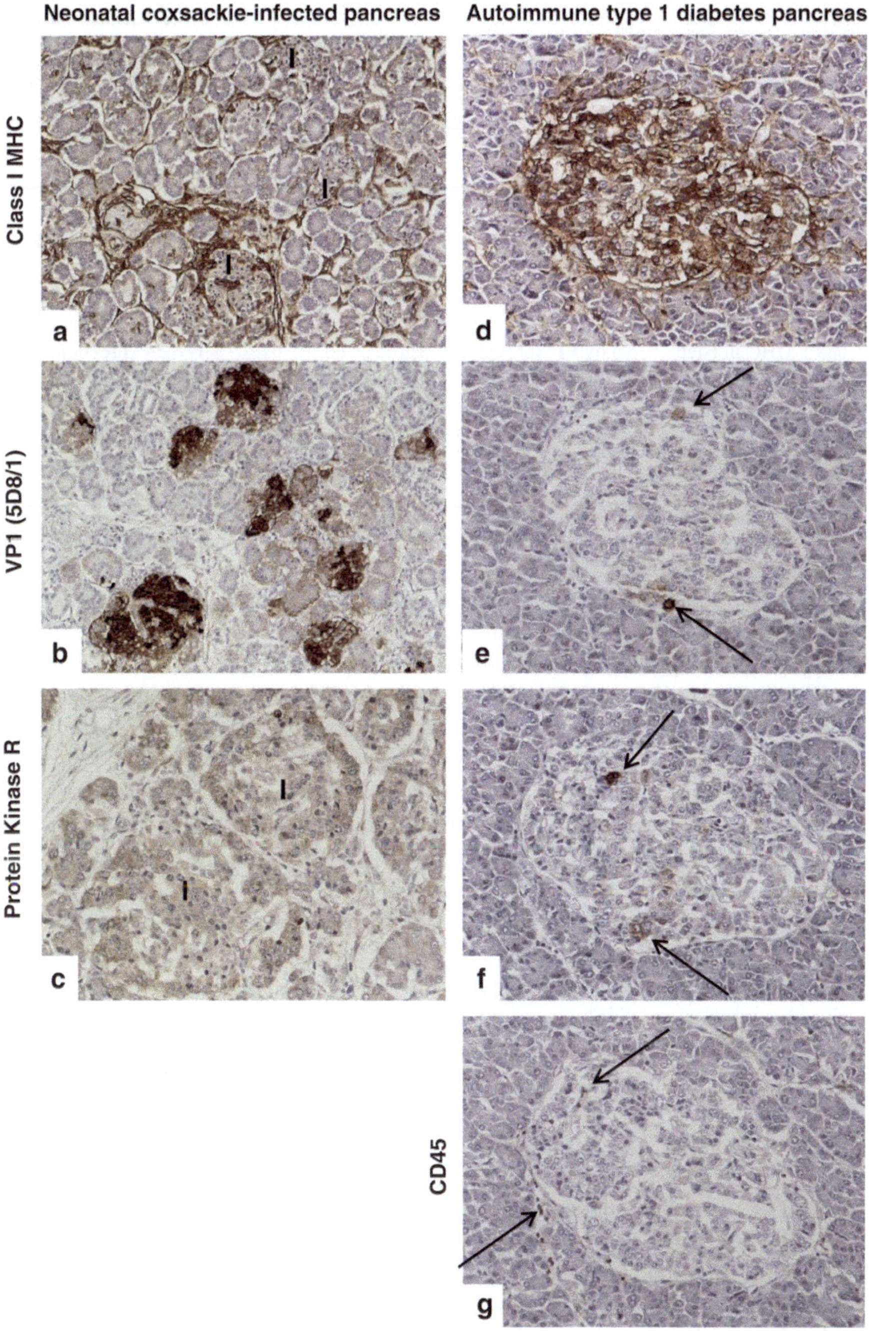

Fig. 17.1 Immunohistochemical staining for Class I MHC (**a** and **d**), VP1 (mAb 5D8/1; **b** and **e**) and PKR (**c** and **f**) in neonatal coxsackie-infected pancreas (**a**–**c**) and serial sections of an autoimmune type 1 diabetes pancreas (**d**–**f**). Further staining for CD45 (**g**) is demonstrated in autoimmune type 1 diabetes. Although both the neonatal coxsackie-infected pancreas and the autoimmune type 1 diabetes pancreas express enteroviral VP1, the islet-specific hyperexpression of Class I MHC and the beta cell selective up-regulation of PKR only occur in autoimmune type 1 diabetes. Endothelial cell and lymphocyte-specific Class I MHC staining is observed in the coxsackie-infected pancreas (**a**), but the endocrine cells remain largely unstained

Table 17.1 A comparison of pancreas pathology between neonatal coxsackie infection, fulminant diabetes and autoimmune type 1 diabetes

	Neonatal CVB infection	Fulminant diabetes	Autoimmune T1D
Endocrine cell loss	Alpha and beta cells	Alpha and beta cells	Beta cells
Insulitis	+	100% +++ but by 1 month negative	++ but selected islets only
Predominant immune cells	Unknown	Macrophage > CD8	CD8 > B cells > Macrophage > CD4
Class I MHC in islets	No hyperexpression	Little or no hyperexpression	Hyperexpression
Evidence of islet VP1 staining	+++	+++	Occasional positive cells
IFN alpha expression by beta cells	++	+	++

much lower in patients with autoimmune type 1 diabetes (Fig. 17.1e) than in those with acute systemic enterovirus infection (Fig. 17.1b). In the autoimmune type 1 cases, VP1 strongly correlates with PKR expression (Fig. 17.1e–f; Richardson unpublished results), which is unlike the situation in the acutely infected pancreas where more VP1$^+$ cells are observed and there appears to be no correlation with PKR (Fig. 17.1b, c).

Interferon-alpha and hyperexpression of Class I MHC are not present in the islets of non-diabetic controls but are each observed at clinical onset and years after the onset of autoimmune type 1 diabetes, and these markers may form part of a "viral footprint" (Bottazzo et al. 1985; Foulis et al. 1987a; Oikarinen et al. 2008) present in infected islets. However, only a small number of the beta cells within such islets also express enteroviral VP1, suggesting that there may be an underlying persistent enteroviral infection in cells that express little or no VP1 protein, but which can still drive the expression of interferon-alpha and the hyperexpression of Class I MHC. In support of this, enteroviruses can establish persistent infections in human tissues and, when this happens, it is thought that the virus exists in a different but stable molecular form, possibly as double-stranded RNA (Cunningham et al. 1990; Tam and Messner 1999). Under these circumstances VP1 is less likely to be expressed (Klingel et al. 1992), but it seems probable that the occasional islet cells which do express VP1 may represent only the "tip of the iceberg" amongst numerous adjacent beta cells that harbour a more persistent infection.

In fulminant type 1 diabetes, high levels of expression of VP1 were detected in the islets at early time points following disease onset suggesting that the induction of this form of diabetes may result from an acute lytic infection of islet cells leading to damage of the surrounding tissue and promoting a generalised inflammatory response. Thus in this regard, fulminant type 1 diabetes has strong similarities to the acutely infected neonates who died of viral myocarditis (Table 17.1, Fig. 17.1). It is important to emphasise, however, that the number of fulminant diabetes patients examined for VP1 expression remains small with viral protein having been detected in three of the three cases studied (Aida et al. 2011; Tanaka et al. 2009) while enteroviral RNA was found in one of the three cases (Shibasaki et al. 2010); so further studies are necessary to confirm these findings.

It's All in the Order

As noted previously, it is not known whether an enteroviral infection precedes the onset of autoimmunity or whether it accelerates the disease process once autoimmunity is established. The latter appears to be the case in the NOD mouse model, where a threshold of established insulitis must be present in order that enteroviral infection can hasten the onset of diabetes (Drescher et al. 2004). In humans the immunopathological evidence seems to suggest that the order is subtly different as islets from patients with recent-onset autoimmune type 1 diabetes can be found that are completely devoid of immune cell infiltration, but with evidence of hyperexpression of Class I MHC, expression of interferon-alpha and VP1 protein (Foulis et al. 1987b; Richardson et al. 2009). The detection of VP1 protein in islets in the absence of any signs of insulitis in an autoantibody positive case has also been documented (Oikarinen et al. 2008). Therefore, more research is required in humans to determine whether a viral infection is the "chicken or the egg" in terms of immune-mediated beta cell destruction.

There is still a long way to go in order to elucidate the precise role of enteroviral infections in the aetiology of type 1 diabetes. It appears though that different host responses, and perhaps different genetic variants among the viruses involved, may determine whether an enteroviral infection of beta cells causes (1) rapid lytic cell death, (2) a persistent infection which evokes an autoimmune reaction to beta cells, eventually resulting in their destruction, or (3), a persistent infection with little host response and little damage done.

References

Aida K, Nishida Y, Tanaka S, Maruyama T, Shimada A, Awata T, Suzuki M, Shimura H, Takizawa S, Ichijo M, Akiyama D, Furuya F, Kawaguchi A, Kaneshige M, Itakura J, Fujii H, Endo T, Kobayashi T (2011) RIG-I- and MDA5-initiated innate immunity linked with adaptive immunity accelerates beta-cell death in fulminant type 1 diabetes. Diabetes 60:884–889

Andreoletti L, Hober D, Hober-Vandenberghe C, Belaich S, Vantyghem MC, Lefebvre J, Wattre P (1997) Detection of coxsackie B virus RNA sequences in whole blood samples from adult patients at the onset of type I diabetes mellitus. J Med Virol 52:121–127

Bottazzo GF, Dean BM, McNally JM, MacKay EH, Swift PG, Gamble DR (1985) In situ characterization of autoimmune phenomena and expression of HLA molecules in the pancreas in diabetic insulitis. N Engl J Med 313:353–360

Chehadeh W, Kerr-Conte J, Pattou F, Alm G, Lefebvre J, Wattre P, Hober D (2000) Persistent infection of human pancreatic islets by coxsackievirus B is associated with alpha interferon synthesis in beta cells. J Virol 74:10153–10164

Clements GB, Galbraith DN, Taylor KW (1995) Coxsackie B virus infection and onset of childhood diabetes. Lancet 346:221–223

Coutant R, Carel JC, Lebon P, Bougneres PF, Palmer P, Cantero-Aguilar L (2002) Detection of enterovirus RNA sequences in serum samples from autoantibody-positive subjects at risk for diabetes. Diabet Med 19:968–969

Cunningham L, Bowles NE, Lane RJ, Dubowitz V, Archard LC (1990) Persistence of enteroviral RNA in chronic fatigue syndrome is associated with the abnormal production of equal amounts of positive and negative strands of enteroviral RNA. J Gen Virol 71:1399–1402

Dotta F, Censini S, van Halteren AG, Marselli L, Masini M, Dionisi S, Mosca F, Boggi U, Muda AO, Prato SD, Elliott JF, Covacci A, Rappuoli R, Roep BO, Marchetti P (2007) Coxsackie B4 virus infection of beta cells and natural killer cell insulitis in recent-onset type 1 diabetic patients. Proc Natl Acad Sci USA 104:5115–5120

Drescher KM, Kono K, Bopegamage S, Carson SD, Tracy S (2004) Coxsackievirus B3 infection and type 1 diabetes development in NOD mice: insulitis determines susceptibility of pancreatic islets to virus infection. Virology 329:381–394

Foulis AK, Stewart JA (1984) The pancreas in recent-onset type 1 (insulin-dependent) diabetes mellitus: insulin content of islets, insulitis and associated changes in the exocrine acinar tissue. Diabetologia 26:456–461

Foulis AK, Farquharson MA, Hardman R (1987a) Aberrant expression of class II major histocompatibility complex molecules by B cells and hyperexpression of class I major histocompatibility complex molecules by insulin containing islets in type 1 (insulin-dependent) diabetes mellitus. Diabetologia 30:333–343

Foulis AK, Farquharson MA, Meager A (1987b) Immunoreactive alpha-interferon in insulin-secreting beta cells in type 1 diabetes mellitus. Lancet 2:1423–1427

Foulis AK, Francis ND, Farquharson MA, Boylston A (1988) Massive synchronous B-cell necrosis causing type 1 (insulin-dependent) diabetes—a unique histopathological case report. Diabetologia 31:46–50

Foulis AK, Farquharson MA, Cameron SO, McGill M, Schonke H, Kandolf R (1990) A search for the presence of the enteroviral capsid protein VP1 in pancreases of patients with type 1 (insulin-dependent) diabetes and pancreases and hearts of infants who died of coxsackieviral myocarditis. Diabetologia 33:290–298

Foulis AK, McGill M, Farquharson MA, Hilton DA (1997) A search for evidence of viral infection in pancreases of newly diagnosed patients with IDDM. Diabetologia 40:53–61

Gladisch R, Hofmann W, Waldherr R (1976) Myocarditis and insulitis following coxsackie virus infection. Z Kardiol 65:837–849

Klingel K, Hohenadl C, Canu A, Albrecht M, Seemann M, Mall G, Kandolf R (1992) Ongoing enterovirus-induced myocarditis is associated with persistent heart muscle infection: quantitative analysis of virus replication, tissue damage, and inflammation. Proc Natl Acad Sci USA 89:314–318

Lonnrot M, Salminen K, Knip M, Savola K, Kulmala P, Leinikki P, Hyypia T, Akerblom HK, Hyoty H (2000) Enterovirus RNA in serum is a risk factor for beta-cell autoimmunity and clinical type 1 diabetes: a prospective study. Childhood Diabetes in Finland (DiMe) Study Group. J Med Virol 61:214–220

Moya-Suri V, Schlosser M, Zimmermann K, Rjasanowski I, Gurtler L, Mentel R (2005) Enterovirus RNA sequences in sera of schoolchildren in the general population and their association with type 1-diabetes-associated autoantibodies. J Med Microbiol 54:879–883

Nairn C, Galbraith DN, Taylor KW, Clements GB (1999) Enterovirus variants in the serum of children at the onset of type 1 diabetes mellitus. Diabet Med 16:509–513

Oikarinen M, Tauriainen S, Honkanen T, Vuori K, Karhunen P, Vasama-Nolvi C, Oikarinen S, Verbeke C, Blair GE, Rantala I, Ilonen J, Simell O, Knip M, Hyoty H (2008) Analysis of pancreas tissue in a child positive for islet cell antibodies. Diabetologia 51:1796–1802

Pujol-Borrell R, Todd I, Doshi M, Gray D, Feldmann M, Bottazzo GF (1986) Differential expression and regulation of MHC products in the endocrine and exocrine cells of the human pancreas. Clin Exp Immunol 65:128–139

Richardson SJ, Willcox A, Bone AJ, Foulis AK, Morgan NG (2009) The prevalence of enteroviral capsid protein VP1 immunostaining in pancreatic islets in human type 1 diabetes. Diabetologia 52:1143–1151

Richardson SJ, Willcox A, Bone AJ, Morgan NG, Foulis AK (2011) Immunopathology of the human pancreas in type-I diabetes. Semin Immunopathol 33:9–21

Sarmiento L, Cabrera-Rode E, Lekuleni L, Cuba I, Molina G, Fonseca M, Heng-Hung L, Borroto AD, Gonzalez P, Mas-Lago P, Diaz-Horta O (2007) Occurrence of enterovirus RNA in serum of children with newly diagnosed type 1 diabetes and islet cell autoantibody-positive subjects in a population with a low incidence of type 1 diabetes. Autoimmunity 40:540–545

Schulte BM, Bakkers J, Lanke KH, Melchers WJ, Westerlaken C, Allebes W, Aanstoot HJ, Bruining GJ, Adema GJ, Van Kuppeveld FJ, Galama JM (2010) Detection of enterovirus RNA in peripheral blood mononuclear cells of type 1 diabetic patients beyond the stage of acute infection. Viral Immunol 23:99–104

Shibasaki S, Imagawa A, Tauriainen S, Iino M, Oikarinen M, Abiru H, Tamaki K, Seino H, Nishi K, Takase I, Okada Y, Uno S, Murase-Mishiba Y, Terasaki J, Makino H, Shimomura I, Hyoty H, Hanafusa T (2010) Expression of Toll-like receptors in the pancreas of recent-onset fulminant type 1 diabetes. Endocr J 57:211–219

Tam PE, Messner RP (1999) Molecular mechanisms of coxsackievirus persistence in chronic inflammatory myopathy: viral RNA persists through formation of a double-stranded complex without associated genomic mutations or evolution. J Virol 73:10113–10121

Tanaka S, Nishida Y, Aida K, Maruyama T, Shimada A, Suzuki M, Shimura H, Takizawa S, Takahashi M, Akiyama D, Arai-Yamashita S, Furuya F, Kawaguchi A, Kaneshige M, Katoh R, Endo T, Kobayashi T (2009) Enterovirus infection, CXC chemokine ligand 10 (CXCL10), and CXCR3 circuit: a mechanism of accelerated beta-cell failure in fulminant type 1 diabetes. Diabetes 58:2285–2291

Tauriainen S, Oikarinen S, Oikarinen M, Hyoty H (2011) Enteroviruses in the pathogenesis of type 1 diabetes. Semin Immunopathol 33:45–55

Tracy S, Drescher KM, Jackson JD, Kim K, Kono K (2010) Enteroviruses, type 1 diabetes and hygiene: a complex relationship. Rev Med Virol 20:106–116

Willcox A, Richardson SJ, Bone AJ, Foulis AK, Morgan NG (2009) Analysis of islet inflammation in human type 1 diabetes. Clin Exp Immunol 155:173–181

Willcox A, Richardson SJ, Bone AJ, Foulis A, Morgan NG (2011) Immunohistochemical analysis of the relationship between islet cell proliferation and the production of the enteroviral capsid protein, VP1, in the islets of patients with recent-onset type 1 diabetes. Diabetologia 54:2417–2420

Yin H, Berg AK, Tuvemo T, Frisk G (2002) Enterovirus RNA is found in peripheral blood mononuclear cells in a majority of type 1 diabetic children at onset. Diabetes 51:1964–1971

Ylipaasto P, Klingel K, Lindberg AM, Otonkoski T, Kandolf R, Hovi T, Roivainen M (2004) Enterovirus infection in human pancreatic islet cells, islet tropism in vivo and receptor involvement in cultured islet beta cells. Diabetologia 47:225–239

Yoon JW, Austin M, Onodera T, Notkins AL (1979) Isolation of a virus from the pancreas of a child with diabetic ketoacidosis. N Engl J Med 300:1173–1179

Chapter 18
Rotavirus and Type 1 Diabetes

Margo C. Honeyman and Leonard C. Harrison

Abstract A strong and temporal association exists between infections of infants and children with the major childhood gastroenteritis virus, RV, and the appearance, or exacerbation, of antibodies to the islet antigens glutamic acid decarboxylase 65 (GAD 65) and islet antigen 2 (IA-2). There is no association with thyroid autoantibodies, and islet autoantibodies do not occur prior to RV infections. Epidemics peak each winter, as do diagnoses of type 1 diabetes. The dominant, HLA-DR4-binding, $CD4^{+}$ T-cell autoepitopes in GAD65 and IA2 are strongly similar to sequences in the RV surface antigen, VP7. These VP7 sequences are also $CD4^{+}$ T-cell epitopes for T cells from children with islet autoimmunity. The VP7-similar GAD65 and IA2 epitopes sequences also contain CD8 T-cell epitopes which are dominant at the onset of clinical type 1 diabetes, suggesting that molecular mimicry between RV sequences and islet antigens could lead to type 1 diabetes. T cells important in diabetes in NOD mice bear the gut-associated alpha4beta7 integrin, while RV becomes infectious in the duodenum by the action of pancreatic trypsin. The fact that RV can infect pancreatic beta cells in vitro and can accelerate the onset of diabetes in NOD mice with islet inflammation strengthens the case for the involvement of RV in the initiation of islet autoimmunity and type 1 diabetes.

Viruses in Type 1 Diabetes

The view that viruses have an etiological role in type 1 diabetes emerged gradually as the immune mechanisms underlying type 1 diabetes were understood better. Initially, the observation that congenital rubella virus infection was associated with

M.C. Honeyman, M.Sc. Syd, Ph.D. UNSW
L.C. Harrison, MB, BS UNSW, M.D. Melb, D.Sc. Melb, FRACP, FRCPA (✉)
Director, Burnet Clinical Research Unit, Walter and Eliza Hall Institute of Medical Research, Melbourne Area, Australia
e-mail: Harrison@wehi.edu.au

K.W. Taylor, H. Hyoty, A. Toniolo, A.J. Zuckerman (eds.), *Diabetes and Viruses*,
DOI 10.1007/978-1-4614-4051-2_18, © Springer Science+Business Media New York 2013

type 1 diabetes (Forrest et al. 1969) and the discovery that the HLA-A1-B8 haplotype was more common in individuals with congenital rubella who developed type 1 diabetes (Menser et al. 1974) and also in type 1 diabetes generally (Nerup et al. 1974) implied that both virus infection and an immune response were required to develop type 1 diabetes. The pathology of the islets in type 1 diabetes indicated that adaptive T-cell immunity, dependent on HLA molecules to present specific virus or autoantigen peptides to T cells or on other immune response genes linked with HLA, was critical.

However, the introduction of vaccination against rubella did not diminish the incidence of type 1 diabetes and evidence for other viruses was then sought, in particular enteroviruses that infect children commonly. Recent studies continue to suggest that enteric viruses are associated with the etiology of type 1 diabetes (Roivainen and Klingel 2010). DNA/RNA studies reveal that enteroviruses are more likely to be detected around the onset of type 1 diabetes (Craig et al. 2003), but in only a small number of cases they have been associated with the initiation of islet autoimmunity (Roivainen and Klingel 2010).

Studies searching for acute or chronic virus infection at the time of symptomatic onset of type 1 diabetes were necessarily cross-sectional and measured antiviral antibodies (Frisk and Tuvemo 2004), viral DNA/RNA (Craig et al. 2003; Elshebani et al. 2007), or attempted to isolate virus from the blood (Elshebani et al. 2007) or in a few cases directly from postmortem pancreas (Foulis 1989). Results were inconsistent but support for the viral hypothesis was sustained by the intriguing finding of Foulis et al. (1987) that interferon (IFN)-alpha, a signature of virus infection, could be detected by immunostaining in the islets of postmortem pancreata in type 1 diabetes. In addition, 2′, 5′ oligoadenylate synthase activity, a marker of viral RNA-induced IFN, was shown to be increased in blood mononuclear cells of individuals with type 1 diabetes (Petrovsky et al. 1997).

Longitudinal studies of "at-risk" children with a first-degree relative with type 1 diabetes revealed that the symptomatic onset of the disease was usually preceded by a clinically silent period of months to years (Powers and Eisenbarth 1985). This preclinical phase is defined by circulating autoantibodies to islet antigens—insulin, glutamic acid decarboxylase 65 (GAD65), tyrosine phosphatase IA-2, and beta-cell-specific zinc (Zn) transporter 8. Autoantibody titer and the number of target specificities correlate with risk for type 1 diabetes (reviewed in Harrison 2001). Importantly, the preclinical phase of type 1 diabetes provides an opportunity to detect and track virus infection closer to the onset of islet pathology, and is a window for intervention–prevention. By extrapolation from the NOD mouse model of type 1 diabetes, beta-cell destruction is mediated by both $CD4^+$ T cells activated by peptides in HLA class II molecules and $CD8^+$ T cells activated by peptides in HLA class I molecules and ultimately kill beta cells. The identification of islet autoantigen T-cell epitopes "restricted by" specific HLA molecules that confer susceptibility to type 1 diabetes is a clue to mimicry with viral peptides as a potential etiological mechanism.

Interest in viral mechanisms of type 1 diabetes was rekindled with the identification of toll-like receptors (TLRs) and cytoplasmic retinoic acid-inducible gene I (RIG-I)-like helicases (RLHs) involved in innate immune recognition of RNA viruses. TLR3 is highly expressed in both pancreatic beta cells and the gut epithelium (Rasschaert et al. 2005; Wen et al. 2004). TLR3 binds double-stranded (ds) RNA, present in dsRNA and dividing single-stranded RNA viruses but not mammals. Binding of dsRNA to TLR3 triggers type 1 IFN and other cytokine responses. Moreover, the dsRNA mimic, poly I:C, was shown to induce beta-cell apoptosis in C57/Bl6 mouse islet cells in vitro (Dogusan et al. 2008). Intracellular IFN-induced helicase C domain 1 (IFIH1), also known as MDA-5, mediates the intracellular response to viral RNA (Smyth et al. 2006). *IFIH1* is a susceptibility gene for type 1 diabetes (Nejentsev et al. 2009). The role of RNA viruses in type 1 diabetes which could replicate in the gut or pancreas months to years before the clinical onset of disease is, therefore, of interest.

Rotaviruses: A Genus of the Reoviruses

The *Reoviridae* family of double-stranded RNA viruses contains strains that are diabetogenic in infant mice (Onodera et al. 1978) and can infect islet beta cells in vitro leading to their destruction in a cytokine-dependent manner (Campbell and Harrison 1989). The ability of reoviruses to upregulate class I histocompatibility molecules on beta cells (Campbell and Harrison 1989) would render beta cells susceptible to attack by CD8 T cells. Rotaviruses (RVs) are in a genus of the *Reoviridae* that infect young mammals and birds by fecal-oral contamination. In humans, RVs are the major cause of acute gastroenteritis in young children, responsible globally for greater than 500,000 deaths annually (Parashar et al. 2009). Epidemics occur in winter, peaking in February/August in the Northern/Southern hemispheres, and most human infections are by RV group A. Until 50 years ago, the initial RV infection in developed communities was most likely to occur either in neonatal hospital nurseries or during breast-feeding. Symptoms were attenuated due to binding of virus to lactadherin in human milk and the presence of maternal milk IgA and transplacental serum IgG to RV—nature's vaccine regimen. In recent times, however, the first infection is more likely to occur in the second year of life and be more severe, at entry to day-care usually after breast-feeding has ceased and protection from maternal antibodies has waned. Within the gut, RV is not infectious until it is activated by exocrine pancreas-derived trypsin in the duodenum. As RV infection spreads down the gut, intestinal permeability increases (Zhang et al. 2000) and a celiac disease-like pathology occurs transiently (Snodgrass et al. 1977). Following the primary infection, clinically silent reinfections occur at least annually and herd immunity is complete by 5 years of age.

T-Cell Epitopes in Islet Autoantigens and Rotavirus

The first hint of a link between type 1 diabetes and RV came serendipitously from studies to determine the dominant epitopes recognized by islet-reactive T cells from individuals with preclinical type 1 diabetes (Honeyman et al. 1998). Overlapping

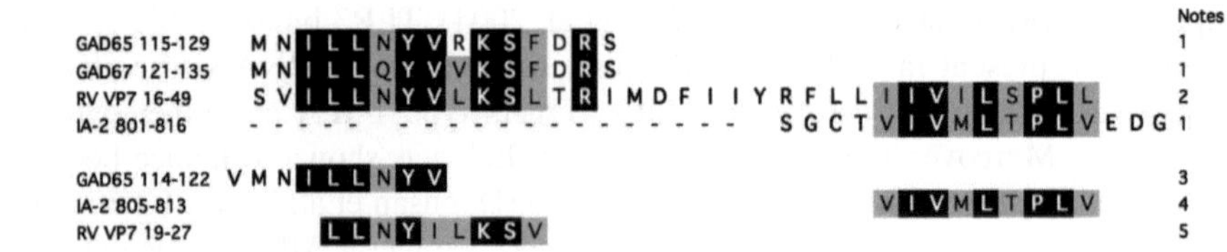

Notes
1 HLA-DR4 CD4 T-cell epitope (Honeyman, 1998)
2 HLA-DR4 epitope predicted (Honeyman, 1998), then established (Honeyman, 2010)
3 HLA-A2 CD8 T-cell epitope (Mallone, 2007)
4 HLA-A2 CD8 T-cell epitope (Martinuzzi, 2008)
5 HLA-A2 CD8 T-cell epitope (Wei, 2009)

Fig. 18.1 T-cell epitopes in islet autoantigens and rotavirus VP7

peptides encompassing IA-2 and GAD65 were assessed for their capacity to bind to the HLA susceptibility molecules HLA-DR3 and -4 and stimulate T-cell proliferation. In HLA DR4 homozygous individuals the dominant DR4-binding epitope in IA-2 was aa 40–52 and in GAD65 aa 121–135 (Fig. 18.1). The IA-2 epitope has 56% identity and 100% similarity with RV VP7 aa 41–49 (Honeyman et al. 1998). The GAD65 epitope has 75% identity and 92% similarity with RV VP7 aa 18–29. HLA DR4 binding studies revealed that identical amino acids in both islet autoantigen and RV VP7 peptides faced the T-cell receptor (Honeyman et al. 1998). Both RV sequences were within the H2 signal sequence of VP7 (aa 1–50) conserved between different strains of RV (Franco et al. 1993). A more recent study established that the RV peptides were indeed CD4 T-cell epitopes in the context of HLA DR4, to which they bound identically to the similar autoepitopes (Honeyman et al. 2010) and elicited $CD4^+$ T-cell proliferation equivalent to the homologous IA-2 and GAD65 epitopes (Honeyman et al. 2010). Two clones that proliferated in response to the IA-2 epitope also proliferated to the similar RV peptide. Furthermore, $CD4^+$ T cells secreted IFN-gamma after stimulation by the IA-2 epitope and when restimulated by same IA-2 epitope or the similar RV epitope. These results are consistent with molecular mimicry, whereby RV peptides could elicit cross-reactive $CD4^+$ T-cell responses to islet autoepitopes.

The HLA class I molecules A2 and A24 also confer susceptibility to type 1 diabetes (Honeyman et al. 1995; Watts et al. 1992) and islet autoepitope peptides that elicit class I-restricted CD8 T-cell responses have recently been reported (Mallone et al. 2007). Martinuzzi and colleagues used an enhanced IFN-gamma-based ELISpot assay (Martinuzzi et al. 2008) to show that GAD65 aa 114–122 and IA-2 aa 805–814, both with high identity and similarity to RV VP7 HLA class II epitopes, were the dominant HLA-A2 restricted CD8 T-cell autoepitopes at diagnosis of type 1 diabetes, in 53% and 42% of individuals, respectively. After 7–16 months, reactivity waned but the IA-2 autoepitope became relatively the more dominant (GAD 19%, IA-2 31%) in the same individuals.

HLA-A2 transgenic mice infected with the human Wa strain of RV demonstrated that RV VP7 aa 19–27 was an epitope (Wei et al. 2009). This epitope is contained within the class II GAD65-similar RV epitope RV VP7 aa 18–29. RV VP7 aa 19–27 also overlaps GAD65 aa 114–122 defined as a CD8 epitope with 60% identity and 100% similarity in the 5 aa overlap (Fig. 18.1). These highly similar regions in GAD65 and RV VP7 contain both CD4 and CD8 epitopes presented by HLA-DR4 and HLA-A2, respectively, which occur frequently on the same haplotype (Alper et al. 2006) associated with susceptibility to type 1 diabetes.

Adaptive autoimmune responses are conditioned by innate immunity. The innate response could be triggered by the interaction of virally derived dsRNA and TLR3 with consequent beta-cell apoptosis mediated via the IFN regulatory-3 pathway (Dogusan et al. 2008) in individuals who are genetically susceptible (Downes et al. 2010). Apoptotic beta cells could release islet autoantigens, which may also be targets of mimicry, thus perpetuating the destructive process. Multiple innate immune insults and/or molecular mimicry between RV and islet autoepitopes could be mechanisms that intensify islet autoimmunity.

Temporal Association of Rotavirus Infection with Islet Autoimmunity

Longitudinal studies of children at risk for type 1 diabetes provide the opportunity to investigate a relationship between markers of RV infection and islet autoimmunity. In the Australian BabyDiab Study (Couper et al. 1999), sera from children followed 6 monthly from birth were tested for islet autoantibodies and for IgG and IgA antibodies to RV (Honeyman et al. 2000b). A specific and highly significant association was found between RV seroconversion and the first appearance of or an increase in the concentration of autoantibodies to either IA-2 (in 86%), insulin (in 62%), and/or GAD65 (in 50%). Subsequently, this relationship was not found in at-risk children in Finland (Blomqvist et al. 2002). The Finnish study differed however in that it did not test for antibodies to a RV G3 serotype found commonly in humans, but rather to a G6 serotype, an uncommon cause of infection in humans. Moreover, the Finnish study did not test for RV IgA which although short-lived is the most important protective antibody. It is therefore possible that RV infections were missed. Indeed, in a later study (Makela et al. 2004), the same senior investigators found that some RV infections could only be defined serologically by a rise in IgA assayed now on both G1 and G6 serotypes. Children in the Finnish study were selected for HLA-DQ susceptibility to type 1 diabetes, whereas the Australian children were selected by relationship to a first-degree relative with type 1 diabetes. Virus infection might be more important in driving islet autoimmunity in individuals who lack the strongest HLA susceptibility genes. This is suggested by the finding that at the onset of type 1 diabetes, children positive for enteroviral RNA were less likely to have DR3-DQ2 haplotypes (OR 0.46) than those without enteroviral RNA, and non-DR3, DR4 children had a higher frequency of enteroviral RNA positivity (OR 2.52) (Craig et al. 2003). Further studies are needed to resolve this issue. In ten children without a family history of type 1 diabetes who were admitted to hospital for acute RV infection, acute and convalescent sera were tested for autoantibodies to insulin, GAD65 and IA-2, thyroid peroxidase and nuclei. Three patients had antibodies to both GAD65 and IA-2, but not to insulin, thyroid peroxidase, or nuclei (Honeyman et al. 2000b). Although only a small sample, a frequency of 30% for two autoantibodies greatly exceeds that of 0.1% in the population at large.

Rotavirus in the Pancreas

Only recently has it been appreciated that RV spreads beyond the gut to cause a viremia and enter other organs (Mossel and Ramig 2003). In case reports, RV infection has been associated with acute pancreatitis, e.g., (Nigro 1991; Parri et al. 2010). Virus infection including by RV has been proposed as the cause of acute fulminant type 1 diabetes (Honeyman et al. 2000a; Imagawa and Hanafusa 2006). However, direct evidence for a pathological effect of RV infection on the pancreas has not yet been reported. In vitro, pancreatic islets from mice, fetal pigs, and macaque monkeys support various degrees of RV growth, while human RVs only replicate in monkey islets (Coulson et al. 2002). After rhesus RV infection by oral inoculation of young, autoimmune diabetes-prone NOD mice, RV antigen was detected in pancreatic macrophages outside islets, as well as in the pancreas itself, spleen, and liver (Graham et al. 2007). However, the onset of spontaneous diabetes in NOD mice was actually delayed by oral inoculation with rhesus RV at birth (Harrison et al. 2008). The latter is in keeping with the "hygiene hypothesis," in which neonatal infection is postulated to be protective against immuno-inflammatory disease, while infection at a later stage may accelerate disease (Graham et al. 2008). In "adolescent" NOD mice established islet inflammation (a model of children with islet autoimmunity) was accelerated by rhesus RV (Graham et al. 2008). The timing of RV infection relative to developmental stage may thus be critical in relation to its role in type 1 diabetes.

Rotavirus and Celiac Disease

The HLA haplotype DR3-DQ2 confers susceptibility not only to type 1 diabetes but also to celiac disease, the prevalence of which is 20 times higher in type 1 diabetes than in the general population (Barera et al. 2002). Blood samples were obtained from 1,900 children with HLA susceptibility to celiac disease and type 1 diabetes, at 9, 15, and 24 months of age and annually thereafter. Of these, 54 developed biopsy-confirmed celiac disease and were tested blindly with two controls for RV antibodies to determine if there was an association between RV infection and the development of celiac diagnostic anti-transglutaminase antibodies. The relative risk in children who developed celiac disease was 1.94 for one infection and 3.76 for ≥ 2 infections, after adjustment for gender, ethnic group, maternal education, breast-feeding, day-care attendance, number of sibs, season of birth, and number of HLA-DR3-DQ2 haplotypes (Stene et al. 2006). Thus a high frequency of RV infections appears to increase the risk of an inflammatory disease associated with type 1 diabetes (Rewers et al. 2004). Further possible evidence for an association between RV infection and celiac disease was the finding that a subset of anti-transglutaminase IgA antibodies is directed against RV VP7 (Zanoni et al. 2006).

Causality

Nine criteria proposed by Hill (1965) for disease causality can be applied to RV and type 1 diabetes.

1. Strength of association: demonstrated by an odds ratio of 7.91 for the temporal association of RV infection with the first appearance of or increase in concentration of islet autoantibodies ($p < 0.02$).
2. Consistency (replication of association): not demonstrated between Australia and Finland but assays and subject populations differed as discussed.
3. Specificity: demonstrated by the association of RV infection and autoantibodies to islet, not thyroid or nuclear, antigens.
4. Temporality (the one essential criterion): demonstrated in the 6 monthly analysis of RV IgA and IgG from birth in relation to the first appearance of or an increase in the concentration of islet autoantibodies. Islet autoantibodies were not detected before RV infection.
5. Dose–response (increased effect with increasing dose or exposure): demonstrated by repeated RV infection being associated with islet autoantibodies.
6. Plausibility (observed associations explained by the biology): supported by the demonstration that $CD4^+$ T-cell autoepitope peptides in GAD65 and IA-2 have strong sequence similarities to T-cell epitopes in RV-VP7, that T-cell clones bearing one T-cell receptor recognize both the IA-2 and RV-VP7 epitopes, and that RV-similar $CD4^+$ T-cell autoepitopes encompass autoepitopes recognized by $CD8^+$ T cells at the onset of type 1 diabetes. Further evidence is the activation of RV in the duodenum by pancreas-derived trypsin.
7. Coherence (should not contradict present substantive knowledge): demonstrated by the increase in incidence of type 1 diabetes in 0–4-year-old children, in whom RV infections now occur more frequently.
8. Rejection of alternatives: supported by the finding that Coxsackie B1-5 infections, in contrast to RV infections, were not associated with islet autoantibodies.
9. Experiments: demonstrated by RV infection of pancreatic islet beta cells in vitro and by the effect of RV infection to accelerate diabetes onset in NOD mice with islet inflammation. An animal model in which RV infection, e.g., at weaning to mimic the human situation, triggers beta-cell dysfunction leading to hyperglycemia, would greatly strengthen the case for causality.

Causality in humans will only be established by either eradicating RV or, more practically, by vaccination to prevent active infection. It will be of interest therefore to monitor whether the incidence of type 1 diabetes decreases following the recent introduction of RV vaccines.

References

Alper CA, Larsen CE, Dubey DP, Awdeh ZL, Fici DA, Yunis EJ (2006) The haplotype structure of the human major histocompatibility complex. Hum Immunol 67:73–84

Barera G, Bonfanti R, Viscardi M, Bazzigaluppi E, Calori G, Meschi F, Bianchi C, Chiumello G (2002) Occurrence of celiac disease after onset of type 1 diabetes: a 6-year prospective longitudinal study. Pediatrics 109:833–838

Blomqvist M, Juhela S, Erkkila S, Korhonen S, Simell T, Kupila A, Vaarala O, Simell O, Knip M, Ilonen J (2002) Rotavirus infections and development of diabetes-associated autoantibodies during the first 2 years of life. Clin Exp Immunol 128:511–515

Campbell IL, Harrison LC (1989) Viruses and cytokines: evidence for multiple roles in pancreatic beta cell destruction in type 1 insulin-dependent diabetes mellitus. J Cell Biochem 40:57–66

Coulson BS, Witterick PD, Tan Y, Hewish MJ, Mountford JN, Harrison LC, Honeyman MC (2002) Growth of rotaviruses in primary pancreatic cells. J Virol 76:9537–9544

Couper JJ, Steele C, Beresford S, Powell T, McCaul K, Pollard A, Gellert S, Tait B, Harrison LC, Colman PG (1999) Lack of association between duration of breast-feeding or introduction of cow's milk and development of islet autoimmunity. Diabetes 48:2145–2149

Craig ME, Howard NJ, Silink M, Rawlinson WD (2003) Reduced frequency of HLA DRB1*03-DQB1*02 in children with type 1 diabetes associated with enterovirus RNA. J Infect Dis 187:1562–1570

Dogusan Z, Garcia M, Flamez D, Alexopoulou L, Goldman M, Gysemans C, Mathieu C, Libert C, Eizirik DL, Rasschaert J (2008) Double-stranded RNA induces pancreatic beta-cell apoptosis by activation of the toll-like receptor 3 and interferon regulatory factor 3 pathways. Diabetes 57:1236–1245

Downes K, Pekalski M, Angus KL, Hardy M, Nutland S, Smyth DJ, Walker NM, Wallace C, Todd JA (2010) Reduced expression of IFIH1 is protective for type 1 diabetes. PLoS One 5(9)

Elshebani A, Olsson A, Westman J, Tuvemo T, Korsgren O, Frisk G (2007) Effects on isolated human pancreatic islet cells after infection with strains of enterovirus isolated at clinical presentation of type 1 diabetes. Virus Res 124:193–203

Forrest JM, Menser MA, Harley JD (1969) Diabetes mellitus and congenital rubella. Pediatrics 44:445–447

Foulis AK (1989) In type 1 diabetes, does a non-cytopathic viral infection of insulin-secreting B-cells initiate the disease process leading to their autoimmune destruction? Diabet Med 6:666–674

Foulis AK, Farquharson MA, Meager A (1987) Immunoreactive alpha-interferon in insulin-secreting beta cells in type 1 diabetes mellitus. Lancet 2:1423–1427

Franco MA, Prieto I, Labbe M, Poncet D, Borras-Cuesta F, Cohen J (1993) An immunodominant cytotoxic T cell epitope on the VP7 rotavirus protein overlaps the H2 signal peptide. J Gen Virol 74:2579–2586

Frisk G, Tuvemo T (2004) Enterovirus infections with beta-cell tropic strains are frequent in siblings of children diagnosed with type 1 diabetes children and in association with elevated levels of GAD65 antibodies. J Med Virol 73:450–459

Graham KL, O'Donnell JA, Tan Y, Sanders N, Carrington EM, Allison J, Coulson BS (2007) Rotavirus infection of infant and young adult nonobese diabetic mice involves extraintestinal spread and delays diabetes onset. J Virol 81:6446–6458

Graham KL, Sanders N, Tan Y, Allison J, Kay TW, Coulson BS (2008) Rotavirus infection accelerates type 1 diabetes in mice with established insulitis. J Virol 82:6139–6149

Harrison LC (2001) Risk assessment, prediction and prevention of type 1 diabetes. Pediatr Diabetes 2:71–82

Harrison LC, Honeyman MC, Morahan G, Wentworth JM, Elkassaby S, Colman PG, Fourlanos S (2008) Type 1 diabetes: lessons for other autoimmune diseases? J Autoimmun 31:306–310

Hill AB (1965) The environment and disease: association or causation? Proc R Soc Med 58: 295–300

Honeyman MC, Harrison LC, Drummond B, Colman PG, Tait BD (1995) Analysis of families at risk for insulin-dependent diabetes mellitus reveals that HLA antigens influence progression to clinical disease. Mol Med 1:576–582

Honeyman MC, Stone NL, Harrison LC (1998) T-cell epitopes in type 1 diabetes autoantigen tyrosine phosphatase IA-2: potential for mimicry with rotavirus and other environmental agents. Mol Med 4:231–239

Honeyman MC, Coulson BS, Harrison LC (2000a) A novel subtype of type 1 diabetes mellitus. N Engl J Med 342:1835 [author reply 1837]

Honeyman MC, Coulson BS, Stone NL, Gellert SA, Goldwater PN, Steele CE, Couper JJ, Tait BD, Colman PG, Harrison LC (2000b) Association between rotavirus infection and pancreatic islet autoimmunity in children at risk of developing type 1 diabetes. Diabetes 49:1319–1324

Honeyman MC, Stone NL, Falk BA, Nepom G, Harrison LC (2010) Evidence for molecular mimicry between human T cell epitopes in rotavirus and pancreatic islet autoantigens. J Immunol 184:2204–2210

Imagawa A, Hanafusa T (2006) Pathogenesis of fulminant type 1 diabetes. Rev Diabet Stud 3:169–177

Makela M, Marttila J, Simell O, Ilonen J (2004) Rotavirus-specific T-cell responses in young prospectively followed-up children. Clin Exp Immunol 137:173–178

Mallone R, Martinuzzi E, Blancou P, Novelli G, Afonso G, Dolz M, Bruno G, Chaillous L, Chatenoud L, Bach JM, van Endert P (2007) CD8+ T-cell responses identify beta-cell autoimmunity in human type 1 diabetes. Diabetes 56:613–621

Martinuzzi E, Novelli G, Scotto M, Blancou P, Bach JM, Chaillous L, Bruno G, Chatenoud L, van Endert P, Mallone R (2008) The frequency and immunodominance of islet-specific CD8+ T-cell responses change after type 1 diabetes diagnosis and treatment. Diabetes 57: 1312–1320

Menser MA, Forrest JM, Honeyman MC, Burgess JA (1974) Letter: diabetes, HL-A antigens, and congenital rubella. Lancet 2:1508–1509

Mossel EC, Ramig RF (2003) A lymphatic mechanism of rotavirus extraintestinal spread in the neonatal mouse. J Virol 77:12352–12356

Nejentsev S, Walker N, Riches D, Egholm M, Todd JA (2009) Rare variants of IFIH1, a gene implicated in antiviral responses, protect against type 1 diabetes. Science 324:387–389

Nerup J, Platz P, Andersen OO, Christy M, Lyngsoe J, Poulsen JE, Ryder LP, Nielsen LS, Thomsen M, Svejgaard A (1974) HL-A antigens and diabetes mellitus. Lancet 2:864–866

Nigro G (1991) Pancreatitis with hypoglycemia-associated convulsions following rotavirus gastroenteritis. J Pediatr Gastroenterol Nutr 12:280–282

Onodera T, Jenson AB, Yoon JW, Notkins AL (1978) Virus-induced diabetes mellitus: reovirus infection of pancreatic beta cells in mice. Science 201:529–531

Parashar UD, Burton A, Lanata C, Boschi-Pinto C, Shibuya K, Steele D, Birmingham M, Glass RI (2009) Global mortality associated with rotavirus disease among children in 2004. J Infect Dis 200(Suppl 1):S9–S15

Parri N, Innocenti L, Collini S, Bechi F, Mannelli F (2010) Acute pancreatitis due to rotavirus gastroenteritis in a child. Pediatr Emerg Care 26:592–593

Petrovsky N, Harrison LC, Kyvik KO, Beck-Nielsen H, Green A, Bonnevie-Nielsen V (1997) Evidence for the viral aetiology of IDDM. Autoimmunity 25:251–252

Powers AC, Eisenbarth GS (1985) Autoimmunity to islet cells in diabetes mellitus. Annu Rev Med 36:533–544

Rasschaert J, Ladriere L, Urbain M, Dogusan Z, Katabua B, Sato S, Akira S, Gysemans C, Mathieu C, Eizirik DL (2005) Toll-like receptor 3 and STAT-1 contribute to double-stranded RNA+ interferon-gamma-induced apoptosis in primary pancreatic beta-cells. J Biol Chem 280:33984–33991

Rewers M, Liu E, Simmons J, Redondo MJ, Hoffenberg EJ (2004) Celiac disease associated with type 1 diabetes mellitus. Endocrinol Metab Clin North Am 33:197–214

Roivainen M, Klingel K (2010) Virus infections and type 1 diabetes risk. Curr Diab Rep 10:350–356

Smyth DJ, Cooper JD, Bailey R, Field S, Burren O, Smink LJ, Guja C, Ionescu-Tirgoviste C, Widmer B, Dunger DB, Savage DA, Walker NM, Clayton DG, Todd JA (2006) A genome-wide association study of nonsynonymous SNPs identifies a type 1 diabetes locus in the interferon-induced helicase (IFIH1) region. Nat Genet 38:617–619

Snodgrass DR, Angus KW, Gray EW (1977) Rotavirus infection in lambs: pathogenesis and pathology. Arch Virol 55:263–274

Stene LC, Honeyman MC, Hoffenberg EJ, Haas JE, Sokol RJ, Emery L, Taki I, Norris JM, Erlich HA, Eisenbarth GS, Rewers M (2006) Rotavirus infection frequency and risk of celiac disease autoimmunity in early childhood: a longitudinal study. Am J Gastroenterol 101:2333–2340

Watts GF, Taub N, Gant V, Wilson I, Shaw KM (1992) The immunogenetics of early nephropathy in insulin-dependent diabetes mellitus: association between the HLA-A2 antigen and albuminuria. Q J Med 83:461–471

Wei J, Li J, Zhang X, Tang Y, Wang J, Wu Y (2009) A naturally processed epitope on rotavirus VP7 glycoprotein recognized by HLA-A2.1-restricted cytotoxic CD8+ T cells. Viral Immunol 22:189–194

Wen L, Peng J, Li Z, Wong FS (2004) The effect of innate immunity on autoimmune diabetes and the expression of Toll-like receptors on pancreatic islets. J Immunol 172:3173–3180

Zanoni G, Navone R, Lunardi C, Tridente G, Bason C, Sivori S, Beri R, Dolcino M, Valletta E, Corrocher R, Puccetti A (2006) In celiac disease, a subset of autoantibodies against transglutaminase binds toll-like receptor 4 and induces activation of monocytes. PLoS Med 3:e358

Zhang Y, Lee B, Thompson M, Glass R, Cama RI, Figueroa D, Gilman R, Taylor D, Stephenson C (2000) Lactulose-mannitol intestinal permeability test in children with diarrhea caused by rotavirus and cryptosporidium. Diarrhea Working Group, Peru. J Pediatr Gastroenterol Nutr 31:16–21

Chapter 19
Viruses, Diabetes, and Autoimmunity: Studies of Subjects at Genetic Risk for Type 1 Diabetes

Sabina Resic Lindehammer and Åke Lernmark

Abstract The possible importance of virus infections before the development of islet autoimmunity and then for the appearance of clinical type 1 diabetes is reviewed. There is a lack of specific data on the role of virus to induce islet autoimmunity. There is also a paucity of data to demonstrate that virus infections may contribute to an accelerated disease process resulting in clinical onset of type 1 diabetes. In contrast, there is a plethora of studies on virus infections at the time of clinical onset. However, these studies have made the understanding of virus in type 1 diabetes less easy. Future studies need to address further gestational infections and the risk of the offspring for islet autoimmunity. Such studies should also investigate the mechanisms by which gestational infections may alter the ability of the offspring to respond to future virus infections related to the development of islet autoimmunity or accelerator of the clinical onset of diabetes.

Enterovirus: A Possible Association Between Infections During Pregnancy and Type 1 Diabetes in the Offspring

Associations between gestational enterovirus (EV) infection during pregnancy and type 1 diabetes in the child have been reported (Dahlquist et al. 1995a; Hyoty et al. 1995). Therefore, gestational EV infection needs to be considered as a possible contributor to type 1 diabetes in the offspring (Table 19.1). Despite reports of intrauterine transmission of EVs in animals (Modlin and Bowman 1987), transplacental infection in humans is not well understood. However, a number of case reports have concluded: EV may be detected in stillbirth infants at autopsy and in the placental

S.R. Lindehammer, Ph.D. • Å. Lernmark, Ph.D. (✉)
Department of Clinical Sciences, Lund University/CRC, Skåne University Hospital SUS, SE-20502 Malmö, Sweden
e-mail: Ake.Lernmark@med.lu.se

K.W. Taylor, H. Hyoty, A. Toniolo, A.J. Zuckerman (eds.), *Diabetes and Viruses*, DOI 10.1007/978-1-4614-4051-2_19, © Springer Science+Business Media New York 2013

Table 19.1 Positive associations between gestational enterovirus infections and type 1 diabetes in the offspring

Pregnancy/delivery	Virus/method	Reference
Early pregnancy		
Swedish Childhood Diabetes register	CBV by PCR and IgA, IgM, and IgG by ELISA	Dahlquist et al. (1999)
Late pregnancy		
DiMe study	IgG, IgM, and IgA capture radioimmunoassay	Hyöty et al. (1995)
Delivery		
Swedish Childhood Diabetes register	CBV 2–4 by IgM capture radioimmunoassay	Dahlquist et al. (1995a)
Swedish Childhood Diabetes register	IgG and IgM towards echo9, echovirus 30, CBV 5	Dahlquist et al. (1995b)
DISS study	EV by PCR and IgM capture immunoassay	Elfving et al. (2008)

tissues of newborns with neuro-developmental delays (Euscher et al. 2001), Coxsackie B virus (CBV 3) has been isolated from a stillborn infant with myocarditis and hydrops fetalis (Bates 1970), postnatal diagnosis of congenital skin lesions may be caused by intrauterine CBV 3 infection (Sauerbrei et al. 2000), CBV 3 may be detected in a stillborn hydropic fetus (Smalling et al. 2002) and present a case of intrauterine CBV 3 infection diagnosed during the second trimester with nonimmune fetal hydrops, resulting in a live birth and subsequent early neonatal death (Ouellet et al. 2004). These studies all support intrauterine transmission of EV. A number of studies have investigated gestational EV infections leading to development of type 1 diabetes in the child. The first report used the nationwide Swedish Childhood diabetes register and published by Dahlquist et al. (1995b). Maternal serum from 57 mothers at delivery of children who developed type 1 diabetes before 15 years of age were compared with serum from 203 mothers of control subjects who were delivered at the same hospital during the same time period. A group-specific enzyme-linked immunosorbent assay for enteroviral IgG and IgM antibodies in this study showed that EV-IgM and IgG antibodies against enteroviral antigens (echovirus 9 and 30, CBV 5) were significantly higher among mothers whose children later developed type 1 diabetes. The second report was conducted by the prospective population-based Childhood Diabetes in Finland study (DiMe) (Hyoty et al. 1995). This study showed an association between EV infections in the mother and type 1 diabetes in the child when presented with disease before 3 years of age, but not in group of children being 4–6 years of age. The serum samples had been collected from mothers at the end of the third month of pregnancy. The control mothers were matched for the time of delivery (±1 day) and gender of offspring. Samples were analyzed for IgG, IgM, and IgA virus antibodies by a capture-based radioimmunoassay. By the end of the same year, the Swedish Childhood diabetes register published a similar paper (Dahlquist et al. 1995a) with approximately the same number of mothers, matched to their control mothers no differently from the previous study but the IgM analysis this time included CBV 2–4. Dahlquist et al. 1995a reported, compared to matched controls, a significantly higher frequency of CBV 3 IgM but not of CBV 2 and CBV 4 at delivery in mothers whose children later developed type 1 diabetes. An additional report on samples collected during early pregnancy was conducted 4 years later, once more by the Swedish Childhood diabetes register. In a case–control study serum samples were collected during the first trimester from 85 mothers whose children developed type 1 diabetes before 15 years of age and compared to 172 controls of mothers whose children did not develop type 1 diabetes (Dahlquist et al. 1999). The samples were tested for CBV with RNA by reverse transcription-polymerase chain reaction (RT-PCR) and EV IgA, IgM, and IgG by ELISA. It was found that three mothers of type 1 diabetes children were CBV RNA positive and another three were positive for CBV IgM antibodies. Thus 6 out of 85 mothers of type 1 diabetes children had signs of EV infection in early pregnancy compared with 1 mother among 172 controls. The Diabetes Incidence in Sweden Study (DISS) published recently a report (Elfving et al. 2008) in which serum samples were examined at delivery from 30 nondiabetic mothers whose offspring developed type 1 diabetes between 15 and 25 years of age and compared them to 90 maternal serum samples matched by date of delivery. The samples were tested for EV-IgM and EV-RNA. Among the 30 nondiabetic mothers, 30% were EV-IgM positive, and none was positive for EV-RNA. In the control group, 16% were EV-IgM positive and 4% were positive for EV-RNA. None of these differences in frequencies were statistically significant, however boys of EV-IgM positive mothers had approximately five times greater risk of developing type 1 diabetes as compared to boys of IgM negative mothers. The German multicenter BABYDIAB found no evidence for an association of EV infections during pregnancy and early childhood with development of islet autoantibodies in offspring (Fuchtenbusch et al. 2001). The study estimated EV infections from birth, prior to and in parallel with the appearance of islet autoantibodies in offspring of parents with type 1 diabetes by IgG antibodies against a panel of CBVs, and IgG and IgM antibodies to CBV 3, CBV 4, and CBV 5. Another study evaluated the role of first trimester EV infections in two series of pregnant women (Viskari et al. 2002). The first series of mothers were 948 women whose children developed clinical diabetes before the age of 15 and the second series 680 women whose child developed clinical diabetes before 7 years of age. IgM class antibodies against CBV 5 and a mixture of CBV 3, CAV 16, and echovirus 11 antigens were analyzed. The results suggested that infection with EV during the first trimester of pregnancy was not a risk factor for type 1 diabetes in the offspring.

Enterovirus: A Possible Trigger of Islet Autoimmunity

A possible triggering time point has been observed between EV infections and the initiation of the autoimmune process as infections appeared to coincide with seroconversion to islet autoantibodies (Hiltunen et al. 1997; Lonnrot et al. 2000a, b; Salminen et al. 2003). Association between EV infections and seroconversion of islet autoantibodies was reported as early as 1982 (Asplin et al. 1982). In this report a family was described in whom serial measurements of ICA and CBV 3–5 titers were determined for 3 years before one of the children developed type 1 diabetes (Asplin et al. 1982). A few years later a fatal case of CBV 6 infection with characteristic features of a viral meningoencephalitis also demonstrated ICA seroconversion during hospitalization in a young child (Nigro et al. 1986). The DiMe study (Hyoty et al. 1995) not only showed an association between EV infections in the mother and type 1 diabetes in the offspring but also demonstrated concurrent infections associated with an ICA response in at least one-third of the prediabetic children, while they were rare in the control subjects. In this study controls were matched for age, gender, and the time period during which samples were collected. The same study followed siblings of type 1 diabetes patients who seroconverted to ICA during a prospective follow-up (Hiltunen et al. 1997). The cohort consisted of 765 nondiabetic siblings and comprised blood sampling every 6 months. IgG, IgM, and IgA class antibodies were analyzed for a panel of EV antigens. Increase in EV antibody levels were significantly more frequent in sample intervals in which ICA first appeared than in sample intervals in ICA negative control sibling. The children who converted to ICA during an EV infection more often had the HLA-DQ 2/8 genotype than children who stayed ICA negative. The Finnish Diabetes Prediction and Prevention (DIPP) study investigated the role of EV infections in children who have tested positive for islet autoantibodies in a prospectively starting at birth (Lonnrot et al. 2000a). Samples were drawn at birth and subsequently every 3–6 months interval. IgG and IgA class antibodies against purified CBV 4, purified echovirus 11, and an EV peptide antigen derived from an immunodominant region of capsid protein VP1 were measured by enzyme immunoassay (EIA). The presence of EV-RNA was also examined by RT-PCR. The cohort consisted of serum from 21 children who developed and retained islet autoantibodies and 104 control subjects matched for the time of birth, gender, and HLA. An association between EV infections and the induction of autoimmunity was found as EV infections were detected in 57% of the case subjects during a 6-month follow-up period preceding the first appearance of islet autoantibodies compared with 31% of the matched control children. It was concluded that EV infections were associated with the development of beta-cell autoimmunity and would provide evidence for the role of EVs in the initiation of beta-cell destruction.

The international study Trial to Reduce Insulin Dependent Diabetes Mellitus in the Genetically at Risk (TRIGR) reported the risk effect of EV infections in children who were followed prospectively from birth and subsequently developed signs of progressive beta-cell autoimmunity, i.e., positivity for type 1 diabetes-associated autoantibodies by the age of 2 years (Sadeharju et al. 2003). The 103 children included 19 cases and 84 control subjects. Their serum samples were analyzed for IgG and IgA toward CBV 4 and echovirus 11 as well as for EV-RNA. The result showed that children with islet autoantibody had more EV infections than autoantibody negative children before the appearance of autoantibodies, which suggested an association between EV infections and induction of beta-cell autoimmunity. Three years after the DIPP study reported EV infections to be associated with beta-cell autoimmunity (Lonnrot et al. 2000a), these authors conducted a similar study but with twice the number of cases as well as controls (Salminen et al. 2003). The second study showed comparable results but with stronger statistical significance.

The first study to evaluate the role of viral infections as accelerating an already initiated disease process in humans and to follow the rate of progression from islet autoimmunity to clinical diabetes was performed in the DAISY study (Stene et al. 2010). The cohort comprised children who tested positive for one or more islet autoantibodies on two or more consecutive clinic visits. Blood samples and rectal swabs were collected every 3–6 months after seroconversion for GADA, IAA, or IA-2A until the diagnosis of diabetes. The rate of progression from islet autoimmunity to diabetes was found to be significantly increased after the detection of EV-RNA in serum but not after detection of EV-RNA in rectal swab samples. The observation led to the conclusion that progression from islet autoimmunity to type 1 diabetes may increase after an EV infection characterized by the presence of viral RNA in the serum. However, conflicting results in observations between EV infections and the initiation of the autoimmune process have been obtained. An earlier DAISY study investigated whether there was an association between EV infections and beta-cell autoimmunity in children at higher risk of developing type 1 diabetes (Graves et al. 2003). A nested matched case–control study of incident cases of beta-cell autoimmunity within two prospective cohorts of genetically high-risk children (cases = 26, controls = 39) used PCR of serum, saliva, and rectal swab samples to detect EV infection. The study showed no evidence that EV infection would be a risk factor for the development of beta-cell autoimmunity. Similar negative findings was reported in the Norwegian MIDIA study that tested whether the frequency of EV-RNA was associated with the development of multiple islet autoantibodies in children with the highest HLA risk genotype in fecal samples from early infancy (Tapia et al. 2011a, b). The study included 27 children who developed two or more islet autoantibodies in two or more consecutive samples and two control subjects per case matched by follow-up time, date of birth, and county of residence. An additional MIDIA study published recently a nested case–control study, which included 27 children who developed islet autoimmunity and 53 children matched for age and community of residence (Tapia et al. 2011a, b). The objective of this study was to investigate a possible association between human parechovirus infections in early infancy and the development of islet autoimmunity. When testing monthly stool samples for human parechovirus using a semi-quantitative RT-PCR, they concluded that there was no associations between human parechovirus infections and the signs of islet autoimmunity. It is noted that parechovirus is another genus member of the *Picornaviridae* family.

The Environmental Determinants of Diabetes in the Young Study

A major effort to understand the role of EV infections in type 1 diabetes is made in The Environmental Determinants of Diabetes in the Young (TEDDY) study, which was initiated in 2005 (Group 2008a; Hagopian et al. 2006). After screening more than 440,000 newborns in four different countries (Finland, Germany, Sweden, and the USA), 8,600 children at increased HLA genetic risk for type 1 diabetes were invited to be followed until 2024 for exposures that may explain the development of islet autoantigen-specific autoantibodies (2008b). The second primary aim of the TEDDY study is to identify environmental factors that may increase the rate of the disease process and thereby a clinical onset of type 1 diabetes. Similar efforts are carried out in the DIPP (Lonnrot et al. 2000a; Salminen et al. 2003). MIDIA (Tapia et al. 2011a, b) or DiPiS studies (Larsson et al. 2005; Lynch et al. 2008). It is important in this respect that the TEDDY study is designed to meet the requirement of matched HLA genotypes as well as intensive follow-up during the first 15 years of life (Group 2008b; Hagopian et al. 2006).

Future Perspectives

It will be important to investigate pregnant women and to follow not only these women but their offspring if they are born with high-risk HLA for type 1 diabetes. The TEDDY study is lacking information other than questionnaire data to evaluate the possible importance of gestational events for the children developing islet autoantibodies. The data so far indicate that the TEDDY children have comparable birth size (Sterner et al. 2011). Longitudinal follow-up of these children should reveal whether the birth size differences that were observed between participating countries may contribute to the risk for islet autoimmunity and type 1 diabetes (Sterner et al. 2011). However, future studies of mothers in early pregnancy should also investigate the mechanisms by which gestational infections may alter the ability of the offspring to respond to future virus infections that may trigger islet autoimmunity.

Acknowledgments The authors wish to thank the European Foundation for the Study of Diabetes (EFSD), Clinical Research in Diabetes Program, the Swedish Research Council (K2008-55X-15312-04-3), the Swedish Diabetes Association, Knut & Alice Wallenberg Foundation, UMAS Funds, the Skane County Council of Research and Development, the Novo Nordisk Foundation, and the National Institutes of Health (NIH) for support.

References

Asplin CM, Cooney MK, Crossley JR, Dornan TL, Raghu P, Palmer JP (1982) Coxsackie B4 infection and islet cell antibodies three years before overt diabetes. J Pediatr 101:398–400

Bates HR Jr (1970) Coxsackie virus B3 calcific pancarditis and hydrops fetalis. Am J Obstet Gynecol 106:629–630

Dahlquist G, Frisk G, Ivarsson SA, Svanberg L, Forsgren M, Diderholm H (1995a) Indications that maternal coxsackie B virus infection during pregnancy is a risk factor for childhood-onset IDDM. Diabetologia 38:1371–1373

Dahlquist GG, Boman JE, Juto P (1999) Enteroviral RNA and IgM antibodies in early pregnancy and risk for childhood-onset IDDM in offspring. Diabetes Care 22:364–365

Dahlquist GG, Ivarsson S, Lindberg B, Forsgren M (1995b) Maternal enteroviral infection during pregnancy as a risk factor for childhood IDDM. A population-based case–control study. Diabetes 44:408–413

Elfving M, Svensson J, Oikarinen S, Jonsson B, Olofsson P, Sundkvist G, Lindberg B, Lernmark A, Hyoty H, Ivarsson SA (2008) Maternal enterovirus infection during pregnancy as a risk factor in offspring diagnosed with type 1 diabetes between 15 and 30 years of age. Exp Diabetes Res 2008:271958

Euscher E, Davis J, Holzman I, Nuovo GJ (2001) Coxsackie virus infection of the placenta associated with neurodevelopmental delays in the newborn. Obstet Gynecol 98:1019–1026

Fuchtenbusch M, Irnstetter A, Jager G, Ziegler AG (2001) No evidence for an association of coxsackie virus infections during pregnancy and early childhood with development of islet autoantibodies in offspring of mothers or fathers with type 1 diabetes. J Autoimmun 17:333–340

Graves PM, Rotbart HA, Nix WA, Pallansch MA, Erlich HA, Norris JM, Hoffman M, Eisenbarth GS, Rewers M (2003) Prospective study of enteroviral infections and development of beta-cell autoimmunity. Diabetes autoimmunity study in the young (DAISY). Diabetes Res Clin Pract 59:51–61

Group Ts (2008) The Environmental Determinants of Diabetes in the Young (TEDDY) Study. Ann N Y Acad Sci 1150:1–13

Group TTs (2008) The Environmental Determinants of Diabetes in the Young (TEDDY) Study. Ann N Y Acad Sci 1150:1–13

Hagopian WA, Lernmark A, Rewers MJ, Simell OG, She JX, Ziegler AG, Krischer JP, Akolkar B (2006) TEDDY—The Environmental Determinants of Diabetes in the Young: an observational clinical trial. Ann N Y Acad Sci 1079:320–326

Hiltunen M, Hyoty H, Knip M, Ilonen J, Reijonen H, Vahasalo P, Roivainen M, Lonnrot M, Leinikki P, Hovi T, Akerblom HK (1997) Islet cell antibody seroconversion in children is temporally associated with enterovirus infections. Childhood Diabetes in Finland (DiMe) Study Group. J Infect Dis 175:554–560

Hyoty H, Hiltunen M, Knip M, Laakkonen M, Vahasalo P, Karjalainen J, Koskela P, Roivainen M, Leinikki P, Hovi T, Åkerblom HK (1995) A prospective study of the role of coxsackie B and other enterovirus infections in the pathogenesis of IDDM. Childhood Diabetes in Finland (DiMe) Study Group. Diabetes 44:652–657

Larsson HE, Lynch K, Lernmark B, Nilsson A, Hansson G, Almgren P, Lernmark A, Ivarsson SA (2005) Diabetes-associated HLA genotypes affect birthweight in the general population. Diabetologia 48:1484–1491

Lonnrot M, Korpela K, Knip M, Ilonen J, Simell O, Korhonen S, Savola K, Muona P, Simell T, Koskela P, Hyoty H (2000a) Enterovirus infection as a risk factor for beta-cell autoimmunity in a prospectively observed birth cohort: the Finnish Diabetes Prediction and Prevention Study. Diabetes 49:1314–1318

Lonnrot M, Salminen K, Knip M, Savola K, Kulmala P, Leinikki P, Hyypia T, Akerblom HK, Hyoty H (2000b) Enterovirus RNA in serum is a risk factor for beta-cell autoimmunity and clinical type 1 diabetes: a prospective study. Childhood Diabetes in Finland (DiMe) Study Group. J Med Virol 61:214–220

Lynch KF, Lernmark B, Merlo J, Cilio CM, Ivarsson SA, Lernmark A (2008) Cord blood islet autoantibodies and seasonal association with the type 1 diabetes high-risk genotype. J Perinatol 28:211–217

Modlin JF, Bowman M (1987) Perinatal transmission of coxsackievirus B3 in mice. J Infect Dis 156:21–25

Nigro G, Pacella ME, Patane E, Midulla M (1986) Multi-system coxsackievirus B-6 infection with findings suggestive of diabetes mellitus. Eur J Pediatr 145:557–559

Ouellet A, Sherlock R, Toye B, Fung KF (2004) Antenatal diagnosis of intrauterine infection with coxsackievirus B3 associated with live birth. Infect Dis Obstet Gynecol 12:23–26

Sadeharju K, Hamalainen AM, Knip M, Lonnrot M, Koskela P, Virtanen SM, Ilonen J, Akerblom HK, Hyoty H (2003) Enterovirus infections as a risk factor for type I diabetes: virus analyses in a dietary intervention trial. Clin Exp Immunol 132:271–277

Salminen K, Sadeharju K, Lonnrot M, Vahasalo P, Kupila A, Korhonen S, Ilonen J, Simell O, Knip M, Hyoty H (2003) Enterovirus infections are associated with the induction of beta-cell autoimmunity in a prospective birth cohort study. J Med Virol 69:91–98

Sauerbrei A, Gluck B, Jung K, Bittrich H, Wutzler P (2000) Congenital skin lesions caused by intrauterine infection with coxsackievirus B3. Infection 28:326–328

Smalling TW, Sefers SE, Li H, Tang YW (2002) Molecular approaches to detecting herpes simplex virus and enteroviruses in the central nervous system. J Clin Microbiol 40:2317–2322

Stene LC, Oikarinen S, Hyoty H, Barriga KJ, Norris JM, Klingensmith G, Hutton JC, Erlich HA, Eisenbarth GS, Rewers M (2010) Enterovirus Infection and Progression from Islet Autoimmunity to Type 1 Diabetes: The Diabetes and Autoimmunity Study in the Young (DAISY). Diabetes 59:3174–3180

Sterner Y, Torn C, Lee HS, Larsson H, Winkler C, McLeod W, Lynch K, Simell O, Ziegler A, Schatz D, Hagopian W, Rewers M, She JX, Krischer JP, Akolkar B, Lernmark A (2011) Country-specific birth weight and length in type 1 diabetes high-risk HLA genotypes in combination with prenatal characteristics. J Perinatol 31:764–769.

Tapia G, Cinek O, Rasmussen T, Grinde B, Stene LC, Ronningen KS (2011a) Longitudinal study of parechovirus infection in infancy and risk of repeated positivity for multiple islet autoantibodies: the MIDIA study. Pediatr Diabetes 12:58–62

Tapia G, Cinek O, Rasmussen T, Witso E, Grinde B, Stene LC, Ronningen KS (2011b) Human Enterovirus RNA in Monthly Fecal Samples and Islet Autoimmunity in Norwegian Children With High Genetic Risk for Type 1 Diabetes: The MIDIA study. Diabetes Care 34:151–155

Viskari HR, Roivainen M, Reunanen A, Pitkaniemi J, Sadeharju K, Koskela P, Hovi T, Leinikki P, Vilja P, Tuomilehto J, Hyoty H (2002) Maternal first-trimester enterovirus infection and future risk of type 1 diabetes in the exposed fetus. Diabetes 51:2568–2571

Chapter 20
Type 1 Diabetes in the Tropics: A Link with Enterovirus Infections

Eduardo Cabrera-Rode, Oscar Díaz-Horta, Antonio Toniolo, and Luis Sarmiento

Abstract The global distribution of the new cases of type 1 diabetes is influenced by genetic and environmental factors. The genetic component is not restricted to HLA, as other loci participate as risk factors. The incidence of type 1 diabetes in tropical/subtropical regions is low, whereas helminth and enterovirus infections are common and widespread. Three Echovirus epidemics have been documented in Cuba over the last two decades. Echovirus infections were associated with seroconversion to diabetes-related autoantibodies. Data from Cuba indicate enteroviruses of the B species as the agents linked most frequently to type 1 diabetes. Further research is necessary to define the enterovirus types linked to type 1 diabetes, which will pave the way for designing novel preventive approaches.

E. Cabrera-Rode, Ph.D. (✉)
Department of Immunology of Diabetes, National Institute of Endocrinology, Havana, Cuba
e-mail: diabetes@infomed.sld.cu

O. Díaz-Horta, Ph.D.
Department of Immunology of Diabetes, National Institute of Endocrinology, Havana, Cuba

Department of Experimental Medicine, University of Insubria and Ospedale di Circolo e Fondazione Macchi, Varese, Italy

A. Toniolo, M.D.
Laboratory of Microbiology, University of Insubria and Ospedale di Circolo, Varese, Italy
e-mail: antonio.toniolo@uninsubria.it

L. Sarmiento, Ph.D.
Department of Virology, "Pedro Kouri" Tropical Medicine Institute, Havana, Cuba

K.W. Taylor, H. Hyoty, A. Toniolo, A.J. Zuckerman (eds.), *Diabetes and Viruses*,
DOI 10.1007/978-1-4614-4051-2_20,

Introduction

Analysis of cumulative data indicates that detection of human enterovirus (EV) genomes is particularly frequent in subjects with newly diagnosed type 1 diabetes and in individuals in the prediabetic stages (Yeung et al. 2011). The isolation and replication in vitro of EVs from the blood of these subjects have been rarely possible in practice, challenging the hypothesis of a viral etiology of type 1 diabetes. However, other lines of evidence support this conjecture.

The incidence of type 1 diabetes, especially in childhood, has been increasing rapidly in the western world over the last 30 years. Interestingly, the global distribution of new cases of type 1 diabetes is not homogeneous, the highest incidence rate being found in Northern countries, the lowest incidence in tropical and subtropical regions. In spite of many studies, the reasons for the temporal increase and marked geographic variations of the disease remain enigmatic. Whether the particular distribution of type 1 diabetes is linked to genetic and/or environmental differences among populations is unknown, but it is likely that both components play a role. The prevalence of type 1 diabetes susceptibility/protective haplotypes varies among different populations, and the incidence of type 1 diabetes is negatively related to the circulation of EVs. In this respect, tropical regions are characterized by an elevated frequency of EV infections.

Genetic and environmental studies undertaken in Cuba are reviewed with special reference to the putative involvement of EVs as environmental factors linked to type 1 diabetes.

Genetic Background and Type 1 Diabetes

The presence of a non-aspartic amino acid in the position 57 of the DQ beta chain was related to the incidence of type 1 diabetes across populations (Dorman et al. 1990). Subsequently, it became clear that this marker was not accurate as expected (Dorman and Bunker 2000). The presence of polar residues at beta 7 and beta 37 in the DR and DQ chains appeared to reflect more closely the risk of developing type 1 diabetes (Parry and Brooks 2008). In comparison, the absence of aspartic acid at DQ beta 57 represents a minor risk factor.

It is important to note that the risk conferred by HLA DR/DQ two-locus haplotypes is dependent on the investigated population. For instance, DRB1*0301-DQB1*0201 and DRB1*0401-DQB1*0302 are associated with the highest risk for type 1 diabetes in European populations. That is not the case for East Asian populations in which DRB1*0405-DQB1*0401 and DRB1*0901-DQB1*0303 are the most important risk factors (Ikegami et al. 2008). In a recent genetic study performed in the tropics it was shown that Cubans share with Europeans the same risk HLA

DR/DQ haplotypes. That was not a surprise according to the genetic structure of the Cuban population (Cintado et al. 2009) which is composed mainly of descendants of European ancestors. In fact, the Cuban population is genetically close to Mediterranean populations and well separated from Amerindian and Oriental populations as assessed by estimated genetic distances of HLA-DRB1 alleles (Alegre et al. 2007). Comparable results were obtained using HLA-DRB1-DQB1 allele frequencies which locate Cubans in close proximity to the French, Berbers, and Spaniards (Alegre et al. 2007). Other major ancestral component in the genetic repertoire of Cubans is that inherited from West African ancestors. However, admixture or stratification of the Cuban population appears not to influence HLA allele- and haplotype-associations with type 1 diabetes (Diaz-Horta et al. 2010).

With regard to of HLA haplotypes, other tropical populations like those of Central America (with high frequency of Amerindians) or Asia are genetically distant to those of European origin.

The data suggest that the differences in type 1 diabetes incidence between tropical and temperate regions could find HLA-based genetic explanations in some countries, but not in others.

The genetic elements associated with type 1 diabetes are not limited to the HLA region. Other genes, most of them participating in the immune modulation, appear to be involved in type 1 diabetes (Liston 2010). The study of the genetic polymorphism for each gene previously associated with type 1 diabetes in every population will be useful to understand their individual contributions to the incidence of type 1 diabetes. In a recent study, the genetic contribution of non-HLA genes was evaluated using estimates of population admixtures (Diaz-Horta et al. 2010). In the case of the Cuban population, the admixture process occurred between Spanish, West African, and—to a lesser degree—Amerindian tribes that inhabited the island (Cintado et al. 2009). Using ancestry informative markers, we found that individuals carrying high European ancestry proportions in their genome appeared to be more prone to develop the disease than those with high African ancestry. The size of this effect was estimated at 5.7 odds ratio (95% CI 1.2–36). The association between particular ancestry informative markers and disease was investigated and corrected by individual ancestry. A correlation was found for three ancestry informative markers: the first situated near the IDDM2 locus which includes the promoter of the insulin gene, the second near the IDDM3 locus and beta2-microglobulin, the third near the interleukin 2 gene.

These results from Cuba show that also in admixed populations some genes which are not limited to those coding for the major histocompatibility complex (MHC) do participate in conferring susceptibility or protection for type 1 diabetes. The results also demonstrate that ethnicity can influence susceptibility to type 1 diabetes. This supports epidemiological findings like those showing the large difference of type 1 diabetes incidence among ethnic groups living in the USA (Borchers et al. 2010). Similarly, a correlation has been reported between the incidence of type

1 diabetes and the proportion of not admixed Amerindian populations among countries of Latin America (Collado-Mesa et al. 2004).

Protective Infections and Type 1 Diabetes

From animal models of type 1 diabetes we learned that the frequency of disease is much greater in those animals bred in specific pathogen-free environments than under conventional conditions (Bach 2002). After decontaminating the food of animals of the latter group an increase in disease frequency has been observed. Prevention of type 1 diabetes in NOD mice has been achieved by exposure to a variety of molecules or organisms (Christen et al. 2004; Cooke 2009; Martins and Aguas 1996; Matsuzaki et al. 1997; McInerney et al. 1991; Nomaguchi et al. 2002; Vaysburd et al. 1995). Interestingly, not all infecting agents, parasites, or related products were able to induce this effect (Cooke 2009). The prevention of type 1 diabetes by stimulation of the immune system in the NOD mouse is similar to that which possibly occurs in human populations. From epidemiological studies of poliomyelitis the term "hygiene hypothesis" was originated. For instance, the circulation of polioviruses (enterovirus genus of *Picornaviridae*) started to decrease at the beginning of the last century in countries with high standards of hygiene (coincidentally, countries with temperate climates), while the incidence of paralytic polio began to increase. The phenomenon was the consequence of the shift of the initial poliovirus infection to older children, when an increasing proportion of children started acquiring infections at 1 year of age or more (i.e., when the protective effect of maternal poliovirus antibodies was no longer present). It is likely that decreasing concentrations of poliovirus antibodies in pregnant women did reduce infants' protection. When maternal antibodies are missing, EVs are able to spread from the intestine to the blood, then to target organs.

Similar observations confirm that over the last 50 years, the decreasing incidence of infectious diseases such as hepatitis A, measles, mumps, rheumatic fever, and tuberculosis was linked to the increasing incidence of immune-mediated disorders (e.g., asthma, multiple sclerosis, Crohn's disease) (Bach 2002). In addition, recognized EV infections appear to be infrequent in countries with an increased incidence of type 1 diabetes (Viskari et al. 2004, 2005).

As mentioned above, type 1 diabetes is rare in most Asian and Latin-American populations. However, its frequency increases noticeably when persons migrate to a northern setting (Karvonen et al. 2000). This might be related to a lower enterovirus circulation, but also to a lower incidence of helminthic infections in these countries (Zaccone et al. 2006). A possible mechanism for the protection elicited by parasites has been suggested. For instance, helminth antigens are able to induce interleukin-10 production by dendritic cells and B cells as well as capable of inducing activated macrophages, invariant natural killer T cells, and regulatory T cells

Table 20.1 Age-standardized incidence of type 1 diabetes in children <15 years of age (per 100,000 per year) and incidence rate of infections by helminths (HEL) and enteroviruses (EV) in Tropical/Subtropical Regions and Europe[a]

Region and country	Incidence of type 1 diabetes 1990–1994 (95% CI)[b]	Incidence of type 1 diabetes 2010[d]	HEL (%)[e]	EV (%)	Reference
Tropical/subtropical					
Cuba	2.9 (2.63–3.24)	2.3	<16.6–50.0	30.0	Sarmiento (2004)
Brazil	8.0 (5.53–11.14)	7.7	28.9–13.10	15.8	Castro et al. (2009)
Colombia	3.8 (2.88–4.93)	1.3	15.8–39.5	13.3	González et al. (2011)
Argentina	6.4	6.8	23.5–8.82	19.5	Cisterna et al. (2007)
Iran	NA[c]	3.7	7.5–2.0	57.0	Kargar et al. (2009)
India	NA	4.2	Endemic	67.0	Dhole et al. (2009)
Afghanistan	NA	1.2	Frequent	Polio endemic	CDC (2011)
Pakistan	NA	0.5	14–1	Polio endemic	CDC (2011)
Thailand	NA	0.3	Frequent	58.3	Chatproedprai et al. (2010)
Malaysia	NA	0.3	38.0–49.0	44.4	Podin et al. (2006)
South Korea	NA	1.1	NA	55.2	Lee and Kim (2008)
Japan	1.6 (0.81–2.65)	2.4	NA	35.0	Yamazaki et al. (2009)
Europe					
Austria	9.6 (8.84–10.31)	13.3	Absent	14.0	Ortner et al. (2009)
Estonia	10.5 (9.05–12.20)	15.3	NA	Frequent[f]	Viskari et al. (2004)
Finland	36.5 (34.83–38.26)	57.4	Absent	Not frequent[f]	Viskari et al. (2004)
France	8.5 (7.86–9.12)	12.2	Absent	5.0	Antona et al. (2007)
Germany	11.0 (10.25–11.69)	18.0	Absent	12.6	Roth et al. (2007)
Greece	9.7 (8.55–10.92)	9.9	NA	6.7	Mavrouli et al. (2007)
Hungary	9.1 (8.43–9.81)	11.3	Absent	Frequent[f]	Viskari et al. (2004)

(continued)

Table 20.1 (continued)

Region and country	Incidence of type 1 diabetes 1990–1994 (95% CI)[b]	Incidence of type 1 diabetes 2010[d]	HEL (%)[e]	EV (%)	Reference
Italy	NA	8.4	Rare/Absent	4.64	Patti et al. (2000)
Lithuania	7.4 (6.57–8.25)	7.8	NA	Not frequent[f]	Viskari et al. (2004)
Poland	NA	12.9	Absent	13.5	Witek et al. (2010)
Spain	12.5 (11.55–13.50)	13.0	Absent	7.9	Trallero et al. (2010)
Sweden	27.5 (26.36–28.67)	41.0	Absent	Not frequent[f]	Viskari et al. (2004)

[a]Data are incidence rates or incidence rates (95% CI) unless otherwise noted

[b]Primary source only. Data from Karvonen et al. 2000

[c]Data not available

[d]Data from the International Diabetes Federation (2010)

[e]Helminth infections: frequency of cases of ascariasis and trichuriasis. Data from de Silva et al. (2003), Zaccone et al. (2006), Hotez et al. (2008) and Hotez (2008, 2009, 2011)

[f]Data based on the relationships between the frequency of enterovirus infections and the incidence of type 1 diabetes in European countries

(Cooke 2009). Table 20.1 summarizes recent data on the prevalence of helminthic and EV infections as well as the negative correlation with the incidence of type 1 diabetes. A recent study (von Herrath 2009) suggests the possible underlying mechanism explaining these epidemiological findings. Regulatory T cells (CD4+ CD25+ Tregs) appear to be invigorated by infection in a toll-like receptor (TLR)-2 dependent fashion. Tregs can turn the immune responses off, thus impeding the cell-mediated autoimmune destruction of beta cells. An increment in TGF-β release induced by viral infection is also associated with Treg invigoration (Aumeunier et al. 2010; von Herrath 2009). Other cytokines induced during infection are the programmed cell death ligand 1 (PD-1 L) and the tumor necrosis factor (TNF)-α which, in turn, mediate the *bystander* death of auto-aggressive T cells. The protective effect of viral infections and other immune modulators has also been associated with the induction of interferon gamma-induced protein 10 kDa (IP-10) and other pro-inflammatory cytokines (Christen et al. 2004).

Taken together, experimental and epidemiological data indicate that the global distribution of the *incidence* of type 1 diabetes is linked, at least in part, to the differential frequency of multiple nonspecific stimuli acting on the immune system during the early periods of life.

Evidence Linking Enteroviruses and Type 1 Diabetes in the Tropics

As reported above, the high circulation of enteroviruses is correlated negatively with the incidence of type 1 diabetes. Thus, it is somehow paradoxical to propose enterovirus as diabetogenic, particularly in the tropics where exposure to these agents is common and occurs year-round with possible peaks in the rainy season. In addition, not all EV types may be linked to type 1 diabetes. Unfortunately, the properties associated with the diabetogenic effect of these viruses remain obscure (Sarmiento and Cabrera-Rode 2007). It seems that the emergence of a diabetogenic EV strain and its interaction with a genetically susceptible individual is a rare event. Otherwise, a wave of type 1 diabetes new cases should be expected after large EV epidemics. In Cuba, meningitis epidemics due to echovirus type 4, type 16, and type 30 have been documented in 1986, 2000, and 2001, respectively (Uriarte et al. 1991; Diaz-Horta et al. 2001; Cabrera-Rode et al. 2003, 2005; Sarmiento 2004). Analysis of these epidemics was particularly relevant since it was demonstrated that some children infected with echovirus either seroconverted to diabetes-related autoantibodies or progressed toward overt type 1 diabetes (Table 20.2). Interestingly, unpublished results indicate that EVs isolated from Cuban meningitis epidemics also display a remarkable tropism for insulin-producing beta cells. Table 20.2 summarizes the evidence linking EV infections with type 1 diabetes in Cuba. In this country, the incidence of type 1 diabetes is particularly low and EV infections are common and widespread.

Table 20.2 Evidence from Cuba associating enterovirus infections with type 1 diabetes

Evidence	References
Year 1986. Association between the ICA seroconversion and a meningoencephalitis epidemic caused by Echovirus type 4. Temporary glucose intolerance was documented in infected children	Uriarte et al. (1991)
Impairment of insulin secretion and protein synthesis in mouse beta cells infected with Echovirus type 4	Szopa et al. (1992)
High frequency of neutralizing antibodies against Echovirus type 4 in newly diagnosed type 1 diabetes patients as compared to matched controls	Diaz-Horta et al. (2001)
Year 2000. Association between the occurrence of a large-scale Echovirus type 16 epidemic and the appearance of diabetes-related autoantibodies (ICA, GADA, IAA)	Cabrera-Rode et al. (2003)
Year 2001. ICA seroconversion during the convalescent phase of infection during a meningitis outbreak due to Echovirus type 30	Cabrera-Rode et al. (2005)
Emergence of pancreatic autoimmunity (ICA and IA-2A) and type 1 diabetes after meningitis due to Echovirus type 30 (case report)	Cabrera-Rode et al. (2005)
Appearance of GADA in serum of rabbits injected with selected enteroviruses of the B species	Sarmiento et al. (2007b)
High frequency of enterovirus RNA in serum of children at the clinical onset of type 1 diabetes and in ICA positive first-degree relatives of the index cases	Cubas-Dueñas et al. (2011), Sarmiento et al. (2007a)

ICA islet cell antibodies, *IAA* insulin autoantibodies, *GADA* glutamic acid decarboxylase autoantibodies, *IA-2A* autoantibodies to the intracellular portion of the protein tyrosine phosphatase-related IA2 molecule

References

Alegre R, Moscoso J, Martinez-Laso J, Martin-Villa M, Suarez J, Moreno A, Serrano-Vela JI, Vargas-Alarcon G, Pacheco R, Arnaiz-Villena A (2007) HLA genes in Cubans and the detection of Amerindian alleles. Mol Immunol 44:2426–2435

Antona D, Lévêque N, Chomel J, Dubrou S, Lévy-Bruhl D, Lina B (2007) Surveillance of enteroviruses in France, 2000–2004. Eur J Clin Microbiol Infect Dis 26:403–412

Aumeunier A, Grela F, Ramadan A, Van Pham L, Bardel E, Gomez Alcala A, Jeannin P, Akira S, Bach JF, Thieblemont N (2010) Systemic toll-like receptor stimulation suppresses experimental allergic asthma and autoimmune diabetes in NOD mice. PLoS One 5:e11484

Bach JF (2002) The effect of infections on susceptibility to autoimmune and allergic diseases. N Engl J Med 347:911–920

Borchers AT, Uibo R, Gershwin ME (2010) The geoepidemiology of type 1 diabetes. Autoimmun Rev 9:A355–A365

Cabrera-Rode E, Sarmiento L, Tiberti C, Molina G, Barrios J, Hernandez D, Diaz-Horta O, Di Mario U (2003) Type 1 diabetes islet associated antibodies in subjects infected by echovirus 16. Diabetologia 46:1348–1353

Cabrera-Rode E, Sarmiento L, Molina G, Perez C, Arranz C, Galvan JA, Prieto M, Barrios J, Palomera R, Fonseca M, Mas P, Diaz-Diaz O, Diaz-Horta O (2005) Islet cell related antibodies and type 1 diabetes associated with echovirus 30 epidemic: a case report. J Med Virol 76: 373–377

Castro CM, Oliveira DS, Macedo O, Lima MJ, Santana MB, Wanzeller AL, Silveira E, Gomes ML (2009) Echovirus 30 associated with cases of aseptic meningitis in state of Pará, Northern Brazil. Mem Inst Oswaldo Cruz 104:444–450

Centers for Disease Control and Prevention (2011) Progress toward interrupting wild poliovirus circulation in countries with reestablished transmission—Africa, 2009–2010. MMWR Morb Mortal Wkly Rep 60:306–311

Chatproedprai S, Theanboonlers A, Korkong S, Thongmee C, Wananukul S, Poovorawan Y (2010) Clinical and molecular characterization of hand-foot-and-mouth disease in Thailand, 2008–2009. Jpn J Infect Dis 63:229–233

Christen U, Benke D, Wolfe T, Rodrigo E, Rhode A, Hughes AC, Oldstone MB, von Herrath MG (2004) Cure of prediabetic mice by viral infections involves lymphocyte recruitment along an IP-10 gradient. J Clin Invest 113:74–84

Cintado A, Companioni O, Nazabal M, Camacho H, Ferrer A, De Cossio ME, Marrero A, Ale M, Villarreal A, Leal L, Casalvilla R, Benitez J, Novoa L, Diaz-Horta O, Duenas M (2009) Admixture estimates for the population of Havana City. Ann Hum Biol 36:350–360

Cisterna DM, Palacios G, Rivero K, Girard D, Lema C, Freire MC (2007) Epidemiology of enterovirus associated with neurologic diseases. Medicina (B Aires) 67:113–119

Collado-Mesa F, Barcelo A, Arheart KL, Messiah SE (2004) An ecological analysis of childhood-onset type 1 diabetes incidence and prevalence in Latin America. Rev Panam Salud Publica 15:388–394

Cooke A (2009) Review series on helminths, immune modulation and the hygiene hypothesis: how might infection modulate the onset of type 1 diabetes? Immunology 126:12–17

Cubas-Dueñas I, Cabrera-Rode E, Sarmiento L, Molina G, Fonseca M, Arranz C, Dominguez E, Gonzalez P, Vera M, Diaz-Horta O (2011) First-degree relatives of persons with type 1 diabetes: insulin resistance and enterovirus infection are associated with different patterns of islet cell autoimmunity. Acta Diabetol. doi:10.1007/s00592-011-0297-6

de Silva NR, Brooker S, Hotez PJ, Montresor A, Engels D (2003) Soil-transmitted helminth infections: updating the global picture. Trends Parasitol 19:547–551

Dhole TN, Ayyagari A, Chowdhary R, Shakya AK, Shrivastav N, Datta T, Prakash V (2009) Non-polio enteroviruses in acute flaccid paralysis children of India: vital assessment before polio eradication. J Paediatr Child Health 45:409–413

Diaz-Horta O, Bello M, Cabrera-Rode E, Suarez J, Mas P, Garcia I, Abalos I, Jofra R, Molina G, Diaz-Diaz O, Di Mario U (2001) Echovirus 4 and type 1 diabetes mellitus. Autoimmunity 34:275–281

Diaz-Horta O, Cintado A, Fernandez-De-Cossio ME, Nazabal M, Ferrer A, Roca J, Camacho H, Benitez J, Ale M, Villarreal A, Molina G, Vera M, Cabrera-Rode E, Novoa L (2010) Relationship of type 1 diabetes to ancestral proportions and HLA DR/DQ alleles in a sample of the admixed Cuban population. Ann Hum Biol 37:778–788

Dorman JS, Bunker CH (2000) HLA-DQ locus of the human leukocyte antigen complex and type 1 diabetes mellitus: a HuGE review. Epidemiol Rev 22:218–227

Dorman JS, LaPorte RE, Stone RA, Trucco M (1990) Worldwide differences in the incidence of type I diabetes are associated with amino acid variation at position 57 of the HLA-DQ beta chain. Proc Natl Acad Sci U S A 87:7370–7374

González M, Castaño J, Giraldo A, Quintero L, Sarmiento L (2011) Prevalence of enterovirus in newborns and infants who consult a first-class Center, Armenia, Colombia 2009. Biomédica 31(4)

Hotez PJ (2008) Holidays in the sun and the Caribbean´s forgotten burden of neglected tropical diseases. PLoS Negl Trop Dis 2:e239. doi:10.1371/journal.pntd.0000239

Hotez PJ (2009) The neglected tropical diseases and their devastating health and economic impact on the member nations of the organization of the Islamic conference. PLoS Negl Trop Dis 3:e539. doi:10.1371/journal.pntd.000539

Hotez PJ (2011) Neglected tropical diseases in the catholic world. PLoS Negl Trop Dis 5:e1132. doi:10.1371/journal.pntd.0001132

Hotez PJ, Bottazzi ME, Franco-Paredes C, Ault SK, Periago MR (2008) The neglected tropical diseases of Latin America and the Caribbean: a review of disease burden and distribution and a roadmap for control and elimination. PLoS Negl Trop Dis 2:e300. doi:10.1371/journal.pntd.0000300

Ikegami H, Noso S, Babaya N, Hiromine Y, Kawabata Y (2008) Genetic basis of type 1 diabetes: similarities and differences between East and West. Rev Diabet Stud 5:64–72

International Diabetes Federation (2010) The diabetes atlas, 4th edn. Brussels. http://www.diabetesatlas.org/map. Accessed 14 Jan 2011

Kargar M, Sadeghipour S, Nategh R (2009) Environmental surveillance of non-polio enteroviruses in Iran. Virol J 6:149. doi:10.1186/1743-422X-6-149

Karvonen M, Viik-Kajander M, Moltchanova E, Libman I, LaPorte R, Tuomilehto J (2000) Incidence of childhood type 1 diabetes worldwide. Diabetes Mondiale (DiaMond) Project Group. Diabetes Care 23:1516–1526

Lee C, Kim SJ (2008) Molecular detection of human enteric viruses in urban rivers in Korea. J Microbiol Biotechnol 18:1156–1163

Liston A (2010) How many genes in type 1 diabetes? Int Diabetes Monitor 22:90–93

Martins TC, Aguas AP (1996) Changes in B and T lymphocytes associated with mycobacteria-induced protection of NOD mice from diabetes. J Autoimmun 9:501–507

Matsuzaki T, Nagata Y, Kado S, Uchida K, Kato I, Hashimoto S, Yokokura T (1997) Prevention of onset in an insulin-dependent diabetes mellitus model, NOD mice, by oral feeding of Lactobacillus casei. Acta Pathol Microbiol Immunol Scand 105:643–649

Mavrouli MD, Spanakis N, Levidiotou S, Politi C, Alexiou S, Tseliou P, Hatzitaki M, Foundouli K, Tsakris A, Legakis NJ, Routsias JG (2007) Serologic prevalence of coxsackievirus group B in Greece. Viral Immunol 20:11–18

McInerney MF, Pek SB, Thomas DW (1991) Prevention of insulitis and diabetes onset by treatment with complete Freund's adjuvant in NOD mice. Diabetes 40:715–725

Nomaguchi H, Yogi Y, Kawatsu K, Okamura H, Ozawa Y, Kasatani T (2002) Prevention of diabetes in non-obese diabetic mice by a single immunization with Mycobacterium leprae. Nihon Hansenbyo Gakkai Zasshi 71:31–38

Ortner B, Huang CW, Schmid D, Mutz I, Wewalka G, Allerberger F, Yang JY, Huemer HP (2009) Epidemiology of enterovirus types causing neurological disease in Austria 1999–2007: detection of clusters of echovirus 30 and enterovirus 71 and analysis of prevalent genotypes. J Med Virol 81:317–324

Parry CS, Brooks BR (2008) A new model defines the minimal set of polymorphism in HLA-DQ and -DR that determines susceptibility and resistance to autoimmune diabetes. Biol Direct 3:42

Patti AM, Santi AL, Fiore L, Vellucci L, De Stefano D, Bellelli E, Study Group (2000) Enterovirus surveillance of Italian healthy children. Eur J Epidemiol 16:1035–1038

Podin Y, Gias EL, Ong F, Leong YW, Yee SF, Yusof MA, Perera D, Teo B, Wee TY, Yao SC, Yao SK, Kiyu A, Arif MT, Cardosa MJ (2006) Sentinel surveillance for human enterovirus 71 in Sarawak, Malaysia: lessons from the first 7 years. BMC Public Health 6:180

Roth B, Enders M, Arents A, Pfitzner A, Terletskaia-Ladwig E (2007) Epidemiologic aspects and laboratory features of enterovirus infections in Western Germany, 2000–2005. J Med Virol 79:956–962

Sarmiento L (2004) Enteroviral meningitis and emergence of rare enterovirus types: Cuban experience. In: Strong Phyllis V (ed) Focus on meningitis research. Nova, New York, pp 1–14

Sarmiento L, Cabrera-Rode E, Lekuleni L, Cuba I, Molina G, Fonseca M, Heng-Hung L, Borroto AD, Gonzalez P, Mas-Lago P, Diaz-Horta O (2007a) Occurrence of enterovirus RNA in serum of children with newly diagnosed type 1 diabetes and islet cell autoantibody-positive subjects in a population with a low incidence of type 1 diabetes. Autoimmunity 40:540–545

Sarmiento L, Cabrera-Rode E, Mas-Lago P, Diaz-Horta O (2007b) Antibodies to human glutamic acid decarboxylase in sera from enterovirus-immunized rabbit. Autoimmunity 40:546–547

Sarmiento-Pérez L, Cabrera-Rode E (2007) Enterovirus infections and type 1 diabetes mellitus. Nova, New York, pp 1–187

Szopa TM, Ward T, Taylor KW (1992) Disturbance of mouse pancreatic beta-cell function following echo 4 virus infection. Biochem Soc Trans 20:315S

Trallero G, Avellon A, Otero A, De Miguel T, Pérez C, Rabella N (2010) Enteroviruses in Spain over the decade 1998–2007: virological and epidemiological studies. J Clin Virol 47:170–176

Uriarte A, Molina G, Cabrera-Rode E, Ventura R, Vargas J, Vera M (1991) Prospective study in children with high risk of type I diabetes as a sequel of ECHO-4 virus infection: 1986–1989. Rev Cubana Endocrinol 2:34–43

Vaysburd M, Lock C, McDevitt H (1995) Prevention of insulin-dependent diabetes mellitus in non-obese diabetic mice by immunogenic but not by tolerated peptides. J Exp Med 182:897–902

Viskari H, Ludvigsson J, Uibo R, Salur L, Marciulionyte D, Hermann R, Soltesz G, Fuchtenbusch M, Ziegler AG, Kondrashova A, Romanov A, Knip M, Hyoty H (2004) Relationship between the incidence of type 1 diabetes and enterovirus infections in different European populations: results from the EPIVIR project. J Med Virol 72:610–617

Viskari H, Ludvigsson J, Uibo R, Salur L, Marciulionyte D, Hermann R, Soltesz G, Fuchtenbusch M, Ziegler AG, Kondrashova A, Romanov A, Kaplan B, Laron Z, Koskela P, Vesikari T, Huhtala H, Knip M, Hyoty H (2005) Relationship between the incidence of type 1 diabetes and maternal enterovirus antibodies: time trends and geographical variation. Diabetologia 48:1280–1287

von Herrath M (2009) Can we learn from viruses how to prevent type 1 diabetes?: the role of viral infections in the pathogenesis of type 1 diabetes and the development of novel combination therapies. Diabetes 58:2–11

Witek A, Wieczorek M, Brzóstkowska M (2010) Enteroviruses serotypes isolated in years 2008–2009 in NIPH-NIH. Przegl Epidemiol 64:509–512

Yamazaki K, Sakon N, Nakata K, Kase T (2009) Prevalence of enterovirus in Osaka, Japan, between April and September 2008. Jpn J Infect Dis 62:80–81

Yeung WC, Rawlinson WD, Craig ME (2011) Enterovirus infection and type 1 diabetes mellitus: systematic review and meta-analysis of observational molecular studies. BMJ 342:d35

Zaccone P, Fehervari Z, Phillips JM, Dunne DW, Cooke A (2006) Parasitic worms and inflammatory diseases. Parasite Immunol 28:515–523

Chapter 21
Diabetes and Viruses in Australia and the Asia-Pacific Region

Myra Poon, William D. Rawlinson, and Maria E. Craig

Abstract Clinical studies from Australia and the Asia-Pacific region support an association between type 1 diabetes and several viruses, including rubella, RV and EV. Meta-analysis has demonstrated a significant association between EV infection and type 1 diabetes globally, as well as in this region. Detailed studies of EV outbreaks have identified genotypes distinct to the region. In particular, an outbreak of EV71 in 1997–1999 was temporally associated with the most common virus isolated from an Australian incident cohort of children with type 1 diabetes. In this study, EV infection was more common in young people without high-risk HLA genotypes, suggesting that there may be a subgroup of virus-induced type 1

M. Poon
Institute of Endocrinology and Diabetes, The Children's Hospital at Westmead, Sydney, NSW, Australia

Discipline of Paediatrics and Child Health, The University of Sydney, Sydney, NSW, Australia

W.D. Rawlinson
Virology Research Laboratory, Prince of Wales Hospital, Sydney, NSW, Australia

School of Biotechnology and Biomolecular Science, University of New South Wales, Sydney, NSW, Australia

School of Medical Sciences, University of New South Wales, Sydney, NSW, Australia

M.E. Craig, M.D. (✉)
Institute of Endocrinology and Diabetes, The Children's Hospital at Westmead, Sydney, NSW, Australia

Discipline of Paediatrics and Child Health, The University of Sydney, Sydney, NSW, Australia

School of Women's and Children's Health, University of New South Wales, Sydney, NSW, Australia
e-mail: m.craig@unsw.edu.au

K.W. Taylor, H. Hyoty, A. Toniolo, A.J. Zuckerman (eds.), *Diabetes and Viruses*,
DOI 10.1007/978-1-4614-4051-2_21, © Springer Science+Business Media New York 2013

diabetes. Fulminant diabetes is a unique form of diabetes found predominantly in Asian patients with rapid onset, ketosis-prone diabetes. Concurrent viral infections, in particular EV, are also associated with this form of diabetes. Further evaluation of the clinical characteristics of young people and adults with virus-induced diabetes in the region, in parallel with examination of molecular characteristics of isolates, may provide innovative insights into the pathogenesis of type 1 diabetes and fulminant diabetes.

Epidemiology of Type 1 Diabetes in the Asia-Pacific Region: Role for Viruses?

The incidence of type 1 diabetes has increased globally in recent decades, by approximately 3% per year (Diamond Project Group 2006). In the Asia-Pacific region, type 1 diabetes has increased in Australia (Catanzariti et al. 2009) and New Zealand (Campbell-Stokes and Taylor 2005), countries with a high incidence of type 1 diabetes (>20 per 100,000 person years), as well as in China (Zhang et al. 2008) and Thailand (Unachak and Tuchinda 2001), where the incidence is relatively low. In contrast, rates of type 1 diabetes have remained steady in Japan (Kawasaki et al. 2006), where the background incidence is also low. Nevertheless, despite a limited body of epidemiological data on type 1 diabetes incidence rates in Asia, the rise in type 1 diabetes incidence across the region has occurred too rapidly to be explained by an increase in population genetic susceptibility for type 1 diabetes. Indeed, there has been an increase in the proportion of individuals with type 1 diabetes without high-risk human leukocyte antigen (HLA) genotypes in Australia in recent years, compared with those diagnosed in earlier decades (Fourlanos et al. 2008). Given environmental factors, in particular viral infections, are strongly associated with the etiology of type 1 diabetes (Yeung et al. 2011); the unprecedented increase in type 1 diabetes suggests that environmental factors are also implicated in the rising incidence.

Regional variation in type 1 diabetes incidence is found across the Asia-Pacific region, as well as within countries (Li et al. 2000). This may be related to variation in the prevalence of type 1 diabetes susceptibility genes, since classic HLA risk genes are generally not found in Asian populations (Ikegami et al. 2006); however, regional and seasonal variation in environmental triggers may also underlie differences in type 1 diabetes incidence rates. Indeed, "epidemics of type 1 diabetes" have been observed in China in association with colder climate and winter months (Yang et al. 2005), while both regional and seasonal variations have been observed in Australia (Chong et al. 2007; Taplin et al. 2005) and New Zealand (Miller et al. 2011). Populations in this region may be more susceptible to infection, due to lower socioeconomic status and/or increased exposure to viruses, particularly Enterovirus (EV) 71 which has caused epidemics across the Asia-Pacific region since 1997 (Mackenzie et al. 2001; Sanders et al. 2006; Solomon et al. 2010).

A causal link between viruses and type 1 diabetes was proposed more than 40 years ago, and some of the early epidemiological studies involved patients from this region. In a case–control study of 122 new onset cases of type 1 diabetes from England, Austria, and Australia, specific IgM responses to coxsackievirus (CV) B1-5 were detected significantly in more children with diabetes compared with controls (Banatvala et al. 1985). In contrast, there was no association with other viruses such as mumps, rubella, or cytomegalovirus (CMV). Following the establishment of molecular methods for viral detection, several studies in the region have similarly demonstrated higher rates of EV infection in cases at onset of type 1 diabetes compared with controls (Table 21.1, Fig. 21.1), as well as an association with islet autoimmunity (Yeung et al. 2011). Viral infections, in particular EVs, are also linked to fulminant diabetes; this specific subgroup of diabetes is associated with the rapid onset of ketosis and hyperglycaemia and is found predominantly in people from Asia.

Viruses and Diabetes in the Asia-Pacific Region: An Historical Perspective

Congenital Rubella

A major outbreak of rubella infection occurred in Australia in the 1940s, promoted by overcrowding in military camps in World War 2. The link between congenital infection and congenital cataracts was first described by an ophthalmic surgeon, Norman Gregg, in 1941 and the congenital rubella syndrome (CRS) was elucidated over the following decades. The first report of an association between maternal rubella and diabetes was made by Hay in 1949; however it was more than 20 years before histopathological evidence confirmed an association between congenital rubella infection and diabetes, including isolation of rubella virus from the pancreas of infants with CRS at autopsy (Cooper et al. 1965) and demonstration of intimal proliferation in pancreatic blood vessels (Rorke and Spiro 1967). Further reports of clinical cases included a 24-year-old Australian man with congenital rubella and diabetes mellitus (Menser et al. 1967) and a case series of four patients, aged between 1.5 and 28 years, with diabetes (Forrest et al. 1969) (Table 21.1). In all four cases, maternal rubella infection occurred in mother during the first trimester of pregnancy. All affected individuals were symptomatic and required insulin treatment; two had diabetic retinopathy. In a follow-up study conducted in 1971, 44 of the original 50 patients with congenital rubella (including the patient originally diagnosed in 1967) underwent oral glucose tolerance tests; 5 of the 44 (11%) had diabetes and a further 4 of the 44 (9%) had evidence of impaired glucose tolerance (Forrest et al. 1971). However, the diagnostic criteria for diabetes (1 h glucose level >8.9 mmol/L and 2 h level >6.7 mmol/L) and impaired glucose tolerance (abnormal

Table 21.1 Summary of studies in the Asia-Pacific region investigating the association between prediabetes or diabetes and viruses

Study	Country	Cases/ controls	Infection rate (%) in cases	Age of cases[a] (years)	Viruses	Method of virus detection	Diabetes type
Enteroviruses							
Akatsuka et al. (2009)	Japan	1/0	–	39	CVB	Serology (neutralizing Ab)	Fulminant
Banatvala et al. (1985)	Australia	12/24	33	2–12	CVB	Serology (IgM)	T1D
Craig et al. (2003)	Australia	206/160	30	0.7–15.7	EV71, CV-B1, CV-B3, ECHOvirus 30	PCR	T1D
Kawashima et al. (2004)	Japan	61/58	38	0.75–40	EV	PCR	T1D
Li et al. (2010)	China	22/30	56	–	CVB	PCR	T1D
Tanaka et al. (2009)	Japan	3/0	–	14–29	EV	Immunostaining of pancreatic tissue	Fulminant
Rubella							
Forrest et al. (1969)	Australia	4	–	1.5–2	Rubella (congenital)	Clinical history	T2D
Forrest et al. (1971)	Australia	5	–	48–52	Rubella (congenital)	Clinical history	T2D
Forrest et al. (2002), McIntosh and Menser (1992), Menser et al. (1967)	Australia	1–7	–		Rubella (congenital)	Clinical history	T2D
Rotavirus							
Honeyman et al. (2000)	Australia	24/17	0.9	Mean 2.5	Rotavirus	Serology (IgA, IgG)	Prediabetes
Human herpesviruses							
Burgess et al. (1974)	Australia	1	–	–	EBV	Serology	Fulminant
Chiou et al. (2006)	Taiwan	1	–	21	HHV-6	Serology (IgG), PCR	Fulminant
Wang et al. (2008)	China	2	–	14–23	EBV	Serology (IgG)	T1D
Other							
Banatvala et al. (1985)	Australia	12/24	0	2–12	CMV, Mumps	Serology (IgM)	T1D
Goto et al. (2008)	Japan	1	–	56	Mumps	Serology (IgG)	Fulminant
Sano et al. (2008)	Japan	1	–	64	Influenza B	Rapid antigen test	Fulminant

CVB Coxsackievirus B, *EBV* Epstein–Barr virus, *EV* Enterovirus, *HHV-6* human herpesvirus-6
[a]Age range (unless otherwise specified)

	Case		Control			Odds Ratio
Study or Subgroup	Events	Total	Events	Total	Weight	M-H, Random, 95% CI
Craig 2003	62	206	6	160	62.2%	11.05 [4.64, 26.33]
Kawashima 2004	23	61	2	58	20.8%	16.95 [3.77, 76.14]
Li 2010	12	22	2	30	17.0%	16.80 [3.19, 88.53]
Total (95% CI)		289		248	100.0%	12.97 [6.54, 25.72]
Total events	97		10			

Heterogeneity: $Tau^2 = 0.00$; $Chi^2 = 0.35$, $df = 2$ ($P = 0.84$); $I^2 = 0\%$

Test for overall effect: $Z = 7.33$ ($P < 0.00001$)

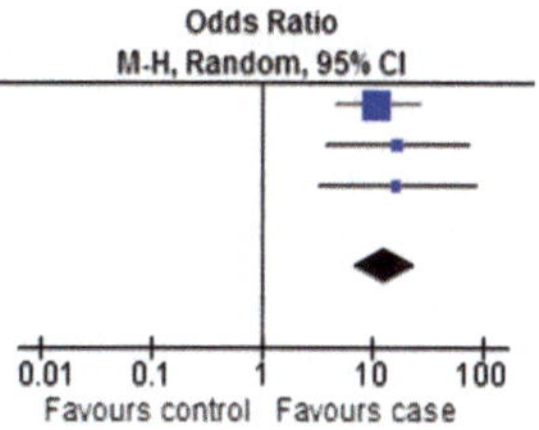

Fig. 21.1 Odds ratios for enterovirus positivity in patients with and without type 1 diabetes from the Asia-Pacific region

insulin responses) were different from those used currently. Six were treated with oral agents and none required insulin therapy (Forrest et al. 1971). In a subsequent follow-up of the cohort in 1991, diabetes persisted in 4 of 5 of the original cases and there was one new case of diabetes (McIntosh and Menser 1992). After 60 years, 7 of the 40 surviving patients had diabetes, of whom two had high levels of glutamic acid decarboxylase (GAD) antibodies, suggesting that they had latent autoimmune diabetes in adulthood. The overall prevalence of diabetes (22%) was higher than the background rate in the Australian population (Forrest et al. 2002). Therefore, congenital rubella appears to be associated with a high prevalence of diabetes, including LADA and type 2 diabetes.

Rotavirus

Rotavirus (RV) is an enteric pathogen that causes gastroenteritis, predominantly in young children. Two case studies reported an association between pancreatitis and acute RV infection (De La Rubia et al. 1996; Nigro 1991). An association between islet autoimmunity and RV infection was demonstrated in a multicentre Australian cohort study of 360 high-risk children (with a parent or sibling with T1D), in which islet autoantibodies and RV IgA or IgG antibodies (RVA or RVG) were measured every 6 months from birth. An increase in RV antibody was associated with the first appearance of tyrosine phosphatase IA-2 autoantibodies (IA-2) in 86% of children, insulin autoantibodies (IAA) in 62%, and GAD in 50%. There was also a significant association between an increase in any islet antibody and an increase in RVA or RVG during the same 6 months period, with a relative risk of ~2 (Honeyman et al. 2000) (Table 21.1). The same group identified T-cell epitope peptides within the intracytoplasmic domain of IA-2 with high sequence homology to the RV VP7 protein, providing molecular evidence that RV may initial β cell autoimmunity by molecular mimicry (Honeyman et al. 1998). Recently, they also reported a significant correlation between proliferative T-cell responses to similar peptides in RV and islet autoantigens, providing further in vitro evidence for molecular mimicry

(Honeyman et al. 2010). In the nonobese diabetic (NOD) mouse, infection with the mouse RV strain RRV exacerbated diabetes once insulitis was established (Graham et al. 2008), while RRV infection in the infant NOD mouse conferred protection from diabetes. However, the association between RV infection and onset of type 1 diabetes in clinical studies is less clear; T-cell responses to RV did not differ between cases of type 1 diabetes ($n=43$), healthy children with multiple islet autoantibodies ($n=36$), or antibody negative controls ($n=104$) (Makela et al. 2006). In our case–control study of 206 children at onset of type 1 diabetes (Craig et al. 2003), we did not find an association between RV infection (using stool EIA) and T1D, with RV detected in no cases of type 1 diabetes and only one control child.

Enteroviruses

Human enteroviruses (HEV), particularly those from HEV group B (Coxsackie B viruses), have been implicated in the pathogenesis of type 1 diabetes on the basis of animal experiments, postmortem studies (Richardson et al. 2009) and clinical studies of individuals with islet autoimmunity and at type 1 diabetes onset, using seroepidemiology and molecular techniques (Yeung et al. 2011). In a multinational study which included children from Australia, Austria, and England, specific IgM responses to CVB 1–5 were higher in 37 of 122 (30%) of the children aged <15 years at type 1 diabetes onset, versus 15 of 204 (6%) of controls ($P=0.001$). Data on the subgroup of Australian children are shown in Table 21.1.

In our Australian incident cohort of 206 young people at onset of T1D, EV RNA was detected in stool or plasma in 30%, compared with 4% of age-matched control children: odds ratio (OR) 11 (95% CI 5–26, $P<0.001$) (Craig et al. 2003). Among children aged ≥5 years, 27% were EV RNA positive versus 1% of controls ($P<0.001$). Cases without high-risk HLA haplotypes (HLA DQB1*02 or *0302) were significantly more likely to be EV RNA positive (OR 2.52, 95% CI 1.02–6.22, $P=0.04$), suggesting that EV infection may be more important at type 1 diabetes onset in those at lower genetic risk. The most common genotype was EV71, which was found in 25% of EV RNA positive cases. These cases were included in a recent systematic review and meta-analysis of observational molecular studies examining the association between EV infection and type 1 diabetes (Yeung et al. 2011). The 24 papers and 2 abstracts identified in the review included 4,448 participants, with 19 of the studies coming from Europe. Meta-analysis showed a significant association between EV infection and type 1 diabetes-related autoimmunity (OR 3.7, 95% CI 2.1–6.8; heterogeneity $\chi^2/\text{df}=1.3$) and clinical type 1 diabetes (9.8, 5.5–17.4; $\chi^2/\text{df}=3.2$) (Yeung et al. 2011). We have updated this review and performed a subgroup analysis of patients from the Asia-Pacific region; three studies involving 537 patients were identified from Australia, (Craig et al. 2003), China (Li et al. 2010), and Japan (Kawashima et al. 2004). The significant association between EV infection and type 1 diabetes remained (OR 13.0, 95% CI 6.5–25.7) (Fig. 21.1).

Molecular Epidemiology of Enteroviruses Associated with Diabetes in the Region

HEV infections in the Asia-Pacific region have a unique epidemiological profile. This was highlighted by an increase in the epidemic activity of EV71, a genetically diverse, rapidly evolving virus (Brown et al. 1999), since 1997 (Mackenzie et al. 2001). While EV71 infection usually manifests as hand, foot, and mouth disease, the recent epidemics have been associated with increased reports of severe sequelae, including meningitis, brainstem encephalitis (Nolan et al. 2003), and acute flaccid paralysis. Outbreaks of EV-71 have occurred in Australia, Japan, the Malay Peninsula, Sarawak, Singapore, and Taiwan since 1997 (Mackenzie et al. 2001). Interestingly, the epidemics of EV71 in Australia (Sanders et al. 2006) occurred in parallel with a significant rise in type 1 diabetes incidence in New South Wales, the most populous state of Australia (Taplin et al. 2005). Indeed, we were the first group to report an association between EV71 infection and onset of type 1 diabetes (Craig et al. 2003). Phylogenetic analysis demonstrated that isolates from Australian children with type 1 diabetes clustered with contemporaneous isolates from South East Asia (Chang et al. 1999). We have subsequently found that EV71 isolates from these cases infect and replicate in rodent beta cell lines. However, there is currently no direct evidence to support a causal relationship between EV71 and type 1 diabetes in humans.

Fulminant Diabetes

This distinct subtype of type 1 diabetes, characterized by rapid onset of β-cell destruction and clinical features of severe hyperglycaemia, ketosis, low HbA1c, and absence of both insulitis and autoimmunity, was first described in 2000 in Japanese adults with clinical type 1 diabetes (Imagawa et al. 2000). Subsequently, cases with a similarly rapid clinical presentation and positive islet autoantibodies were described (Shimada et al. 2002), with approximately 5% being GAD positive (Imagawa et al. 2003). In a nationwide survey in Japan, fulminant diabetes accounted for 20% of type 1 diabetes cases with ketosis at onset (Imagawa et al. 2003). Influenza-like symptoms were typically present, suggesting a viral trigger. Japanese patients with fulminant diabetes have distinct HLA class II susceptibility and protective haplotypes compared with those who have autoimmune type 1 diabetes. Although most reports of fulminant diabetes have been described in Asian populations, including Japanese, Korean (Cho et al. 2007), Chinese (Wang et al. 2008), and Taiwanese (Chiou et al. 2006), these cases were reported in Caucasian French women (Moreau et al. 2008).

There is considerable evidence suggesting that viral infections are implicated in the pathogenesis of fulminant diabetes. As early as 1974, an Australian case report described the onset of fulminant diabetes in association with Epstein–Barr virus

(EBV) infection (Burgess et al. 1974). Fulminant diabetes has been described in association with human herpesvirus 6 (HHV-6) reactivation (Chiou et al. 2006), following influenza B infection (Sano et al. 2008), mumps (Goto et al. 2008) and CVB4 (Akatsuka et al. 2009). Significantly higher titres of EV IgA antibodies have been reported in cases of fulminant diabetes compared with type 1 diabetes or controls (Imagawa et al. 2005), suggesting that fulminant type 1 diabetes is associated with recurrent EV infection. EV capsid protein was detected by immunohistochemistry in the pancreata of patients who died from diabetic ketoacidosis within 2–5 days of onset of fulminant diabetes (Tanaka et al. 2009). Extensive infiltration of CXCR3 receptor-bearing T-cells and macrophages into islets was observed, along with strong expression of interferon gamma (IFN-γ) and CXC chemokine ligand 10 (CXCL10) in all subtypes of islet cells. It was proposed that EV infection initiated co-expression of IFN-γ and the CXCL10 in β-cells, attracting autoreactive T-cells and macrophages to the islets and resulting in β-cell damage via the release of inflammatory cytokines. In an autopsy study of three patients with fulminant diabetes, Toll-like receptor (TLR) 3, a sensor of viral components, was detected in the majority of macrophages infiltrating the islets of all three cases (Shibasaki et al. 2010), and it was concluded that macrophage-dominated insulitis contributed to beta cell destruction in fulminant T1D, rather than T cell autoimmunity. EV RNA was detected by in situ hybridization in β-cell positive islets from one of the three patients. Collectively, these studies provide some evidence that EV infection has an etiological role in fulminant diabetes.

References

Akatsuka H, Yano Y, Gabazza EC, Morser J, Sasaki R, Suzuki T, Fujiwara R, Katsuki A, Takei Y, Sumida Y (2009) A case of fulminant type 1 diabetes with coxsackie B4 virus infection diagnosed by elevated serum levels of neutralizing antibody. Diabetes Res Clin Pract 84:e50–e52

Banatvala JE, Bryant J, Schernthaner G, Borkenstein M, Schober E, Brown D, De Silva LM, Menser MA, Silink M (1985) Coxsackie B, mumps, rubella, and cytomegalovirus specific IgM responses in patients with juvenile-onset insulin-dependent diabetes mellitus in Britain, Austria, and Australia. Lancet 1:1409–1412

Brown BA, Oberste MS, Alexander JP Jr, Kennett ML, Pallansch MA (1999) Molecular epidemiology and evolution of enterovirus 71 strains isolated from 1970 to 1998. J Virol 73: 9969–9975

Burgess JA, Kirkpatrick KL, Menser MA (1974) Fulminant onset of diabetes mellitus during an attack of infectious mononucleosis. Med J Aust 2:706–707

Campbell-Stokes PL, Taylor BJ (2005) Prospective incidence study of diabetes mellitus in New Zealand children aged 0 to 14 years. Diabetologia 48:643–648

Catanzariti L, Faulks K, Moon L, Waters AM, Flack J, Craig ME (2009) Australia's national trends in the incidence of type 1 diabetes in 0–14-year-olds, 2000–2006. Diabet Med 26:596–601

Chang LY, Lin TY, Huang YC, Tsao KC, Shih SR, Kuo ML, Ning HC, Chung PW, Kang CM (1999) Comparison of enterovirus 71 and coxsackie-virus A16 clinical illnesses during the Taiwan enterovirus epidemic, 1998. Pediatr Infect Dis J 18:1092–1096

Chiou CC, Chung WH, Hung SI, Yang LC, Hong HS (2006) Fulminant type 1 diabetes mellitus caused by drug hypersensitivity syndrome with human herpesvirus 6 infection. J Am Acad Dermatol 54(Suppl 2):S14–S17

Cho YM, Kim JT, Ko KS, Koo BK, Yang SW, Park MH, Lee HK, Park KS (2007) Fulminant type 1 diabetes in Korea: high prevalence among patients with adult-onset type 1 diabetes. Diabetologia 50:2276–2279

Chong JW, Craig ME, Cameron FJ, Clarke CF, Rodda CP, Donath SM, Werther GA (2007) Marked increase in type 1 diabetes mellitus incidence in children aged 0–14 yr in Victoria, Australia, from 1999 to 2002. Pediatr Diabetes 8:67–73

Cooper LZ, Green RH, Krugman S, Giles JP, Mirick GS (1965) Neonatal thrombocytopenic purpura and other manifestations of rubella contracted in utero. Am J Dis Child 110:416–427

Craig ME, Howard NJ, Silink M, Rawlinson WD (2003) Reduced frequency of HLA DRB1*03-DQB1*02 in children with type 1 diabetes associated with enterovirus RNA. J Infect Dis 187:1562–1570

De La Rubia L, Herrera MI, Cebrero M, De Jong JC (1996) Acute pancreatitis associated with rotavirus infection. Pancreas 12:98–99

Diamond Project Group (2006) Incidence and trends of childhood type 1 diabetes worldwide 1990–1999. Diabet Med 23:857–866

Forrest JM, Menser MA, Harley JD (1969) Diabetes mellitus and congenital rubella. Pediatrics 44:445–447

Forrest JM, Menser MA, Burgess JA (1971) High frequency of diabetes mellitus in young adults with congenital rubella. Lancet 2:332–334

Forrest JM, Turnbull FM, Sholler GF, Hawker RE, Martin FJ, Doran TT, Burgess MA (2002) Gregg's congenital rubella patients 60 years later. Med J Aust 177:664–667

Fourlanos S, Harrison LC, Colman PG (2008) The accelerator hypothesis and increasing incidence of type 1 diabetes. Curr Opin Endocrinol Diabetes Obes 15:321–325

Goto A, Takahashi Y, Kishimoto M, Nakajima Y, Nakanishi K, Kajio H, Noda M (2008) A case of fulminant type 1 diabetes associated with significant elevation of mumps titers. Endocr J 55:561–564

Graham KL, Sanders N, Tan Y, Allison J, Kay TW, Coulson BS (2008) Rotavirus infection accelerates type 1 diabetes in mice with established insulitis. J Virol 82:6139–6149

Hay DR (1949) The relation of maternal rubella to congenital deafness and other abnormalities in New Zealand. New Zeal Med J 48:604–608

Honeyman MC, Stone NL, Harrison LC (1998) T-cell epitopes in type 1 diabetes autoantigen tyrosine phosphatase IA-2: potential for mimicry with rotavirus and other environmental agents. Mol Med 4:231–239

Honeyman MC, Coulson BS, Stone NL, Gellert SA, Goldwater P, Steele CE, Couper JJ, Tait BD, Colman PG, Harrison LC (2000) Association between rotavirus infection and pancreatic islet autoimmunity in children at risk of developing type 1 diabetes. Diabetes 49:1319–1324

Honeyman MC, Stone NL, Falk BA, Nepom G, Harrison LC (2010) Evidence for molecular mimicry between human T cell epitopes in rotavirus and pancreatic islet autoantigens. J Immunol 184:2204–2210

Ikegami H, Fujisawa T, Kawabata Y, Noso S, Ogihara T (2006) Genetics of type 1 diabetes: similarities and differences between Asian and Caucasian populations. Ann N Y Acad Sci 1079:51–59

Imagawa A, Hanafusa T, Miyagawa J, Matsuzawa Y (2000) A novel subtype of type 1 diabetes mellitus characterized by a rapid onset and an absence of diabetes-related antibodies. Osaka IDDM Study Group. N Engl J Med 342:301–307

Imagawa A, Hanafusa T, Uchigata Y, Kanatsuka A, Kawasaki E, Kobayashi T, Shimada A, Shimizu I, Toyoda T, Maruyama T, Makino H (2003) Fulminant type 1 diabetes: a nationwide survey in Japan. Diabetes Care 26:2345–2352

Imagawa A, Hanafusa T, Makino H, Miyagawa JI, Juto P (2005) High titres of IgA antibodies to enterovirus in fulminant type-1 diabetes. Diabetologia 48:290–293

Kawasaki E, Matsuura N, Eguchi K (2006) Type 1 diabetes in Japan. Diabetologia 49:828–836

Kawashima H, Ihara T, Ioi H, Oana S, Sato S, Kato N, Takami T, Kashiwagi Y, Takekuma K, Hoshika A, Mori T (2004) Enterovirus-related type 1 diabetes mellitus and antibodies to glutamic acid decarboxylase in Japan. J Infect 49:147–151

Li XH, Li TL, Yang Z, Liu ZY, Wei YD, Jin SX, Hong C, Qin RL, Li YQ, Dorman JS, Laporte RE, Wang KA (2000) A nine-year prospective study on the incidence of childhood type 1 diabetes mellitus in China. Biomed Environ Sci 13:263–270

Li Q, Xing H, Zhou Y, Qiu LL, Zhang ZW, Liao L (2010) Association of coxsackie virus infection and T lymphocyte subset changes with type 1 diabetes. Nan Fang Yi Ke Da Xue Xue Bao 30:2699–2701

Mackenzie JS, Chua KB, Daniels PW, Eaton BT, Field HE, Hall RA, Halpin K, Johansen CA, Kirkland PD, Lam SK, McMinn P, Nisbet DJ, Paru R, Pyke AT, Ritchie SA, Siba P, Smith DW, Smith GA, van den Hurk AF, Wang LF, Williams DT (2001) Emerging viral diseases of Southeast Asia and the Western Pacific. Emerg Infect Dis 7(Suppl 3):497–504

Makela M, Oling V, Marttila J, Waris M, Knip M, Simell O, Ilonen J (2006) Rotavirus-specific T cell responses and cytokine mRNA expression in children with diabetes-associated autoantibodies and type 1 diabetes. Clin Exp Immunol 145:261–270

McIntosh ED, Menser MA (1992) A fifty-year follow-up of congenital rubella. Lancet 340:414–415

Menser MA, Dods L, Harley JD (1967) A twenty-five-year follow-up of congenital rubella. Lancet 2:1347–1350

Miller LJ, Willis JA, Pearce J, Barnett R, Darlow BA, Scott RS (2011) Urban-rural variation in childhood type 1 diabetes incidence in Canterbury, New Zealand, 1980–2004. Health Place 17:248–256

Moreau C, Drui D, Arnault-Ouary G, Charbonnel B, Chaillous L, Cariou B (2008) Fulminant type 1 diabetes in Caucasians: a report of three cases. Diabetes Metab 34:529–532

Nigro G (1991) Pancreatitis with hypoglycemia-associated convulsions following rotavirus gastroenteritis. J Pediatr Gastroenterol Nutr 12:280–282

Nolan MA, Craig ME, Lahra MM, Rawlinson WD, Prager PC, Williams GD, Bye AM, Andrews PI (2003) Survival after pulmonary edema due to enterovirus 71 encephalitis. Neurology 60:1651–1656

Richardson SJ, Willcox A, Bone AJ, Foulis AK, Morgan NG (2009) The prevalence of enteroviral capsid protein vp1 immunostaining in pancreatic islets in human type 1 diabetes. Diabetologia 52:1143–1151

Rorke LB, Spiro AJ (1967) Cerebral lesions in congenital rubella syndrome. J Pediatr 70:243–255

Sanders SA, Herrero LJ, McPhie K, Chow SS, Craig ME, Dwyer DE, Rawlinson W, McMinn PC (2006) Molecular epidemiology of enterovirus 71 over two decades in an Australian urban community. Arch Virol 151:1003–1013

Sano H, Terasaki J, Tsutsumi C, Imagawa A, Hanafusa T (2008) A case of fulminant type 1 diabetes mellitus after influenza B infection. Diabetes Res Clin Pract 79:8–9

Shibasaki S, Imagawa A, Tauriainen S, Iino M, Oikarinen M, Abiru H, Tamaki K, Seino H, Nishi K, Takase I, Okada Y, Uno S, Murase-Mishiba Y, Terasaki J, Makino H, Shimomura I, Hyoty H, Hanafusa T (2010) Expression of toll-like receptors in the pancreas of recent-onset fulminant type 1 diabetes. Endocr J 57:211–219

Shimada A, Oikawa Y, Shigihara T, Senda T, Kodama K (2002) A case of fulminant type 1 diabetes with strong evidence of autoimmunity. Diabetes Care 25:1482–1483

Solomon T, Lewthwaite P, Perera D, Cardosa MJ, McMinn P, Ooi MH (2010) Virology, epidemiology, pathogenesis, and control of enterovirus 71. Lancet Infect Dis 10:778–790

Tanaka S, Nishida Y, Aida K, Maruyama T, Shimada A, Suzuki M, Shimura H, Takizawa S, Takahashi M, Akiyama D, Arai-Yamashita S, Furuya F, Kawaguchi A, Kaneshige M, Katoh R, Endo T, Kobayashi T (2009) Enterovirus infection, CXC chemokine ligand 10 (CXCL10), and CXCR3 circuit: a mechanism of accelerated beta-cell failure in fulminant type 1 diabetes. Diabetes 58:2285–2291

Taplin CE, Craig ME, Lloyd M, Taylor C, Crock P, Silink M, Howard NJ (2005) The rising incidence of childhood type 1 diabetes in New South Wales, 1990–2002. Med J Aust 183:243–246

Unachak K, Tuchinda C (2001) Incidence of type 1 diabetes in children under 15 years in northern Thailand, from 1991 to 1997. J Med Assoc Thai 84:923–928

Wang T, Xiao XH, Li WH, Yuan T, Sun XF, Wang H (2008) Fulminant type 1 diabetes: report of two cases. Chinese Med J-Peking 121:181–182

Yang Z, Long X, Shen J, Liu D, Dorman JS, LaPorte RE, Chang YF (2005) Epidemics of type 1 diabetes in China. Pediatr Diabetes 6:122–128

Yeung WC, Rawlinson WD, Craig ME (2011) Enterovirus infection and type 1 diabetes mellitus: systematic review and meta-analysis of observational molecular studies. BMJ 342:d35

Zhang H, Xia W, Yu Q, Wang B, Chen S, Wang Z, Love EJ (2008) Increasing incidence of type 1 diabetes in children aged 0–14 years in Harbin, China (1990–2000). Prim Care Diabetes 2:121–126

Chapter 22
Fulminant Type 1 Diabetes in Japan

Akihisa Imagawa and Toshiaki Hanafusa

Abstract Fulminant type 1 diabetes is a novel subtype within type 1 diabetes characterized by a remarkably abrupt onset, absence of islet-related autoantibodies, and almost no C-peptide secretion even at the onset of the disease. Several lines of evidence suggest that viral infection contributes to the development of fulminant type 1 diabetes: (1) preceding influenza-like symptoms, (2) elevated IgA level to the enterovirus common antigen, (3) association of *HLA-B*4002*, (4) several patients with an elevation of antiviral antibodies, (5) TLR expression in the pancreas of patients, and (6) detection of enterovirus or cytomegalovirus in the pancreas of patients.

Introduction

Fulminant type 1 diabetes was first reported in 2000 as a distinct clinical entity within type 1 diabetes (Imagawa et al. 2000a, b). The clinical characteristics of this subtype have been defined as (1) remarkably abrupt onset; (2) very short duration (usually less than 1 week) of hyperglycemic symptoms, e.g., polyuria, thirst; (3) acidosis at the time of diagnosis; (4) absence of islet-related autoantibodies, such as islet-cell antibodies (ICA), anti-glutamic acid decarboxylase antibodies (GAD antibodies), insulin autoantibodies (IAA), and anti-insulinoma-associated antigen 2 antibodies (IA-2Ab); (5) almost no C-peptide secretion; (6) elevated serum pancreatic enzyme levels. These characteristics are found in approximately 20% of

A. Imagawa, M.D., Ph.D. (✉)
Department of Metabolic Medicine, Graduate School of Medicine,
Osaka University, Suita 565-0871, Japan
e-mail: aimagawa@endmet.med.osaka-u.ac.jp

T. Hanafusa, M.D., Ph.D.
Department of Internal Medicine (I), Osaka Medical College, Takatsuki 569-8686, Japan
e-mail: hanafusa@poh.osaka-med.ac.jp

K.W. Taylor, H. Hyoty, A. Toniolo, A.J. Zuckerman (eds.), *Diabetes and Viruses*,
DOI 10.1007/978-1-4614-4051-2_22, © Springer Science+Business Media New York 2013

Table 22.1 Criteria for diagnosis of fulminant type 1 diabetes mellitus

Fulminant type 1 diabetes mellitus is confirmed when all the following three features are present:
1. Occurrence of diabetic ketosis or diabetic ketoacidosis soon (around 7 days) after the onset of hyperglycemic symptoms (elevation of urine and/or serum ketone bodies at first visit)
2. Plasma glucose level ≥288 mg/dl (16.0 mmol) and glycated hemoglobin level <8.5% (JDS value) at first visit
3. Urinary C-peptide excretion <10 μg/day or fasting serum C-peptide level <0.3 ng/ml (0.10 nmol/l) and <0.5 ng/ml (<0.17 nmol/l) after intravenous glucagon (or after meal) load at onset
Related findings
(a) Islet-related autoantibodies such as antibodies to GAD, IA-2, and insulin are undetectable in general
(b) Duration of the disease before the start of insulin treatment can be 1–2 weeks
(c) Elevation of serum pancreatic enzyme levels (amylase, lipase, or elastase-1) is observed in 98% of the patients
(d) Influenza-like symptoms (e.g., fever, upper respiratory symptoms) or gastrointestinal symptoms (e.g., upper abdominal pain, nausea, and/or vomiting) precede the disease onset in 70% of patients
(e) The disease could occur during pregnancy or just after delivery

acute-onset type 1 diabetes in Japan. A nationwide survey conducted by the Japan Diabetes Society Committee on Fulminant Type 1 Diabetes Mellitus Research recorded 161 patients since 2000 across Japan. This study confirmed the clinical characteristics of fulminant type 1 diabetes listed in the first report and added several new ones such as frequent influenza-like symptoms just before disease onset, association with pregnancy, transient elevation of transaminase, and early progression of diabetic microvascular complications (Imagawa et al. 2003; Hanafusa et al. 2005; Shimizu et al. 2006; Hanafusa and Imagawa 2007; Murase et al. 2007; Takaike et al. 2008). On the basis of these findings, diagnostic criteria were established in 2004; these are listed in Table 22.1 (Hanafusa and Imagawa 2007; Hanafusa et al. 2005). According to these criteria, many cases have been reported in other countries (Chiou et al. 2006; Cho et al. 2007; Moreau et al. 2008; Kim et al. 2009; Zheng et al. 2009, 2011); these cases were reported mainly in Korean and Chinese populations, but they also included other Asians and Caucasians.

Sekine et al. (2001) reported a typical case of fulminant type 1 diabetes in which precise laboratory data related to onset were available. Although blood glucose and C-peptide levels were within normal limits just before the onset of diabetes, the patient's blood glucose level suddenly increased to more than 1,000 mg/dl, and endogenous insulin-secreting capacity was abolished on the day of onset. Hemoglobin A1c level was 5.9% (JDS value, normal range: –5.8%), suggesting the extremely rapid progression of beta cell destruction in fulminant type 1 diabetes. ICA and GAD antibodies were negative, and serum lipase and elastase-1 were elevated in this patient.

Influenza-like symptoms observed frequently around the time of onset prompted us to suspect that fulminant type 1 diabetes could be "virus-induced diabetes." To clarify this issue, we present several lines of evidence that suggest that viral infection contributes to the development of fulminant type 1 diabetes.

Contribution of Viral Infection to the Etiology of Fulminant Type 1 Diabetes

Preceding Influenza-Like Symptoms

Influenza-like symptoms suggest that a virus contributes to the development of the disease. In a nationwide survey, preceding influenza-like symptoms were observed more frequently in persons who presented with fulminant type 1 diabetes (71.7%) than in those who presented with autoimmune type 1A diabetes (26.9%), which was thought previously to make up almost all type 1 diabetes; fever was observed in 60% of patients with fulminant type 1 diabetes, a sore throat in 25.2%, and coughing in 12.0%. These symptoms were similar regardless of the age and gender of those with fulminant type 1 diabetes (Imagawa et al. 2008).

Elevation of Anti-Enterovirus IgA

Anti-enterovirus IgA was elevated in patients with recent-onset fulminant type 1 diabetes compared to those with recent-onset type 1A diabetes and healthy control subjects, as determined by ELISA with a broadly reacting anti-enterovirus antibody (Imagawa et al. 2005a). IgM antibodies specific for enterovirus were not detected in any subject. These results suggest that patients presenting with fulminant type 1 diabetes have had recurrent enterovirus infections, indicating that they are more susceptible to this type of infection than are type 1A diabetic individuals and healthy controls.

Resemblance of EMC Virus-Induced Diabetes in Mice to Fulminant Type 1 Diabetes in Humans

A sudden increase in blood glucose levels has been reported after the injection of some variants of encephalomyocarditis (EMC) virus into several strains of mice, such as DBA/2 or SJL mice (Jun and Yoon 2001). This finding means that beta cell destruction and overt diabetes can be triggered by viral infection. In these mice, the physiological and pathological characteristics are very similar to those in humans with fulminant type 1 diabetes (Shimada and Maruyama 2004; Sano et al. 2011). First, animals suffer from diabetes from 4 or 5 days after injection, and the duration from the preceding symptoms to overt diabetes was 4 days on average in human fulminant type 1 diabetes. Second, in addition to very rapid onset, beta cells are destroyed almost completely in both cases, even at the onset of overt diabetes. Third, high serum amylase levels are observed in both fulminant type 1 diabetes and EMC virus-induced diabetes. Fourth, cellular infiltration of the islets and the exocrine pancreas was observed in both cases. Such inflammation is present around the time of onset but not several weeks after onset in both fulminant type 1 diabetes and EMC virus-induced diabetes. Fifth, autoantibodies were not detectable or detectable at low levels in fulminant type 1 diabetes and EMC virus-induced diabetes. Lastly, individuals with susceptible and resistant genotypes have been reported for fulminant type 1 diabetes (as shown below), and mouse strains that are susceptible or resistant to EMC virus-induced diabetes have also been reported. DBA/2 and SJL mice are susceptible to developing this type of diabetes, whereas C57Bl/6 and CBA mice are resistant.

Association with Class I HLA

In a recent nationwide study, *B*4002*, a class I HLA subtype, was reported to be frequent in individuals with fulminant type 1 diabetes (Kawabata et al. 2009). *B*4002* was observed in 17.4% of those with fulminant type 1 diabetes but just 9.3% of those with acute-onset type 1A diabetes, 6.3% of slowly progressive type 1A diabetes, and 6.7% of healthy controls. This finding suggests that carrying this HLA subtype renders an individual susceptible to developing fulminant type 1 diabetes but not acute-onset type 1A diabetes. Viral antigens are presented to immune cells in combination with class I HLA. Therefore, an association of disease with class I HLA subtype is compatible with the involvement of viral infection in fulminant type 1 diabetes. In addition, two class II HLA genotypes, *DRB1*0405-DQB1*0401* and *DRB1*0901-DQB1*0303*, were found more frequently in individuals with fulminant type 1 diabetes. Both *DRB1*1501-DQB1*0602* and *DRB1*1502-DQB1*0601* were less frequent in individuals with type 1A diabetes, whereas *DRB1*1502-DQB1*0601* but not *DRB1*1501-DQB1*0602* was less frequent in fulminant type 1 diabetes than in control subjects (Imagawa et al. 2005b; Tsutsumi et al. 2009). *CTLA-4 CT60* was also associated with fulminant type 1 diabetes, although the study was performed in a small number of patients (Kawasaki et al. 2008).

Case Reports with Elevation of Antiviral Antibodies

Several cases have been reported to be accompanied by an elevation in the titer of serum antiviral antibody or by reactivation of a latent virus close to the onset of fulminant type 1 diabetes (Sano et al. 2008; Goto et al. 2008; Akatsuka et al. 2009; Hwang et al. 2010). This includes the case shown in the introduction as a typical case of fulminant type 1 diabetes (Sekine et al. 2001). In that case, DNA from human herpes virus 6 (HHV6) was detected in the serum at the onset of overt diabetes. In other cases, the elevation of various antiviral antibodies, such as those specific for herpes simplex viruses, coxsackievirus, Epstein–Barr virus (EBV), and the mumps virus and influenza B virus, has been reported at the onset of fulminant type 1 diabetes.

A nationwide survey has also been performed with regard to antiviral antibody around the time of onset of fulminant type 1 diabetes (Hanafusa et al. 2008). In that study, antiviral antibodies in sera from 55 patients close to the onset of fulminant type 1 diabetes were tested. The analysis included a search for 23 antiviral antibodies to parainfluenza virus types 1–3, coxsackie virus types A2–10, A16 and B1–6, rotavirus, cytomegalovirus, EBV, and HHV6 and HHV7. All 55 participating patients met the criteria for fulminant type 1 diabetes and had already registered with, and provided serum samples to the Japan Diabetes Society Committee on Fulminant Type 1 Diabetes Mellitus Research. As a result, elevation of antiviral antibody was observed for 11 antibodies in 7 of 39 patients for whom paired sera, which were taken at least 2 weeks interval, were available (e.g., antibodies specific for coxsackie virus types A4, A5, A6, and B1 were found in one patient, and antibodies specific for rotavirus were found in two patients). In addition, IgM specific for cytomegalovirus was found in one patient and IgM specific for EBV and HHV6 was found in another patient. Furthermore, IgG specific for HHV6 and HHV7 was increased in paired sera of one patient. An elevation of anti-cytomegalovirus IgM levels was observed in 2 of 16 patients for whom paired sera were not available. It is interesting to note that some individuals had elevated serum concentrations of antibodies specific for more than one type of virus. These results and various case reports suggest that it is not the virus itself but the reaction to viral infection that is essential for beta cell damage and the development of fulminant type 1 diabetes (Table 22.2).

Signature of the Antiviral Immune Response in the Pancreas

It is reasonable to consider that the most important evidence of viral infection would be found in the pancreas, where beta cells are destroyed in patients with diabetes. However, it is difficult to detect dead beta cells because they are phagocytosed rapidly. In such a situation, one of the best methods for detecting in situ beta cell death caused by a virus is to identify the signature of viral infection.

Three of the patients examined died soon after the onset of fulminant type 1 diabetes (Shibasaki et al. 2010). Two patients died 3 days after the onset of diabetic symptoms, and one patient died 5 days after the onset of diabetic symptoms. All of the patients met the criteria for fulminant type 1 diabetes (Hanafusa and Imagawa 2007). As a result, very few insulin-positive cells and only several glucagon-positive cells were observed in the pancreas of all three patients. The mean beta cell area was only 0.00256% in these three patients, whereas that of normal controls was 1.745%. In other words, most islets in the three patients with fulminant type 1 diabetes are so-called glucagon islets. Glucagon islets are observed commonly in the pancreas of patients with long-standing type 1 diabetes. The duration of diabetes was very short in our patients, indicating that the beta cells had been destroyed over a short period of time in these cases of fulminant type 1 diabetes. Cellular infiltration to both the islets and the exocrine pancreas was also common in the pancreas of all three patients. In the serial sections, the infiltrating cells were CD3-positive T cells and CD68-positive macrophages but not CD56-positive

Table 22.2 Viral antibody at disease onset in fulminant type 1 diabetes (adapted from Hanafusa et al. 2008 with minor modifications)

Age	sex	Accompanied symptoms	Coxsackie virus	Rota virus	CMV	EBV	HHV6	HHV7
77	F	Drowsiness, nausea, hypersensitivity syndrome	–	–	1.20 (IgM)	–	10 (IgM)	–
23	M	Drowsiness, nausea, abdominal pain	–	x4/x16	–	–	–	–
45	F	Fever, nausea, abdominal pain	–	–	–	–	–	x10/x80(IgG)
57	M	Fever, nausea, abdominal pain	–	<4/x8	–	3.2(IgM)	x40/x160 (IgG)	–
38	M	Abdominal pain, fever, headache	x16/x64 (A4) x16/x64 (B1)	–	–	–	–	–
34	M	ND	x8/x32(A6)	–	–	–	–	–
39	M	Nausea, headache	x4/x16(A5)	–	–	–	–	–
61	M	Fever, nausea, abdominal pain, diarrhea	ND	ND	1.03 (IgM)	–	–	–
25	F	Fever, nausea, abdominal pain	ND	ND	0.86 (IgM)	–	–	–

ND denotes not determined

natural killer (NK) cells. Although both T cells and macrophages had infiltrated the islets in very large numbers, macrophages dominated islet infiltration and were detected in 92.6% of the islets. To detect the signature of viral infection, the expression of Toll-like receptor (TLR)-3, TLR7, and TLR9 was investigated in the pancreas specimens. Once viruses infect the human body, the innate immune system is activated. In this system, several receptors, which are called TLRs, act as pattern recognition receptors (PRRs). For example, TLR3 recognizes not specific viruses but double-stranded RNA, which is produced in the process of viral replication. Similarly, TLR7 recognizes single-stranded viral RNA, and TLR9 recognizes CpG DNA. Therefore, excessive expression of TLRs is thought to be a signature of viral infection. Expression of TLR3 was identified in islet lesions of the autopsied pancreas in all three patients. Using the double-staining method, TLR3-positive cells were also found to be positive for CD3 or CD68, indicating that TLR3 is expressed by T cells and macrophages in islet infiltrates. TLR3 was detected in 84.7±7.0% of T cells and 62.7±32.3% of macrophages (mean±SD) in all three patients. TLR3 expression was higher in the pancreas of patients than in the spleen cells of normal control subjects. In the same serial sections, TLR7-positive cells and TLR9-positive cells were also detected in all three patients with fulminant type 1 diabetes. Together, these factors represent evidence (a signature) of viral infection of the pancreas in fulminant type 1 diabetes.

Direct Detection of Virus in the Pancreas of Patients

The most direct evidence of viral infection is detection of the virus itself in the pancreas of the patients with fulminant type 1 diabetes. In one of the three patients mentioned above, enterovirus RNA was identified in the islets by using in situ hybridization (Shibasaki et al. 2010; Fig. 22.1).

In another case of fulminant type 1 diabetes for which autopsy samples were available, cytomegalovirus infection was reported (Sakai M et al., unpublished observation). This patient died 31 days after the onset of fulminant type 1 diabetes. At onset, cytomegalovirus antigen was detected in his sera. In his autopsy specimens, inclusion bodies were detected by hematoxylin and eosin staining. This finding was confirmed by immunohistochemistry using anti-cytomegalovirus antibody. These cases suggest that both enterovirus and cytomegalovirus infections may induce beta cell death, thus causing fulminant type 1 diabetes. In other words, together with the elevation of various serum antiviral antibodies, infection with a variety of viruses has been shown to contribute to the development of fulminant type 1 diabetes.

Recently, another research group has reported islet-cell expression of RIG-I and melanoma differentiation-associated gene-5, which are also PRRs present in the cytoplasm, and has also confirmed the expression of TLR3 by mononuclear cells that have infiltrated islets. In addition, this group reported the expression of interferon-alpha and interferon-beta by islet cells (Aida et al. 2011). Those findings further support our concept of "virus-induced diabetes" in the etiology of fulminant type 1 diabetes.

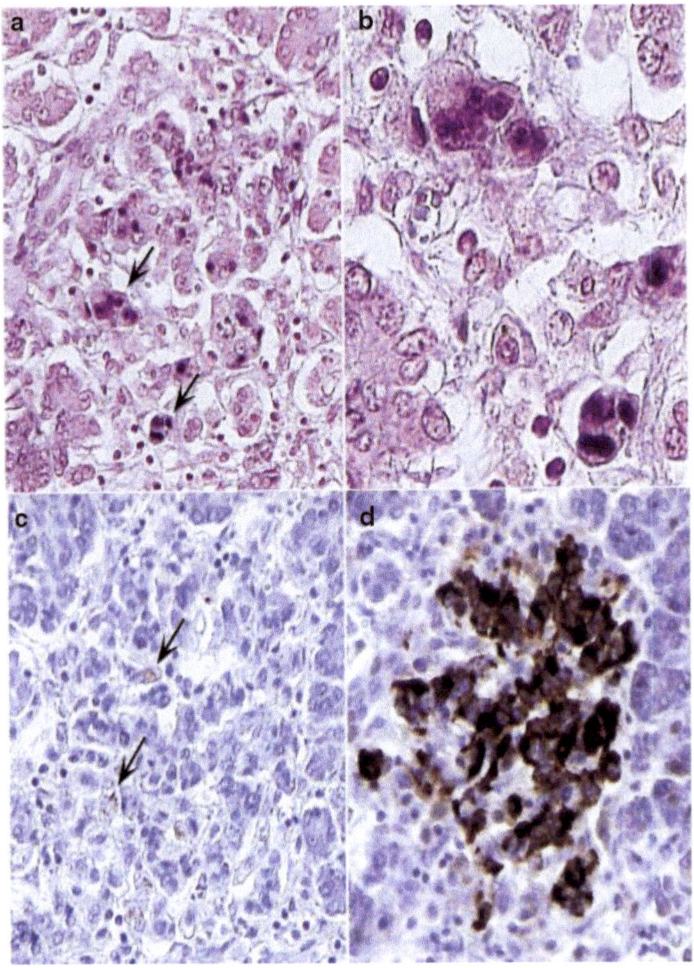

Fig. 22.1 Enterovirus RNA in the pancreas. Enterovirus RNA; (**a**) low magnification; (**b**) high magnification; (**c**) insulin; (**d**) glucagon (adapted from Shibasaki et al. 2010)

A Possible Mechanism of Beta Cell Death

A possible mechanism of beta cell death in fulminant type 1 diabetes is shown in Fig. 22.2. The virus itself might destroy some beta cells, but we suppose that this is unlikely to be the main pathway of beta cell death, since several types of virus contribute to the development of fulminant type 1 diabetes, not one specific virus. Our hypothesis is that viral infection in individuals with genetic factors such as particular alleles of HLA molecules or CTLA-4 will promote the antiviral immune reaction that causes beta cell death. This reaction culminates in macrophage and T-cell infiltration of the islets.

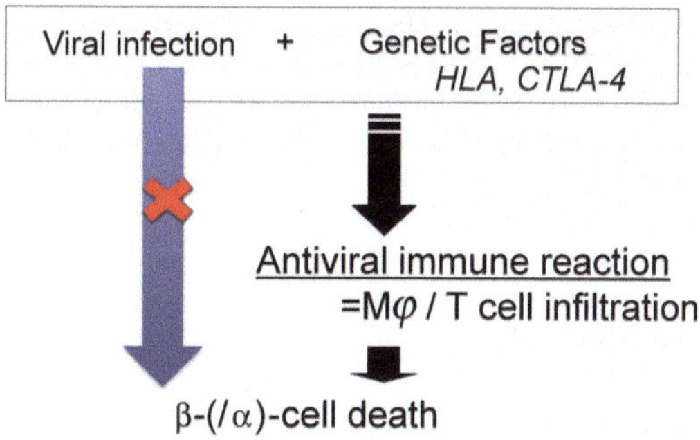

Fig. 22.2 A possible mechanism of beta cell death in fulminant type 1 diabetes

Immunoregulation Mechanisms as the Next Horizon

It should be noted that there are similarities in the etiology of fulminant type 1 diabetes and classical type 1A diabetes. Viral infection has also been discussed as a trigger of classical type 1A diabetes (Jun and Yoon 2001; Hyöty and Taylor 2002). Enteroviruses have been the most likely candidates and have been investigated extensively. In other words, the difference in the mechanism of beta cell destruction in fulminant type 1 and classical type 1A diabetes might be due to quantitative rather than qualitative factors. Therefore, the mechanism of the acceleration of the immune reaction to the beta cells should be investigated in the context of fulminant type 1 diabetes. To answer this question, a study of the factors that affect this immune reaction to beta cells is in progress. First, emphasis was placed on cytotoxic T-lymphocyte antigen 4 (CTLA-4). CTLA-4 is an inhibitory immunoregulatory molecule that plays an important role in adjusting the threshold for T-cell activation and preventing autoimmunity (Haseda et al. 2011). Peripheral $CD4^+$ T cells revealed a significant reduction in CTLA-4 expression in patients with fulminant type 1 diabetes compared to individuals with type 1A diabetes, type 2 diabetes, or normal controls. A significant negative correlation between the expression of CTLA-4 and cellular proliferation was observed. Lower expression of CTLA-4 correlated with higher cell proliferation in individuals with fulminant type 1 diabetes. These findings suggest that reduced expression of CTLA-4 by helper T cells might promote increased proliferation of effector ($CD4^+$ $CD25^-$) T cells and an accelerated immune reaction that leads to beta cell loss and the development of fulminant type 1 diabetes. In this way, the "immunological brake" may not work well in these patients, thus leading to an accelerated immune reaction and fulminant type 1 diabetes.

Closing Remarks

The findings presented above provide strong evidence that viral infection contributes to the development of fulminant type 1 diabetes. Further studies will be required to identify the molecular mechanism underlying beta cell death in this type of diabetes.

Acknowledgments This study was supported in part by a Grant-in Aid from the Japanese Society for the Promotion of Science (Kakenhi), a grant from a Health and Labour Sciences Research Grant on Research on Intractable Diseases from the Japanese Ministry of Health, Labour and Welfare, a grant from the Japan Medical Association and continuous support of the Japan Diabetes Society.

References

Aida K, Nishida Y, Tanaka S, Maruyama T, Shimada A, Awata T, Suzuki M, Shimura H, Takizawa S, Ichijo M, Akiyama D, Furuya F, Kawaguchi A, Kaneshige M, Itakura J, Fujii H, Endo T, Kobayashi T (2011) RIG-I- and MDA5-initiated innate immunity linked with adaptive immunity accelerates beta-cell death in fulminant type 1 diabetes. Diabetes 60:884–889

Akatsuka H, Yano Y, Gabazza EC, Morser J, Sasaki R, Suzuki T, Fujiwara R, Katsuki A, Takei Y, Sumida Y (2009) A case of fulminant type 1 diabetes with coxsackie B4 virus infection diagnosed by elevated serum levels of neutralizing antibody. Diabetes Res Clin Pract 84:e50–e52

Chiou CC, Chung WH, Hung SI, Yang LC, Hong HS (2006) Fulminant type 1 diabetes mellitus caused by drug hypersensitivity syndrome with human herpesvirus 6 infection. J Am Acad Dermatol 54(Suppl 2):S14–S17

Cho YM, Kim JT, Ko KS, Koo BK, Yang SW, Park MH, Lee HK, Park KS (2007) Fulminant type 1 diabetes in Korea: high prevalence among patients with adult-onset type 1 diabetes. Diabetologia 50:2276–2279

Goto A, Takahashi Y, Kishimoto M, Nakajima Y, Nakanishi K, Kajio H, Noda M (2008) A case of fulminant type 1 diabetes associated with significant elevation of mumps titers. Endocr J 55:561–564

Hanafusa T, Imagawa A (2007) Fulminant type 1 diabetes: a novel clinical entity requiring special attention by all medical practitioners. Nat Clin Pract Endocrinol Metab 3:36–45

Hanafusa T, Imagawa A, Iwahashi H, Uchigata Y, Kanatsuka A, Kawasaki E, Kobayashi T, Shimada A, Shimizu I, Toyoda T, Maruyama T, Makino H (2005) Report of Japan Diabetes Society Committee on fulminant type 1 diabetes mellitus research: epidemiological and clinical analysis and proposal of diagnostic criteria. J Japan Diab Soc 48(Suppl 1):A1–A13 (in Japanese)

Hanafusa T, Imagawa A, Iwahashi H, Uchigata Y, Kanatsuka A, Kawasaki E, Kobayashi T, Shimada A, Shimizu I, Toyoda T, Maruyama T, Makino H (2008) Report of the Japan Diabetes Society's Committee on research on fulminant type 1 diabetes mellitus: analysis of antiviral antibodies at disease onset. J Japan Diab Soc 51:531–536 (in Japanese)

Haseda F, Imagawa A, Murase-Mishiba Y, Sano H, Hirano-Kuwata S, Ueda H, Terasaki J, Hanafusa T (2011) Low CTLA-4 expression in CD4(+) helper T-cells in patients with fulminant type 1 diabetes. Immunol Lett 139:80–86

Hwang YC, Jeong IK, Chon S, Oh S, Ahn KJ, Chung HY, Woo JT, Kim SW, Kim JW, Kim YS (2010) Fulminant type 1 diabetes mellitus associated with acute hepatitis A. Diabet Med 27:366–367

Hyöty H, Taylor KW (2002) The role of viruses in human diabetes. Diabetologia 45:1353–1361

Imagawa A, Hanafusa T, Miyagawa J, Matsuzawa Y, Osaka IDDM Study Group (2000a) A novel subtype of type 1 diabetes mellitus characterized by a rapid onset and an absence of diabetes-related antibodies. N Engl J Med 342:301–307

Imagawa A, Hanafusa T, Miyagawa J, Matsuzawa Y (2000b) A proposal of three distinct subtypes of type 1 diabetes mellitus based on clinical and pathological evidence. Ann Med 32:539–543

Imagawa A, Hanafusa T, Uchigata Y, Kanatsuka A, Kawasaki E, Kobayashi T, Shimada A, Shimizu I, Toyoda T, Maruyama T, Makino H (2003) Fulminant type 1 diabetes: a nationwide survey in Japan. Diabetes Care 26:2345–2352

Imagawa A, Hanafusa T, Makino H, Miyagawa JI, Juto P (2005a) High titres of IgA antibodies to enterovirus in fulminant type-1 diabetes. Diabetologia 48:290–293

Imagawa A, Hanafusa T, Uchigata Y, Kanatsuka A, Kawasaki E, Kobayashi T, Shimada A, Shimizu I, Maruyama T, Makino H (2005b) Different contribution of class II HLA in fulminant and typical autoimmune type 1 diabetes mellitus. Diabetologia 48:294–300 [Erratum in: Diabetologia. 2008;51:524–526]

Imagawa A, Hanafusa T, Iwahashi H, Uchigata Y, Kanatsuka A, Kawasaki E, Kobayashi T, Shimada A, Shimizu I, Maruyama T, Makino H (2008) Uniformity in clinical and HLA-DR status regardless of age and gender within fulminant type 1 diabetes. Diabetes Res Clin Pract 82:233–237

Jun HS, Yoon JW (2001) The role of viruses in type I diabetes: two distinct cellular and molecular pathogenic mechanisms of virus-induced diabetes in animals. Diabetologia 44:271–285

Kawabata Y, Ikegami H, Awata T, Imagawa A, Maruyama T, Kawasaki E, Tanaka S, Shimada A, Osawa H, Kobayashi T, Hanafusa T, Tokunaga K, Makino H, Committee on Type 1 Diabetes, Japan Diabetes Society (2009) Differential association of HLA with three subtypes of type 1 diabetes: fulminant, slowly progressive and acute-onset. Diabetologia 52:2513–2521

Kawasaki E, Imagawa A, Makino H, Uga M, Abiru N, Hanafusa T, Uchigata Y, Eguchi K (2008) Differences in the contribution of the CTLA4 gene to susceptibility to fulminant and type 1A diabetes in Japanese patients. Diabetes Care 31:1608–1610

Kim NH, Kim HY, Seo JA, Kim NH, Choi KM, Baik SH, Choi DS, Kim SG (2009) A pooled analysis of 29 patients with fulminant type 1 diabetes in Korea: a comparison with a nationwide survey in Japan. Diabetes Res Clin Pract 86:e43–e45

Moreau C, Drui D, Arnault-Ouary G, Charbonnel B, Chaillous L, Cariou B (2008) Fulminant type 1 diabetes in Caucasians: a report of three cases. Diabetes Metab 34:529–532

Murase Y, Imagawa A, Hanafusa T, Iwahashi H, Uchigata Y, Kanatsuka A, Kawasaki E, Kobayashi T, Shimada A, Shimizu I, Maruyama T, Makino H (2007) Fulminant type 1 diabetes as a high risk group for diabetic microangiopathy–a nationwide 5-year-study in Japan. Diabetologia 50:531–537

Sano H, Terasaki J, Tsutsumi C, Imagawa A, Hanafusa T (2008) A case of fulminant type 1 diabetes mellitus after influenza B infection. Diabetes Res Clin Pract 79:e8–e9

Sano H, Terasaki J, Mishiba Y, Imagawa A, Hanafusa T (2011) Exenatide, a glucagon-like peptide-1 receptor agonist, suppresses pancreatic beta-cell destruction induced by encephalomyocarditis virus. Biochem Biophys Res Commun 404:756–761

Sekine N, Motokura T, Oki T, Umeda Y, Sasaki N, Hayashi M, Sato H, Fujita T, Kaneko T, Asano Y, Kikuchi K (2001) Rapid loss of insulin secretion in a patient with fulminant type 1 diabetes mellitus and carbamazepine hypersensitivity syndrome. J Am Med Assoc 285:1153–1154

Shibasaki S, Imagawa A, Tauriainen S, Iino M, Oikarinen M, Abiru H, Tamaki K, Seino H, Nishi K, Takase I, Okada Y, Uno S, Murase-Mishiba Y, Terasaki J, Makino H, Shimomura I, Hyoty H, Hanafusa T (2010) Expression of toll-like receptors in the pancreas of recent-onset fulminant type 1 diabetes. Endocr J 57:211–219

Shimada A, Maruyama T (2004) Encephalomyocarditis-virus-induced diabetes model resembles "fulminant" type 1 diabetes in humans. Diabetologia 47:1854–1855

Shimizu I, Makino H, Imagawa A, Iwahashi H, Uchigata Y, Kanatsuka A, Kawasaki E, Kobayashi T, Shimada A, Maruyama T, Hanafusa T (2006) Clinical and immunogenetic characteristics of fulminant type 1 diabetes associated with pregnancy. J Clin Endocrinol Metab 91:471–476

Takaike H, Uchigata Y, Iwamoto Y, Imagawa A, Iwahashi H, Kanatsuka A, Kawasaki E, Kobayashi T, Shimada A, Shimizu I, Maruyama T, Hanafusa T, Makino H (2008) Nationwide survey to compare the prevalence of transient elevation of liver transaminase during treatment of diabetic ketosis or ketoacidosis in new-onset acute and fulminant type 1 diabetes mellitus. Ann Med 40:395–400

Tsutsumi C, Imagawa A, Ikegami H, Makino H, Kobayashi T, Hanfusa T (2009) The Research Committee of type 1 diabetes in the Japan diabetes. Class II HLA genotype in fulminant type 1 diabetes. Diabetes 58(Suppl 1):A299

Zheng C, Zhou Z, Zang C, Lin J, Yang L, Wang J, Zhou W, Yang Y, Che Z (2009) Prevalence, clinical and immunological features of fulminant type 1 diabetes among Chinese Hunan Han population. J Japan Diab Soc 51(Suppl 1):S–188

Zheng C, Zhou Z, Yang L, Lin J, Huang G, Li X, Zhou W, Wang X, Liu Z (2011) Fulminant type 1 diabetes mellitus exhibits distinct clinical and autoimmunity features from classical type 1 diabetes mellitus in Chinese. Diabetes Metab Res Rev 27:70–78

Part IV
Evaluation of Causality in Human Studies

Chapter 23
Defining Causal Relationships Between Viral Infections and Human Diabetes

Lars C. Stene and Marian Rewers

Abstract Type 1 diabetes is a multifactorial, chronic disease with a long induction period, and little is known about its etiology. This poses specific challenges to the study of viral infections as a potential causes. Statistical associations may be due to either chance, bias, or causality. Study designs (in combination with methods of viral detection) have important impact on ability to make causal inference under various hypothetical mechanisms of disease induction. Causes of multifactorial disease may be categorized as necessary, sufficient, both, or neither. Most single risk factors for type 1 diabetes is likely to be neither. Causality can never be inferred with certainty, but practical criteria can be employed to assess various aspects of the available evidence. Koch's postulates and Hill's "criteria" are well known, but often misinterpreted. Many established causal relations do not fulfil most of these criteria. Subjective judgment will always be part of the causal inference process, but modern analytical methods may help validate the process. While summarizing existing evidence with respect to potential causality may help to justify specific intervention trials, such as vaccination, the risk of side effects should be seriously considered.

Introduction

The art and science of causal inference has troubled philosophers and scientists for decennia. Despite the impossibility of scientific proof, scientists attempt to make causal inference, and there are numerous examples of successful applications in

L.C. Stene, Ph.D. (✉)
Division of Epidemiology, Norwegian Institute of Public Health,
P.O. Box 4404, Nydalen, NO 0403, Oslo, Norway
e-mail: lars.christian.stene@fhi.no

M. Rewers, M.D., Ph.D.
Barbara Davis Center for Childhood Diabetes, University of Colorado School of Medicine, Aurora, CO 80045, USA

K.W. Taylor, H. Hyoty, A. Toniolo, A.J. Zuckerman (eds.), *Diabetes and Viruses*,
DOI 10.1007/978-1-4614-4051-2_23,

medicine. Both Koch's postulates and Austin Bradford Hill's causal criteria are widely known, but frequently misinterpreted. Throughout history, paradigms that influenced the criteria for evaluating causal relationships in medicine depended on available technologies (Evans 1993; Susser 1973).

Methodological advance has nearly always accompanied important new discoveries. In this chapter, we will review the merits of causal criteria as well as the designs of studies to determine the role of viruses as potential causes of type 1 diabetes. Such studies pose specific challenges (Lipkin 2010; Oberste and Pallansch 2003). However, we will not discuss technologies for viral detection, covered in other chapters in this volume and in the excellent recent review by Lipkin (Lipkin 2010).

Associations, Causes, and Models of Multifactorial Causation

We refer here to (statistical) association as a difference in disease risk between groups exposed or not exposed to a proposed causal factor. The exposure (treatment, infection, etc.) should be clearly specified and contrasted with an equally specific alternative. While a cause may be simply defined as "something that makes a difference," it is difficult to find out whether the observed difference was due to a causal effect, and not chance, bias, or confounding. If a microbial agent is isolated from a single patient, chance is likely. Statistical analysis of groups of individuals helps to estimate the probability that the observed difference could have been due to chance. Multiple testing increases the chance of finding a "false positive" association, while low statistical power may lead to "false negative" lack of associations. In addition to chance, biases and confounding factors may lead to incorrect findings. Even in the presence of biases or confounding, there may still be a component of the association that can be ascribed to causality.

There are many ways to categorize causes, and single causal factors may in theory be necessary, sufficient, neither, or both (Table 23.1). Most diseases are multifactorial where the component causes can be proximal or distal, falling into a "web of causation." The traditional depiction of agent, environment, vector, and host (Fig. 23.1) is an example of such a model. Many diseases are likely influenced by factors at different levels of organization, an illustrative example being the concept of herd immunity for infectious diseases. Although beyond the scope of this chapter, there are formal methods for analysis of data from observational studies to estimate average causal effect (Halloran and Struchiner 1995; Hernán and Robins 2012; Little and Rubin 2000; Pearl 2009). This requires not testable assumptions such as knowledge of the structure of the causal model and error-free measurements of all relevant variables, but this work has nevertheless provided interesting insights of practical relevance (Rothman et al. 2008).

Several general, theoretical causal models exist and they have different strengths and weaknesses (Greenland and Brumback 2002). An interesting conceptualization of multifactorial causation is the "causal pie" model (Rothman et al. 2008). Here, disease is considered as a product of several component causes which together

Table 23.1 Some attributes of idealized single causes of multifactorial disease

Necessary	Sufficient	Occurrence in patients with disease[a]	Occurrence in non-diseased individuals	Comment
Yes	Yes	100%	0%[b]	If studied, would be very strong and perfectly predictive and therefore easily identifiable. Such causes are rare, probably nonexistent in type 1 diabetes
Yes	No	100%	0–100%	Such causes may easily go unnoted in epidemiological studies because the whole healthy population may be exposed. Identification of such causes is the holy grail, as elimination would lead to elimination of disease. Modifiable necessary causes may be few or nonexistent in type 1 diabetes
No	Yes	1 subject to 99%	0%[b]	If rare, low statistical power may lead to inability to demonstrate significant association. See also text discussing implications for strength of causes. Furthermore, such causes may also be difficult to distinguish from heterogeneity (disease due to this cause may be something different to disease due to other causes). There are potentially many such types of causes in type 1 diabetes
No	No	1 subject to 100%	>0%, but fewer than among the diseased	Probably characteristic of most causes in type 1 diabetes and other complex disease. Can be difficult to detect

[a]Occurrence before disease, a necessary condition for a cause
[b]In case of long latency some non-diseased individuals may be exposed, and only later develop disease

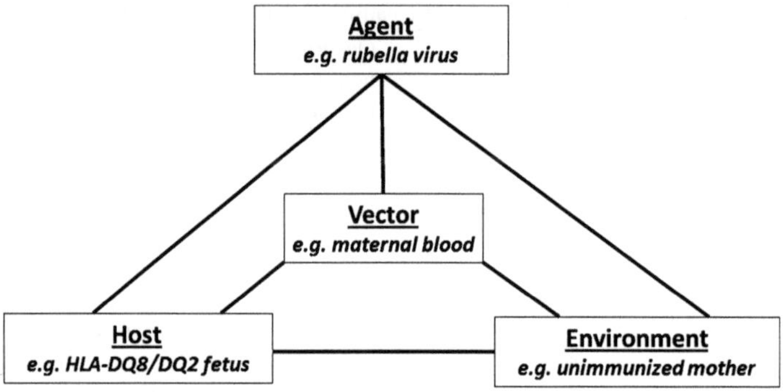

Fig. 23.1 Illustration of the classical causal model involving the agent, environment, vector and host, with the example of congenital rubella

comprise a sufficient cause. Two or more different sets of sufficient causes are viewed as different causal mechanisms. This simple model can incorporate surprisingly many complex concepts. For instance, it helps illustrating interesting points regarding strength of causes (see below) and the fact that the attributable fractions for different causes may add up to more than 100% because of interactions. The sufficient sets of causes for human type 1a (autoimmune) diabetes are still being formulated, but all would need to include as a component cause the HLA Class II genotypes.

The Henle–Koch Postulates

While often referred to as "Koch postulates," credit should also be given to Jacob Henle's important contributions (Evans 1976):

1. The parasite occurs in every case of the disease in question and under circumstances which can account for the pathological changes and clinical course of the disease.
2. It occurs in no other disease as a fortuitous and nonpathogenic parasite.
3. After being fully isolated from the body and repeatedly grown in pure culture, it can induce the disease anew.

It is sometimes also added that the pathogen should be isolated from the target organ in which it "induced disease anew." Koch himself and many others realized that strict application of these postulates is not advisable. There are numerous examples of well-established causal relations which do not satisfy any of the criteria. The first two criteria define a cause as necessary and sufficient (given causality) and do not "allow" pleiotropic effects or multifactorial diseases. Note that the criteria were developed chiefly for diseases caused by bacterial pathogens, thus isolation should be done in "pure culture," meaning without cells. This postulate had to be modified for application in virology. Not all pathogens can be grown in culture, however. The famous study by Yoon et al. (1979) where a Coxsackievirus was isolated from the pancreas of a child with type 1 diabetes may be said to fulfill the third postulate, if we allow for growth in cell culture rather than "pure culture," but postulates 1 and 2 were not fulfilled.

Hill's Viewpoints

Austin Bradford Hill proposed strength, consistency, specificity, temporality, biological gradient, plausibility, coherence, experiment, and analogy as "viewpoints from all of which we should study association before we cry causation" (Hill 1965). While these viewpoints have been later referred to as causal criteria, the original paper is filled with reservations. The shortcomings and difficulty of applying Hill's and other criteria have been extensively discussed (Evans 1976; Evans 1993; Kaufman and Poole 2000; Rothman et al. 2008; Susser 1973).

1. *Strength.* In the causal pie model, the strength of a cause (the magnitude of statistical association) depends on the relative frequency of the different component causes that together constitutes sufficient causes in the study population (Rothman et al. 2008). A single necessary cause may be of paramount importance while demonstrating weak or no statistical association in the population. For instance, intake (yes/no and amount) of wheat products is not strongly associated with the presence of undiagnosed celiac disease, because most people eat wheat products. Hill's main point was that strong associations are less likely the result of bias or confounding.
2. *Consistency.* Has the association been repeatedly observed by different persons, in different places, circumstances and times? In our view this comprises two components, namely reproducible under the *same* (or very similar) conditions and also found under a variety of different conditions. Experience from genetic epidemiology has thought us the importance of distinguishing the two (Chanock et al. 2007). Meta-analysis illustrates the fallacy of inferring lack of consistency from statistical significance alone. It is to be expected that not all studies demonstrate statistically significant results even if the association is truly causal. This has to do with sample variability and statistical power. While repeating results under different circumstances would prove the relationship to be robust, it is perfectly possible that a cause operates only under certain circumstances and not others. Paralytic complications of polio infections exemplify the importance of age as one critical circumstance.
3. *Specificity.* Hill stated that "We must not, however, over-emphasize the importance of the characteristic. We must also keep in mind that diseases may have more than one cause."
4. *Temporality.* Note that this does not refer merely to occurrence at the same time, but to the occurrence of the cause *before* the effect, sometimes called time order. Temporality is a logical, absolute criterion for a cause. We have emphasized that most of the human studies of enterovirus and type 1 diabetes have not

been able to demonstrate time order, even among some of the longitudinal studies (Stene and Rewers 2012). When detecting a virus at or after disease onset one has to rely on the assumption that it was present before disease onset. At diagnosis, even at the true onset of symptoms, disease may have been present for some time while the autoimmune disease process may have been going on for years.

5. *Biological gradient.* This refers to a monotonous dose–response relationship. Apart from the fact that the appearance of an estimated dose–response curve depends on the choice of scale, the relationship may demonstrate a threshold effects or other nonlinear forms.
6. *Plausibility.* Hill noted that "What is biologically plausible depends upon the biological knowledge of the day." This criterion is obviously open to subjective interpretation. Instructive examples abound, such as that of Ignaz Semmelweiss that clarified the transmission of streptococci in puerperal fever as well as the discovery of *Helicobacter pylori* by Marshall and Warren. Furthermore, the idea that a mechanism supported by an in vitro or animal model can be generalized to human disease is indisputably subjective.
7. *Coherence.* Here, Hill meant that the cause-and-effect interpretation of available data should "not seriously conflict with the generally known facts of the natural history and biology of the disease." This is of course important, but this criterion suffers from the same limitations as discussed above for the plausibility criterion.
8. *Experiment.* There are many types of experimental data, including in vitro laboratory experiments, experiments in animal models, randomized clinical trials, or randomized field trials. Today, results from "well-conducted" randomized clinical trials is widely accepted as the strongest form of causal evidence, although such studies are often not feasible and currently not available in the field of virus and diabetes. Furthermore, randomized clinical trials may suffer from a range of potential biases and can often not be generalized.
9. *Analogy.* Are there other similar situations where causality is established? Hill exemplified with the observation that intrauterine exposure to thalidomide and rubella could lead to malformations, and suggested that "we would surely be ready to accept slighter but similar evidence with another drug or another viral disease in pregnancy." We have cited congenital rubella and risk of diabetes as an analogy to justify our studies (Forrest et al. 2002; Stene et al. 2008). This criterion overlaps plausibility, and a problem is that scientists, with their creative minds, find analogies everywhere (Rothman et al. 2008).

Research Designs and Implications for Causal Inference

We cannot provide a full overview of study designs in this chapter, but we will briefly discuss selected topics in human studies. All designs have their strengths and limitations.

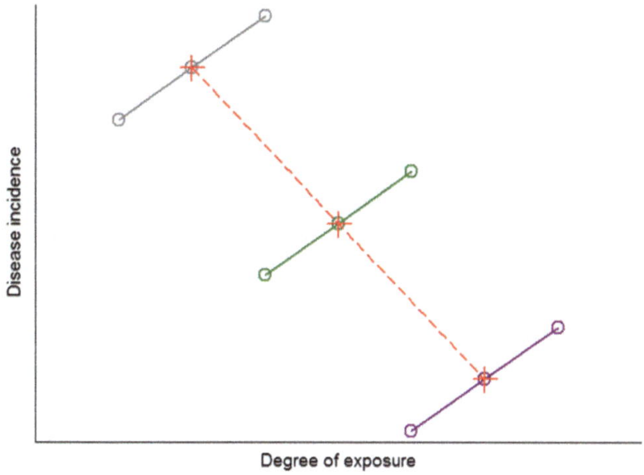

Fig. 23.2 Illustration of the potential fallacy of making inference from ecological, or aggregated data. Degree exposure to a hypothetical factor (e.g., frequency of viral infections) of individuals, may be positively associated with increasing risk of a multifacorial disease such as type 1 diabetes (*open circles*) in each of three groups, marked in *gray*, *green*, and *purple color*, respectively. The groups may be different countries, which for reasons other than virus exposure differ in disease risk. If only ecological data (mean disease incidence and mean degree of exposure in each of the three groups) are available, this would be represented only by the three data points labeled with *red crosses*. The ecological relationship would then appear as the *red dashed line*. The direction of the association at the two levels of organization is opposite. Examples can easily be constructed where no association at the individual level would correspond to a correlation at the aggregated level, either positively or inversely

Ecological Studies

While ecological studies are widely known to be the weakest type of design, "ecological reasoning" is common. For instance, since type 1 diabetes is common in Finland and rare in Paraguay, what is different between the countries? The answer is of course: a lot. Ecological studies here means studies where information on the exposure (infection) and disease is not available for individuals, only as average occurrence of the two across groups (Morgenstern 1995). The ecological study of epidemics of parotitis and mortality of diabetes (Gundersen 1927) was a foray into the studies of the potential infectious causes of type 1 diabetes. Ecological associations can easily occur when there is no statistical association at the individual level at all. Furthermore, an association that is positive at the individual level may appear inverse at the aggregated (ecological) level (Fig. 23.2). This can occur if there are other causes of disease that are responsible for the difference between groups. The currently available data to support the polio model for type 1 diabetes is of ecological nature (Gamble 1980; Viskari et al. 2005).

Case Crossover Studies

We have previously discussed the so-called case crossover studies as applied to viruses and diabetes, which attempt to mimic randomized crossover studies (Stene and Hyöty 2006). Here, only patients are studied, and their exposure to viral infections in a relevant time window before disease is compared with the same exposure in nonrelevant (control) time windows, and each case is his/her own control. Although appealing, it is difficult to apply without bias because of the sensitivity to seasonal variation, age-related factors, and other time trends in exposure. The design is optimal for rapid effects of acute events, and there is limited knowledge about the truly relevant (or nonrelevant) time windows of exposure to causal factors in type 1 diabetes (Stene and Rewers 2012).

Outbreak Workup

Examinations of individuals affected by an outbreak of viral disease can provide important knowledge. Among the many problems is the known long induction period (incubation period) of type 1 diabetes and that it is rare. Most such studies would be expected to detect few or no cases with type 1 diabetes during a few years of follow-up, except under the unrealistic scenario that the causative agent was a nearly sufficient cause of rapidly developing type 1 diabetes. An interesting study was done in an isolated island population after a Coxsackievirus B4 outbreak, and no association with diabetes was observed (Dippe et al. 1975). Due to the small study population and consequent low statistical power and other weaknesses, the only inference that can be drawn from this study is that an extremely strong effect on diabetes can be excluded. Experience with outbreaks of other enterovirus associated diseases (see, e.g., Witsø et al. 2007) have demonstrated resemblance to what Evans referred to as "the five realities of acute respiratory disease" (Evans 1976):

1. The same clinical syndrome may be produced by a variety of agents.
2. The same etiologic agent may produce a variety of clinical syndromes.
3. The predominating agent in a given clinical syndrome may vary according to the age group involved, the year, the geographic location, and the type of population (military or civilian).
4. Diagnosis of the etiological agent is frequently impossible on the basis of the clinical findings alone.
5. The cause (or causes) of a large percentage of common infectious disease syndromes is still unknown.

Patient Control and Prospective Cohort Studies

We have recently reviewed the implications of the prevalent case–control design and longitudinal birth cohorts in etiological studies of enterovirus and type 1 diabetes, and we will not repeat the details here (Stene and Rewers 2012). Briefly, case–control studies cannot demonstrate time order, whether signs of the virus is detected by serology, PCR of biofluids, or virus detected in pancreatic tissue. Prospective studies are costly and time consuming. However, they represent the only study design where time order can potentially be demonstrated. Study design itself is not enough; data analysis must also be properly executed (Stene and Rewers 2012). While cohort studies can contribute to the field, the complete history of infectious exposures cannot be ascertained without obtaining microbiological samples at least every 2 weeks, clearly an impractical task using current technologies.

Randomized Studies

Randomized (experimental) studies are the only ones where unmeasured confounding can be minimized. Randomized clinical trials and randomized preventive "field trials" have many shortcomings and practical obstacles, particularly for a relatively rare disease with long induction period like type 1 diabetes. Trials to prevent type 1 diabetes conducted to date have been costly, time consuming, and without demonstrable effect. In our view, the observational and mechanistic study evidence should be extremely strong to initiate a randomized trial to reduce the risk of type 1 diabetes targeting an infectious agent. Potential side effects and feasibility of "practical" interventions, such as vaccination, must be considered.

Modern Systems for Synthesis of Evidence, and Concluding Remarks

Experience from other fields has shown that demonstration of efficacy in a single randomized study is rarely sufficient for a widespread adoption of the intervention in clinical practice. The GRADE working group and others have developed a system attempting to make the process of causal inference and associated value judgments as transparent as possible (Guyatt et al. 2011; Guyatt et al. 2008). The focus is on the effect of interventions, rather than on etiology. Sophisticated versions of Hill's criteria are incorporated, as well as important aspects of study design and execution. A number of additional, less obvious factors that cannot be covered here do also influence the medical literature (Ioannidis 2005).

Synthesis of quantitative evidence often culminates in a meta-analysis of studies with similar design, preferably randomized studies. A few attempts have been made to systematically review the literature on enterovirus and human type 1 diabetes, which consists only of observational studies (Green et al. 2004; Yeung et al. 2011). Meta-analysis of observational studies may lead to false conclusions as all studies may suffer from bias or confounding. One should be particularly diligent when judging the comparability of the study designs and methods of viral detection

(Egger et al. 1998; Stene and Rewers 2012). Employing GRADE criteria in the field of viruses and type 1 diabetes is probably not feasible today, because of the limited amount of data. However, this should inspire investigators to contribute pieces to this puzzle.

References

Chanock SJ, Manolio T, Boehnke M, Boerwinkle E, Hunter DJ, Thomas G, Hirschhorn JN, Abecasis G, Altshuler D, Bailey-Wilson JE, Brooks LD, Cardon LR, Daly M, Donnelly P, Fraumeni JF Jr, Freimer NB, Gerhard DS, Gunter C, Guttmacher AE, Guyer MS, Harris EL, Hoh J, Hoover R, Kong CA, Merikangas KR, Morton CC, Palmer LJ, Phimister EG, Rice JP, Roberts J, Rotimi C, Tucker MA, Vogan KJ, Wacholder S, Wijsman EM, Winn DM, Collins FS (2007) Replicating genotype-phenotype associations. Nature 447:655–660

Dippe SE, Bennett PH, Miller M, Maynard JE, Berquist KR (1975) Lack of causal association between Coxsackie B4 virus infection and diabetes. Lancet 1:1314–1317

Egger M, Schneider M, Davey Smith G (1998) Spurious precision? Meta-analysis of observational studies. BMJ 316:140–144

Evans AS (1976) Causation and disease: the Henle-Koch postulates revisited. Yale J Biol Med 49:175–195

Evans AS (1993) Causation and disease: a chronological journey. Plenum Press, New York

Forrest JM, Turnbull FM, Sholler GF, Hawker RE, Martin FJ, Doran TT, Burgess MA (2002) Gregg's congenital rubella patients 60 years later. Med J Aust 177:664–667

Gamble DR (1980) The epidemiology of insulin dependent diabetes, with particular reference to the relationship of virus infection to its etiology. Epidemiol Rev 2:49–70

Green J, Casabonne D, Newton R (2004) Coxsackie B virus serology and type 1 diabetes mellitus: a systematic review of published case–control studies. Diabet Med 21:507–514

Greenland S, Brumback B (2002) An overview of relations among causal modelling methods. Int J Epidemiol 31:1030–1037

Gundersen E (1927) Is diabetes of infectious origin? J Infect Dis 41:197–202

Guyatt GH, Oxman AD, Vist GE, Kunz R, Falck-Ytter Y, Alonso-Coello P, Schünemann HJ (2008) GRADE: an emerging consensus on rating quality of evidence and strength of recommendations. BMJ 336:924–926

Guyatt G, Oxman AD, Akl EA, Kunz R, Vist G, Brozek J, Norris S, Falck-Ytter Y, Glasziou P, DeBeer H, Jaeschke R, Rind D, Meerpohl J, Dahm P, Schünemann HJ (2011) GRADE guidelines: 1. Introduction-GRADE evidence profiles and summary of findings tables. J Clin Epidemiol 64:383–394

Halloran ME, Struchiner CJ (1995) Causal inference in infectious diseases. Epidemiology 6:142–151

Hernán MA, Robins JM (2012) Causal inference. (Draft of the book downloadable in 2012 from http://www.hsph.harvard.edu/faculty/miguel-hernan/causal-inference-book/): Chapman & Hall/CRC Press (in press)

Hill AB (1965) The environment and disease: association or causation? Proc R Soc Med 58:295–300

Ioannidis JP (2005) Why most published research findings are false. PLoS Med 2:e124

Kaufman JS, Poole C (2000) Looking back on "causal thinking in the health sciences". Annu Rev Public Health 21:101–119

Lipkin WI (2010) Microbe hunting. Microbiol Mol Biol Rev 74:363–377

Little RJ, Rubin DB (2000) Causal effects in clinical and epidemiological studies via potential outcomes: concepts and analytical approaches. Annu Rev Public Health 21:121–145

Morgenstern H (1995) Ecologic studies in epidemiology: concepts, principles, and methods. Annu Rev Public Health 16:61–81

Oberste MS, Pallansch MA (2003) Establishing evidence for enterovirus infection in chronic disease. Ann N Y Acad Sci 1005:23–31

Pearl J (2009) Causation: models, reasoning, and inference. Cambridge University Press, New York

Rothman KJ, Greenland S, Lash TL (2008) Modern epidemiology. Lippincott, Williams & Wilkins, Philadelphia

Stene LC, Hyöty H (2006) A novel approach to the investigation of potential precipitating factors in type 1 diabetes. Pediatr Diabetes 7:143–145

Stene LC, Rewers M (2012) Immunology in the clinic review series; focus on type 1 diabetes and viruses: the enterovirus link to type 1 diabetes: critical review of human studies. Clin Exp Immunol 168:12–23

Stene LC, Tuomilehto J, Rewers M (2008) Global epidemiology of type 1 diabetes. In: Ekoé J-M, Rewers M, Williams R, Zimmet P (eds) The epidemiology of diabetes mellitus. Wiley-Blackwell, Chichester, pp 355–383

Susser M (1973) Causal thinking in the health sciences: concepts and strategies of epidemiology. Oxford University Press, New York

Viskari H, Ludvigsson J, Uibo R, Salur L, Marciulionyte D, Hermann R, Soltesz G, Füchtenbusch M, Ziegler AG, Kondrashova A, Romanov A, Kaplan B, Laron Z, Koskela P, Vesikari T, Huhtala H, Knip M, Hyöty H (2005) Relationship between the incidence of type 1 diabetes and maternal enterovirus antibodies: time trends and geographical variation. Diabetologia 48:1280–1287

Witsø E, Palacios G, Rønningen KS, Cinek O, Janowitz D, Rewers M, Grinde B, Lipkin WI (2007) Asymptomatic circulation of HEV71 in Norway. Virus Res 123:19–29

Yeung WC, Rawlinson WD, Craig ME (2011) Enterovirus infection and type 1 diabetes mellitus: systematic review and meta-analysis of observational molecular studies. BMJ 342:d35

Yoon JW, Austin M, Onodera T, Notkins AL (1979) Isolation of a virus from the pancreas of a child with diabetic ketoacidosis. N Engl J Med 300:1173–1179

Chapter 24
The JDRF Network for the Pancreatic Organ Donor with Diabetes (nPOD): A novel Resource and Study Approach in Type 1 Diabetes Research

Alberto Pugliese, Martha Campbell-Thompson, Teodora Staeva, and Mark A. Atkinson

Abstract The JDRF nPOD organization has three goals: To establish a biobank from organ donors with T1D. To distribute tissues to nPOD investigators in support of diverse studies of human T1D. To promote collaboration and data sharing among nPOD investigators and achieve a comprehensive understanding of human T1D. The JDRF nPOD supports 71 projects in key research areas. Findings from nPOD investigators include abnormalities of self-antigen expression in the pancreatic lymph node, which could impair self-tolerance, the localization of antigen-specific autoreactive T-cells in the insulitic lesions, evidence of heterogeneity and distinct patterns of beta-cell destruction, lack of insulitis in donors with a single autoantibody, and beta-cell persistence in patients with long standing disease. Programmatic expansions will apply the nPOD model to additional areas of the diabetes spectrum, such as diabetes complications and diabetes in patients after transplantation. The JDRF nPOD is organizing working groups, which apply the concepts of real-time data sharing to accelerate the rate of discovery. The first working group is addressing

A. Pugliese, M.D. (✉)
Diabetes Research Institute, University of Miami Miller School of Medicine,
1450 NW 10th Avenue, Miami, FL 33136, USA
e-mail: apuglies@med.miami.edu

M. Campbell-Thompson
Department of Pathology, Immunology, and Laboratory Medicine,
The University of Florida College of Medicine, PO Box 100275, Room D11-38A,
1600 SW Archer Road, Gainesville, FL 32610-0275, USA
e-mail: mct@ufl.edu

T. Staeva
JDRF, 26 Broadway, 14th Floor, New York, NY 10004, USA
e-mail: tstaeva@jdrf.org

M.A. Atkinson
Departments of Pathology and Pediatrics, The University of Florida College of Medicine,
1600 SW Archer Road, Gainesville, FL 32610-0275, USA
e-mail: atkinson@ufl.edu

K.W. Taylor, H. Hyoty, A. Toniolo, A.J. Zuckerman (eds.), *Diabetes and Viruses*,
DOI 10.1007/978-1-4614-4051-2_24, © Springer Science+Business Media New York 2013

viruses in diabetes. Ultimately, nPOD provides a scientific venue and a "cloud" environment consisting of shared tissues, data sharing and collaborative, coordinated study design developed with collective input.

Introduction

Despite many decades of research and scientific advances, type 1 diabetes remains a disorder for which there is no cure. Many aspects of its pathogenesis remain unclear to a significant extent, reflecting the limitation of clinical research to peripheral blood specimens and the scarce access to the pancreas and other disease-related tissues. Our current views of the disease pathogenesis mostly rely on data from experimental mice and when data from human pancreas are available, these are largely from older studies which did not rely on modern molecular technologies (In't Veld 2011a; Rowe et al. 2011). While animal models are valuable in many ways, these do not allow addressing critical questions about key biological aspects that are unique to the human condition. Addressing these questions should improve our understanding of the disease pathogenesis and lead to the identification of new, T1D-specific and human-specific therapeutic targets. These include but are not limited to: (1) autoreactive T and B cells, which could be targeted by phenotypic and functional features, by antigen specificity and by knowledge of T cell receptor sequences; (2) viral infections, which could perhaps be averted by vaccination if responsible viruses were identified and isolated; (3) pathways of beta-cell regeneration, which could perhaps be stimulated with drugs; (4) potentially, additional pathogenic mechanisms that may have yet to be uncovered and could contribute to disease heterogeneity, and perhaps help explaining the limited success of clinical trials, which mostly have relied on manipulation of single pathways of the immune system (Matthews et al. 2010). Thus, there is clear need for more research that is focused on the study of the human condition through the study of human tissues from patients. Results could inform strategies for combinatorial therapies that target multiple disease pathways, both immune and nonimmune.

The JDRF nPOD and Its Mission

Recognizing that lack of access to human disease-relevant tissues is hampering our ability to understand diabetes and develop better cures, in 2007 the Juvenile Diabetes Research Foundation (JDRF) began supporting the establishment of the Network for the Pancreatic Organ Donors with Diabetes (nPOD; www.JDRFnPOD.org). The JDRF nPOD has three goals: (1) to establish a biobank of tissues (pancreas, spleen, pancreatic lymph nodes, blood, and other) from organ donors with type 1 diabetes (at any point after clinical diagnosis and during the asymptomatic period leading to diagnosis when islet autoimmunity silently leads to beta-cell destruction); (2) to distribute

Table 24.1 nPOD donor groups, with age, as of December 2011

Donor type ID	Number	Age (mean)	Age (SD)	Age range
No diabetes	69	25.99	20.31	0.01–75
Autoantibody positive	16	34.27	21.77	0.17–69.2
Type 1 diabetes	52	28.32	11.58	4.4–50.8
Type 1 diabetes medalist	12	76.00	10.57	59–93
Type 2 diabetes	16	45.79	16.66	18.8–76.3
Other	8	30.08	14.20	15.5–62
Gestational diabetes	2	33.10	0.14	33–33.2
Total	175			

such tissues, worldwide, to nPOD approved investigators for in-depth and diverse studies of human T1D; and (3) to promote collaboration and data sharing among nPOD investigators to achieve a comprehensive understanding of human T1D.

The JDRF nPOD was initially funded as a pilot study to explore the feasibility of recovering organs from deceased donors and to support research projects. Since its launch, the nPOD program has demonstrated that it is feasible to collect high quality tissues for research, including but not limited to pancreas that can be used for histology studies, gene expression, and genetic analysis, as well as protein studies. By working with numerous organ procurement organizations (OPOs), and by applying rigorous laboratory methods for tissue processing and quality control, nPOD has shown that procuring quality organs is possible even across large geographical distances, with organs being sent to the Florida-based nPOD repository from any region in the USA. Since its inception, nPOD has collected tissues from 64 type 1 diabetes patients, 16 donors with T1D-associated autoantibodies and 69 control donors, as well as other donors with related pancreatic disorders such as gestational diabetes (Table 24.1). nPOD has provided and/or currently provides specimens in support of 71 research studies from approved investigators worldwide, many of whom have multiple collaborating investigators engaged in the project. Over 24,611 specimens have been distributed to investigators, supported by a very efficient review and approval process of projects and requests for specimens.

Selected Studies Supported by nPOD

Amongst the studies supported by nPOD, several have been published that describe novel and important findings. Selected examples are summarized below:

1. The finding that the expression of self-molecules genes in the pancreatic lymph node is depressed in type 1 diabetes patients (Yip et al. 2009), as a result of alternative splicing of the Deaf1 transcription factor, which could be driven by inflammation; this finding can help explain loss of self-tolerance at an immune site that is believed to play a critical role in the activation of islet autoimmunity. Genetic regulation of self-tolerance via the expression of self-molecule genes by

a series of specialized cells is a critical aspect of type 1 diabetes pathogenesis, both in thymus and in peripheral lymphoid tissues (Kyewski and Klein 2006; Metzger and Anderson 2011; Pugliese 2004), and this study has identified a mechanism that could be explored for therapeutic manipulation. As a similar abnormality was demonstrated in the NOD mouse (Kodama et al. 2008; Yip et al. 2009), it should be possible to attempt therapeutic modulation in this preclinical model to conduct further mechanistic investigations on a finding that originates from the study of the human disease and not vice versa. This is an example of how this approach modifies the typical research flow in a way that is better geared to address questions that are relevant to the human condition.

2. The demonstration of islet antigen-specific CD8 T cells in the insulitic lesions in the pancreas of an nPOD donor with recent onset diabetes unequivocally links those T cells with disease and validates those cells and antigen specificity as a bona fide therapeutic target (Coppieters et al. 2012).
3. The demonstration that the pathology of the diabetic pancreas may be heterogeneous, with distinct patterns of beta-cell loss, implicates heterogeneity in the disease pathogenesis and also highlights a potential role for the survivin molecule in the possible persistence of beta-cells in the pancreas of patients with long disease duration (Gianani et al. 2010).
4. Related to this, nPOD has supported the Joslin Medalist study which recently reported on the persistence of insulin production and beta-cells in the pancreas of type 1 diabetes patients with disease of long duration (Keenan et al. 2010). Similar to a study by Meier et al. (2005), the findings suggest potential turnover of β-cells, but chronic apoptosis may hamper the regenerative potential, again providing insight for potential therapeutic approaches.
5. Studies have shown that unlike in mouse, human beta-cells express the glucose transporter glut-1 rather than glut-2, which is more abundant in mouse beta-cells. Using nPOD pancreata, it was also found that glut-1 expression remains present in surviving beta-cells at various stages of disease progression (Coppieters et al. 2011), including long-term diabetes and in recent onset diabetes in islets which are severely infiltrated by lymphocytes.

Challenges and Opportunities

While the above are significant success stories, many more projects are ongoing and generating meaningful results. As nPOD activities have progressed from the pilot stage to a full-fledged effort, operational challenges have emerged. For one, the population of organ donors tends to be biased towards the adult age, while type 1 diabetes is more often (but not exclusively) diagnosed in childhood and adolescence. Additionally, most of the type 1 diabetes donors that OPOs recover are those with disease of long duration, which are less likely to have active autoimmunity, at least in the pancreas; however, persistent autoimmune responses and viral infections could be present in lymphoid tissues; immune responses could represent memory responses, which in themselves are of great interest as a therapeutic target (Monti et al. 2007, 2008b; Ziegler and Nepom 2010). The nPOD's autoantibody screening effort to identify donors with islet autoimmunity has so far mostly identified donors with a single autoantibody, and these do not typically show insulitis. However, this is in itself an important finding, given that studies in first-degree relatives and the general population identify those with a single autoantibody has having lower risk of progression (Bingley et al. 1997; Krischer et al. 2003; Maclaren et al. 2003; Orban et al. 2009) but lack an assessment of the pancreatic pathology. Thus, this finding would suggest that single autoantibody positivity may not necessarily reflect active disease in the pancreas. This is consistent with studies of organ donors whose pancreas was used for islet transplantation that were retrospectively assessed for autoantibodies, noting that in those cases insulitis was assessed in a limited amount of tissue that was spared from the islet isolation (In't Veld et al. 2007). Yet the challenge remains to obtain tissues from donors with ongoing insulitis.

The nPOD program is gearing to address these challenges by developing a multi-pronged approach:

1. By adopting an improved autoantibody screening strategy that adds ZnT8 autoantibodies to the panel (Wenzlau et al. 2009), focuses selection on donors with multiple autoantibodies (Maclaren et al. 2003), and adds HLA types as an additional selection criteria that should facilitate finding donors with a higher likelihood of ongoing disease
2. By developing partnering strategies with emergency room departments to improve the likelihood that donors with recent onset diabetes may be found
3. By expanding the scope of operations through the establishment of satellites nPOD screening/procurement sites in Europe (nPOD Europe)
4. By collaborating with other consortia with overlapping interests, such as the Persistent Virus Infection in Diabetes Network (PEVNET), which operates in Europe and adopts strategies similar to nPOD to recover pancreas from patients with type 1 diabetes and/or autoantibodies focusing on the study of viruses in the disease pathogenesis (Tauriainen et al. 2010). These interactions involve sample sharing, data sharing, logistics support, and scientific collaboration
5. By expanding the nPOD model to other clinical settings relevant to T1D, such as the area of transplantation (nPOD-Transplantation). This is in recognition that type 1 diabetes represents a clinical spectrum of conditions that evolve from the preclinical stage to disease onset, often leading to chronic complications and transplantation to reverse kidney failure and/or diabetes (Fig. 24.1)

The nPOD-Transplantation program focuses on studying autoimmunity and beta-cells after transplantation, when changes of interest for potential therapeutic targeting may be brought about by chronic immunosuppression. This effort offers the opportunity to evaluate both the native pancreas and the transplanted pancreas. Studying tissues from pancreas transplant recipients can provide additional insight into type 1 diabetes etiology, autoimmunity and alloimmunity, as shown by earlier, pioneer studies (Dotta et al. 2007; Velthuis et al. 2009). There are critical questions about the long-term persistence of autoimmunity, and its possible reactivation which may lead to recurrent disease (Sutherland et al. 1984; Vendrame et al. 2010) despite

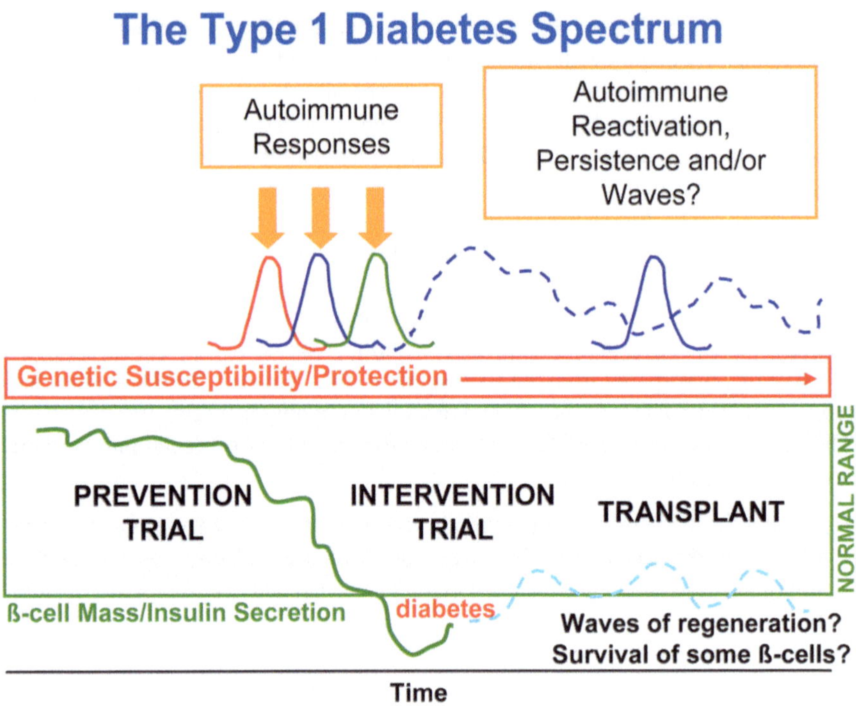

Fig. 24.1 Diagram illustrating various stages of the clinical spectrum of type 1 diabetes, the relationship to clinical events and biological phenomena that are the subject of intense research

chronic immunosuppression. Indeed, biopsies of the transplanted pancreas of patients with diabetes recurrence revealed insulitis and beta-cell loss (Vendrame et al. 2010), which did not appear dissimilar to that seen in patients with spontaneous disease reported in the literature (In't Veld 2011a). Thus, biopsies from transplanted patients could help us better understand type 1 diabetes and could help pinpoint molecular targets expressed by autoreactive T and B cells.

Another important consideration is that patients who develop disease recurrence may have memory autoreactive lymphocytes that are reactivated following transplantation many years after the onset of the original disease (Burke et al. 2011; Monti et al. 2008a; Velthuis et al. 2009). These cells might represent disease-relevant populations; studying their antigen specificity and functional/phenotypic features could help us identify therapeutic targets.

There was evidence of enterovirus infection (Fig. 24.2) in some of the pancreas transplant biopsies from patients with diabetes recurrence (Vendrame et al. 2010). Further studies in transplanted patients could also explore the potential role of viral infections, in relation to persistence and reactivation, or reinfection, which perhaps could trigger disease recurrence in the transplanted pancreas (Vendrame et al. 2010).

Studying tissues from transplanted patients will also help investigating the regenerative potential of the native pancreas. While patients who are immunosuppressed

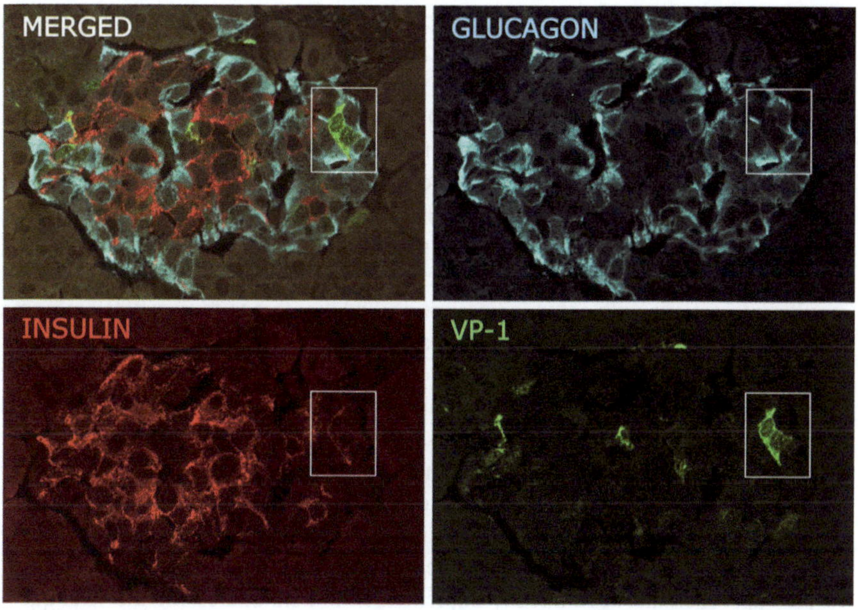

Fig. 24.2 An islet from a transplant pancreas biopsy from a patient with recurrent type 1 diabetes stained for insulin, glucagon, and the enterovirus protein VP-1 demonstrating VP-1 staining in a few beta-cells (highlighted by the *white box*)

(e.g., kidney recipients with T1D) are not cured of their diabetes, it is possible that subclinical changes may occur and could be favored by immunosuppression (Bonner-Weir et al. 2010). Recent studies have been conducted to assess potential beta-cell regeneration in immunosuppressed patients but have been limited by the lack of biopsy data to corroborate metabolic data (Liu et al. 2009; Martin-Pagola et al. 2008; Rother et al. 2009), and conversely there could be subclinical changes that may not be demonstrated by metabolic evaluation. Evidence for changes come from the examination of pancreas transplant biopsies from patients with recurrent autoimmunity, which revealed the presence of ductal cells expressing insulin (Martin-Pagola et al. 2008); these cells stained for the pancreatic-duodenal homeobox-1 transcription factor (PDX-1) and at times co-expressed chromogranin A (suggesting endocrine differentiation) and the proliferation marker Ki-67. Further studies could help advance our knowledge in this area and again guide therapeutic innovations.

The nPOD Research Model

Another key aspect of the nPOD program is the emphasis on identifying synergies among projects; promoting collaboration through the application of the concept of "real-time" data sharing, in the context of coordinated research efforts. This approach should accelerate the rate of discovery and maximize the potential for new and robust advances with the contribution of many investigators to the development of a synergistic research strategy. The nPOD program has indeed built a substantial team of investigators who subscribe to the concepts of data sharing and collaboration, and ongoing efforts capitalize on the collective strength of nPOD as a coordinated, scientific powerhouse. Thus, nPOD functions not just only as a biobank, but also as a synergistic research network with multiple but well-defined goals. Taking advantage of web-based connectivity, including online pathology, nPOD has begun regular interactions with extended groups of investigators to review data emerging from notable nPOD donors. As samples are studied by multiple investigators with a variety of approaches, this will help develop a comprehensive and integrated understanding of the disease pathogenesis. Importantly, the sharing of tissues and analysis by multiple investigators affords a key unifying element in science and the rare opportunity to coordinate studies that take into account multiple approaches and design input from multiple investigators. As the activity of these working groups progresses, findings emerge and are shared, informing the group about any changes in strategies and focus on the next questions.

This approach has been already implemented to help addressing the viral question. As noted earlier, demonstrating a pathogenic role for one or more viruses in type 1 diabetes could have very important therapeutic implications. Several investigators have made key contributions that support a role for viruses, and especially enteroviruses (Dotta et al. 2007; Lempainen et al. 2011; Oikarinen et al. 2008, 2011; Richardson et al. 2009; Salminen et al. 2004; Stene et al. 2010; Tauriainen et al. 2011; Tracy et al. 2011; Willcox et al. 2011). Yet there are outstanding questions, which have been defined collectively by several of these investigators through the research collaborative venue afforded by nPOD. These include whether there is an acute or chronic (persistent) infection, whether there is a replication-defective enterovirus, whether infection affects the beta-cells and or other cell types, not just cells in the pancreas but potentially immune cells, and what functional consequences infection may have on key beta-cell function but also, potentially, the immune response. In addition, emerging data suggest a potential link between beta-cell proliferation and enterovirus infection (In't Veld 2011b; Willcox et al. 2011). Through an nPOD coordinated effort, many leading investigators in the field are applying a multi-pronged approach (immunohistochemistry, RNA sequencing and gene expression, protein analysis, immunology, etc.) to study viruses in diabetes. The group also aims at cross-validating reagents and standardizing readouts, which was also identified as a critical need for this area of research, through the results exchange that is being made possible by the nPOD sample/data sharing framework. Thus, nPOD activities supplement research ongoing in many individual laboratories; a working group such as this one should facilitate progress towards addressing key questions about the role of viruses in T1D.

Closing Remarks

Human type 1 diabetes is a complex and heterogeneous disease. A multi-disciplinary and unbiased approach is required if the mysteries of diabetes are to be unraveled and new, directly relevant therapeutic targets identified. Recognizing this, the JDRF nPOD approaches the study of diabetes with a focus on human samples and a strong emphasis on collaborative research that is uniquely centered on the study of common samples and, whenever applicable, the promotion of real-time data sharing. While these are relevant, pragmatic considerations, the operative philosophy implemented by nPOD and its investigators largely stems from compassion and respect for the loss of human lives, and from full awareness of the moral responsibility that comes with acceptance of organ donation, the supreme gift that defeats death and disease by embracing life. None of this would be possible without the encouragement, thrust and support unreservedly given by patients and their families, both individually and through the JDRF.

Acknowledgements The authors wish to acknowledge the critical contributions to JDRF nPOD of the nPOD staff, the OPOs and the Investigators. Support for nPOD has been provided by JDRF.

References

Bingley PJ, Bonifacio E, Williams AJ, Genovese S, Bottazzo GF, Gale EA (1997) Prediction of IDDM in the general population: strategies based on combinations of autoantibody markers. Diabetes 46:1701–1710

Bonner-Weir S, Li WC, Ouziel-Yahalom L, Guo L, Weir GC, Sharma A (2010) Beta-cell growth and regeneration: replication is only part of the story. Diabetes 59:2340–2348

Burke III GW, Vendrame F, Pileggi A, Ciancio G, Reijonen H, Pugliese A (2011) Recurrence of autoimmunity following pancreas transplantation. Curr Diab Rep 11(5):413–419

Coppieters KT, Wiberg A, Amirian N, Kay TW, von Herrath MG (2011) Persistent glucose transporter expression on pancreatic beta cells from longstanding type 1 diabetic individuals. Diabetes Metab Res Rev 27:746–754

Coppieters KT, Dotta F, Amirian N, Campbell PD, Kay TW, Atkinson MA, Roep BO, von Herrath MG (2012) Demonstration of islet-autoreactive CD8 T cells in insulitic lesions from recent onset and long-term type 1 diabetes patients. J Exp Med 209(1):51–60

Dotta F, Censini S, van Halteren AG, Marselli L, Masini M, Dionisi S, Mosca F, Boggi U, Muda AO, Prato SD, Elliott JF, Covacci A, Rappuoli R, Roep BO, Marchetti P (2007) Coxsackie B4 virus infection of beta cells and natural killer cell insulitis in recent-onset type 1 diabetic patients. Proc Natl Acad Sci USA 104:5115–5120

Gianani R, Campbell-Thompson M, Sarkar SA, Wasserfall C, Pugliese A, Solis JM, Kent SC, Hering BJ, West E, Steck A, Bonner-Weir S, Atkinson MA, Coppieters K, von Herrath M, Eisenbarth GS (2010) Dimorphic histopathology of long-standing childhood-onset diabetes. Diabetologia 53:690–698

In't Veld P (2011a) Insulitis in human type 1 diabetes: the quest for an elusive lesion. Islets 3:131–138

In't Veld P (2011b) Insulitis in the human endocrine pancreas: does a viral infection lead to inflammation and beta cell replication? Diabetologia 54:2220–2222

In't Veld P, Lievens D, De Grijse J, Ling Z, Van der Auwera B, Pipeleers-Marichal M, Gorus F, Pipeleers D (2007) Screening for insulitis in adult autoantibody-positive organ donors. Diabetes 56:2400–2404

Keenan HA, Sun JK, Levine J, Doria A, Aiello LP, Eisenbarth G, Bonner-Weir S, King GL (2010) Residual insulin production and pancreatic ss-cell turnover after 50 years of diabetes: Joslin Medalist Study. Diabetes 59:2846–2853

Kodama K, Butte AJ, Creusot RJ, Su L, Sheng D, Hartnett M, Iwai H, Soares LR, Fathman CG (2008) Tissue- and age-specific changes in gene expression during disease induction and progression in NOD mice. Clin Immunol 129:195–201

Krischer JP, Cuthbertson DD, Yu L, Orban T, Maclaren N, Jackson R, Winter WE, Schatz DA, Palmer JP, Eisenbarth GS (2003) Screening strategies for the identification of multiple antibody-positive relatives of individuals with type 1 diabetes. J Clin Endocrinol Metab 88:103–108

Kyewski B, Klein L (2006) A central role for central tolerance. Annu Rev Immunol 24:571–606

Lempainen J, Tauriainen S, Vaarala O, Makela M, Honkanen H, Marttila J, Veijola R, Simell O, Hyoty H, Knip M, Ilonen J (2012) Interaction of enterovirus infection and Cow's milk-based formula nutrition in type 1 diabetes-associated autoimmunity. Diabetes Metab Res Rev 10 28(2):177–185

Liu EH, Digon BJ III, Hirshberg B, Chang R, Wood BJ, Neeman Z, Kam A, Wesley RA, Polly SM, Hofmann RM, Rother KI, Harlan DM (2009) Pancreatic beta cell function persists in many patients with chronic type 1 diabetes, but is not dramatically improved by prolonged immunosuppression and euglycaemia from a beta cell allograft. Diabetologia 52:1369–1380

Maclaren NK, Lan MS, Schatz D, Malone J, Notkins AL, Krischer J (2003) Multiple autoantibodies as predictors of type 1 diabetes in a general population. Diabetologia 46:873–874

Martin-Pagola A, Sisino G, Allende G, Dominguez-Bendala J, Gianani R, Reijonen H, Nepom GT, Ricordi C, Ruiz P, Sageshima J, Ciancio G, Burke GW, Pugliese A (2008) Insulin protein and proliferation in ductal cells in the transplanted pancreas of patients with type 1 diabetes and recurrence of autoimmunity. Diabetologia 51:1803–1813

Matthews JB, Staeva TP, Bernstein PL, Peakman M, von Herrath M (2010) Developing combination immunotherapies for type 1 diabetes: recommendations from the ITN-JDRF type 1 diabetes combination therapy assessment group. Clin exp Immunol 160:176–184

Meier JJ, Bhushan A, Butler AE, Rizza RA, Butler PC (2005) Sustained beta cell apoptosis in patients with long-standing type 1 diabetes: indirect evidence for islet regeneration? Diabetologia 48:2221–2228

Metzger TC, Anderson MS (2011) Control of central and peripheral tolerance by Aire. Immunol Rev 241:89–103

Monti P, Scirpoli M, Rigamonti A, Mayr A, Jaeger A, Bonfanti R, Chiumello G, Ziegler AG, Bonifacio E (2007) Evidence for in vivo primed and expanded autoreactive T cells as a specific feature of patients with type 1 diabetes. J Immunol 179:5785–5792

Monti P, Scirpoli M, Maffi P, Ghidoli N, De TF, Bertuzzi F, Piemonti L, Falcone M, Secchi A, Bonifacio E (2008) Islet transplantation in patients with autoimmune diabetes induces homeostatic cytokines that expand autoreactive memory T cells. J Clin Invest 118:1806–1814

Oikarinen M, Tauriainen S, Honkanen T, Oikarinen S, Vuori K, Kaukinen K, Rantala I, Maki M, Hyoty H (2008) Detection of enteroviruses in the intestine of type 1 diabetic patients. Clin exp Immunol 151:71–75

Oikarinen S, Martiskainen M, Tauriainen S, Huhtala H, Ilonen J, Veijola R, Simell O, Knip M, Hyoty H (2011) Enterovirus RNA in blood is linked to the development of type 1 diabetes. Diabetes 60:276–279

Orban T, Sosenko JM, Cuthbertson D, Krischer JP, Skyler JS, Jackson R, Yu L, Palmer JP, Schatz D, Eisenbarth G (2009) Pancreatic islet autoantibodies as predictors of type 1 diabetes in the Diabetes Prevention Trial-Type 1. Diabetes Care 32:2269–2274

Pugliese A (2004) Central and peripheral autoantigen presentation in immune tolerance. Immunology 111:138–146

Richardson SJ, Willcox A, Bone AJ, Foulis AK, Morgan NG (2009) The prevalence of enteroviral capsid protein vp1 immunostaining in pancreatic islets in human type 1 diabetes. Diabetologia 52:1143–1151

Rother KI, Spain LM, Wesley RA, Digon BJ III, Baron A, Chen K, Nelson P, Dosch HM, Palmer JP, Brooks-Worrell B, Ring M, Harlan DM (2009) Effects of exenatide alone and in combination with daclizumab on beta-cell function in long-standing type 1 diabetes. Diabetes Care 32:2251–2257

Rowe PA, Campbell-Thompson ML, Schatz DA, Atkinson MA (2011) The pancreas in human type 1 diabetes. Semin Immunopathol 33:29–43

Salminen KK, Vuorinen T, Oikarinen S, Helminen M, Simell S, Knip M, Ilonen J, Simell O, Hyoty H (2004) Isolation of enterovirus strains from children with preclinical type 1 diabetes. Diabet Med 21:156–164

Stene LC, Oikarinen S, Hyoty H, Barriga KJ, Norris JM, Klingensmith G, Hutton JC, Erlich HA, Eisenbarth GS, Rewers M (2010) Enterovirus infection and progression from islet autoimmunity to type 1 diabetes: the Diabetes and Autoimmunity Study in the Young (DAISY). Diabetes 59:3174–3180

Sutherland DE, Sibley R, Xu XZ, Michael A, Srikanta AM, Taub F, Najarian J, Goetz FC (1984) Twin-to-twin pancreas transplantation: reversal and reenactment of the pathogenesis of type I diabetes. Trans Assoc Am Phys 97:80–87

Tauriainen S, Salmela K, Rantala I, Knip M, Hyoty H (2010) Collecting high-quality pancreatic tissue for experimental study from organ donors with signs of beta cell autoimmunity. Diabetes Metab Res Rev 26:585–592

Tauriainen S, Oikarinen S, Oikarinen M, Hyoty H (2011) Enteroviruses in the pathogenesis of type 1 diabetes. Semin Immunopathol 33:45–55

Tracy S, Drescher KM, Chapman NM (2011) Enteroviruses and type 1 diabetes. Diabetes Metab Res Rev 27:820–823

Velthuis JH, Unger WW, Van Der Slik AR, Duinkerken G, Engelse M, Schaapherder AF, Ringers J, van Kooten C, de Koning EJ, Roep BO (2009) Accumulation of autoreactive effector T cells and allo-specific regulatory T cells in the pancreas allograft of a type 1 diabetic recipient. Diabetologia 52:494–503

Vendrame F, Pileggi A, Laughlin E, Allende G, Martin-Pagola A, Molano RD, Diamantopoulos S, Standifer N, Geubtner K, Falk BA, Ichii H, Takahashi H, Snowhite I, Chen Z, Mendez A, Chen L, Sageshima J, Ruiz P, Ciancio G, Ricordi C, Reijonen H, Nepom GT, Burke GW III, Pugliese A (2010) Recurrence of type 1 diabetes after simultaneous pancreas-kidney transplantation, despite immunosuppression, is associated with autoantibodies and pathogenic autoreactive CD4 T-cells. Diabetes 59:947–957

Wenzlau JM, Frisch LM, Gardner TJ, Sarkar S, Hutton JC, Davidson HW (2009) Novel antigens in type 1 diabetes: the importance of ZnT8. Curr Diab Rep 9:105–112

Willcox A, Richardson SJ, Bone AJ, Foulis AK, Morgan NG (2011) Immunohistochemical analysis of the relationship between islet cell proliferation and the production of the enteroviral capsid protein, VP1, in the islets of patients with recent-onset type 1 diabetes. Diabetologia 54:2417–2420

Yip L, Su L, Sheng D, Chang P, Atkinson M, Czesak M, Albert PR, Collier AR, Turley SJ, Fathman CG, Creusot RJ (2009) Deaf1 isoforms control the expression of genes encoding peripheral tissue antigens in the pancreatic lymph nodes during type 1 diabetes. Nat Immunol 10:1026–1033

Ziegler AG, Nepom GT (2010) Prediction and pathogenesis in type 1 diabetes. Immunity 32:468–478

Part V
Possible Mechanisms

Chapter 25
Virus-Induced Models for Type 1 Diabetes in Mice

Urs Christen and Matthias G. von Herrath

Abstract Epidemiological and experimental data strongly suggest that type 1 diabetes occurs as a detrimental combination of genetic predisposition and environmental triggering factors. Viruses have been implicated in the etiology of type 1 diabetes as environmental triggers due to epidemiological association as well as by direct isolation from pancreatic type 1 diabetes patients. The concept of virus-induced type 1 diabetes has been used as an experimental basis in several mouse models. Studies are reviewed on how virus-induced animal models have helped to understand the mechanisms of the pathogenesis of autoimmune-mediated diabetes. We intend to review how such findings obtained from animal models have helped to design novel therapeutic interventions for type 1 diabetes.

Introduction

It is the current point of view that autoimmune diseases, such as type 1 diabetes, arise as a detrimental combination of genetic predisposition and environmental triggering factors. In the case of type 1 diabetes, besides the predominant HLA class I and II susceptibility loci, several other diabetes loci have been identified, and polymorphisms within these loci have been associated with the changes in the risk of developing the disease (for review see Todd 2010; Ziegler and Nepom 2010). However, several observations indicate that additional factors contribute to the clinical manifestation and the etiology of type 1 diabetes. For example, there is no strict

U. Christen, Ph.D. (✉)
Pharmazentrum Frankfurt/ZAFES, Goethe University Hospital, Frankfurt am Main, Germany
e-mail: christen@med.uni-frankfurt.de

M.G. von Herrath, M.D.
La Jolla Institute for Allergy and Immunology, La Jolla, CA, USA
e-mail: matthias@liai.org

K.W. Taylor, H. Hyoty, A. Toniolo, A.J. Zuckerman (eds.), *Diabetes and Viruses*,
DOI 10.1007/978-1-4614-4051-2_25,

concordance between homozygous twins (Redondo et al. 1999). Further, not all individuals that carry a predominance of susceptibility genes develop type 1 diabetes and even individuals with protective genetic loci develop disease (Ziegler and Nepom 2010). Human microbial pathogens are environmental factors that have been associated with autoimmune diseases in the past. In type 1 diabetes evidence has accumulated over the last decades that in particular enteroviruses might play a key role in the initiation and/or propagation etiology of the disease (Tracy et al. 2010). Very recently a large meta-analysis confirmed that there is a significant association between enterovirus infection and the development of type 1 diabetes (Yeung et al. 2011). Yeung et al. (2011) performed a systematic review of 33 type 1 diabetes prevalence studies from 1990 to 2010 involving 1,931 type 1 diabetes patients and 2,517 control individuals and found odds ratios of 3.7 between enterovirus infection and type 1 diabetes-related autoimmunity and 9.8 between enterovirus infection and clinical type 1 diabetes.

Mechanisms of How Viruses Could Trigger Autoimmune Diabetes

There are several mechanisms by which viruses might provoke an autodestructive immune response. First, viruses might infect and damage the target cell directly resulting in spontaneous (virus-induced) or an immune-mediated cell lysis. Target cell antigen presentation will be massively elevated including presentation of determinants of normally sequestered antigens that have not yet been recognized by the immune system. Second, the strong inflammatory response caused in the infected tissue might generate an environment that allows further attraction and activation of aggressive immune cells that would under normal conditions not have migrated in critical number to the site of inflammation. Viruses might therefore generate a "fertile field" for subsequent autoimmune damage (von Herrath et al. 2003). It is important to note that the presence of cytokines and other inflammatory factors per se might impair function of β-cells (Eizirik et al. 2009; Rhode et al. 2005). Third, viruses might carry determinants with structural similarity to components of the host. Thereby, an immune response directed against the virus might in addition attack the similar structure in the host as well. This concept has been termed "molecular mimicry" (Christen et al. 2010; Christen and von Herrath 2004a; Damian 1964; Oldstone 1989).

Animal Models for Virus-Induced Type 1 Diabetes

The RIP-LCMV model uses a viral infection to initiate an aggressive immune response to a transgenically expressed β-cell target antigen derived from the initiating virus. RIP-LCMV mice express the glycoprotein (GP) or the nucleoprotein (NP) of the lymphocytic choriomeningitis virus (LCMV) under control of the rat insulin promoter (RIP) in the β-cells of the islets of Langerhans (Ohashi et al. 1991; Oldstone et al. 1991). Infection of such mice with LCMV expressing the identical antigens as environmental triggers initiates an immune response that is directed against LCMV including the transgenically expressed GP or NP in the β-cells and subsequently type 1 diabetes ensues (Ohashi et al. 1991; Oldstone et al. 1991). Importantly, without viral infection or following immunization with peptides, such mice rarely develop diabetes (Martinic et al. 2010). Thus, the design of the RIP-LCMV model is based on the molecular antigenic recognition of a transgenically expressed viral protein in β-cells as well as the "fertile field" theory, which implies that viral infections induce inflammation that accelerates autoimmunity by recruitment of excessive aggressive lymphocytes and augmented presentation of islet autoantigens. The RIP-LCMV mouse model comes in at least three "flavors." All variants use proteins of LCMV, expressed specifically in the pancreatic β-cells, as target antigens to direct an aggressive immune response to the islets of Langerhans. The models have been developed in the late 1980s in the laboratory of Hans Hengartner and Rolf Zinkernagel in Zürich, Switzerland (Ohashi et al. 1991) and at the same time in Michael Oldstone's laboratory in La Jolla, CA (Oldstone et al. 1991). The RIP-LCMV-GP model developed in Zürich uses the GP of the LCMV stain WE as target antigen. Such mice do not develop type 1 diabetes unless infected with LCMV-WE (Ohashi et al. 1991). In contrast, the transgenic mouse lines generated in La Jolla express the GP or the NP of the LCMV strain Armstrong (Arm) (Oldstone et al. 1991). The fast-onset RIP-LCMV-GP line that expresses the target antigen exclusively in the β-cells develops type 1 diabetes within 10–14 days after infection with LCMV-Arm, whereas the slow-onset RIP-LCMV-NP mice express the LCMV-NP in the pancreas as well as in the thymus and develop type 1 diabetes within 1–6 months after infection (von Herrath et al. 1994). It is important to note that the immunodominant CD8 and CD4 epitopes (GP_{33-41} and GP_{64-80}) are identical in LCMV-WE and LCMV-Arm. Interestingly, the level of viral antigen expression in the pancreatic β-cells affects the development of type 1 diabetes significantly (Martinic et al. 2010). RIP-LCMV-WE mice express higher levels of LCMV-GP in the β-cells than RIP-LCMV-Arm mice and developed more severe type 1 diabetes characterized by a higher incidence, slightly faster onset and more rapid destruction of islets (Martinic et al. 2010).

Antigen-induced models can add in many ways to the insight which can be obtained from spontaneous models, such as the non-obese diabetes (NOD) mouse (Makino et al. 1980) or the biobreeding (BB) rat (Nakhooda et al. 1977). The NOD mouse is definitely the most popular animal model for type 1 diabetes and develops spontaneously autoimmune disorders including diabetes and thyroiditis (for a detailed review see for example: Driver et al. 2010). Described initially by Makino et al. (1980), the NOD mouse has been used widely since then and not surprisingly searching for the term "NOD mouse" results in more than 6,700 hits in the US Library of Medicine (PubMed) and millions of hits in any internet search engine. Research on the pathogenesis and genetics of type 1 diabetes in the NOD mouse over the last three decades led to the discovery of genetic susceptibility loci as well as mechanisms of islet infiltration, destruction, and regeneration (Roep et al. 2004;

Shoda et al. 2005). Although the NOD mouse model is used widely and its type 1 diabetes pathogenesis resembles human type 1 diabetes in some aspects, such spontaneous type 1 diabetes models have some disadvantages. In particular, the precise time of disease initiation is not defined or known. The onset of disease varies from 12 to 30 weeks of age between individual mice, depending on several factors including breeding facilities and animal housing, and it is unfortunately all too easy to prevent diabetes in the NOD (Atkinson and Leiter 1999; Roep and Atkinson 2004). The lack of a defined initiating event and the asynchronized disease progress impedes investigations for prevention and also of pathogenic mechanisms occurring early during the disease process.

In contrast to spontaneous models, inducible mouse models have the advantage of knowing the precise beginning of the immunopathogenic processes that result in the destruction of the β-cells and subsequent type 1 diabetes. Several models have been developed, such as the Ins-HA (Lo et al. 1992), the RIP-mOVA (Kurts et al. 1996), and the RIP-LCMV (Ohashi et al. 1991; Oldstone et al. 1991) model. These mouse models have in common that they express a model target antigen specifically in the β-cells of the pancreatic islets of Langerhans. However, whereas the RIP-LCMV model uses infection by LCMV alone as a trigger for disease, the Ins-HA and the RIP-mOVA model require transfer of effector T cells in the form of T cell clones or as isolated from TcR transgenic mice. The Ins-HA mouse expresses the influenza virus hemagglutinin (HA) under control of the insulin (Ins) promoter specifically by the pancreatic β-cells (Lo et al. 1992). In contrast to the RIP-LCMV model, infection of mice with the corresponding virus (i.e., influenza virus) is not sufficient to induce type 1 diabetes and transfer of HA-specific TcR transgenic T cells is necessary. Alternatively, double transgenic mice expressing both HA and HA-specific TcR have been generated that develop spontaneously type 1 diabetes within 2 weeks of age due to the extensive destruction of HA-bearing β-cells by TcR transgenic T cells (Lo et al. 1992). The RIP-mOVA mouse expresses ovalbulin under control of the rat insulin promoter as target antigen (Kurts et al. 1996). Diabetes is induced by transfer of Ova-specific OT-I CD8 T cells that are activated by cross-presentation of Ova and migrated to islets where they destroy the β-cells (Kurts et al. 1996). Crossing RIP-mOVA mice with DO11.10 TcR transgenic mice that express Ova-specific CD4 T cells results in the development of type 1 diabetes around 10 weeks of age (Clough et al. 2008). It is important to note that the spontaneous type 1 diabetes in such TcR-double transgenic models that carry a high frequency of β-cell-specific T cells is accelerated considerably when compared to the spontaneous NOD mouse model, in which the autodestructive T cell response is being generated over a longer period of time from a low frequency of precursor T cells.

Similar accelerations of the immunopathogenesis can be achieved by transferring β-cells-specific T cells to the NOD mouse. For example, in the BDC2.5/NOD model the onset of type 1 diabetes is decreased dramatically by transfer of BDC2.5 TcR transgenic CD4 T cells with specificity to a β-cell antigen (Haskins and McDuffie 1990). The transfer of such diabetogenic anti-β-cell antigen T cell clones was shown to induce massive insulitis and type 1 diabetes within 2 week in NOD neonates (Katz et al. 1995).

The Immunopathogenesis of Virus-Induced Type 1 Diabetes

The RIP-LCMV model helped to identify potentially critical factors of acute and chronic inflammation that could drive the destruction of β-cells by autoaggressive T cells. Further, several mechanisms of β-cell destruction as well as the activation, proliferation, and migration of autoreactive T cells have been described in detail in this model. The identification of such pathogenic mechanisms has helped to evaluate therapeutic interventions that prevent the development of disease or block the ongoing autoimmune destruction of the β-cells and to abrogate type 1 diabetes. Figure 25.1 displays schematically how the infection of RIP-LCMV mice with a single dose of LCMV subsequently leads to the destruction of most of the β-cells. It has been shown that blockade of some of the steps in the immunopathogenesis of type 1 diabetes in the RIP-LCMV model results in partial or complete abrogation of the disease. One of the most interesting inflammatory factors is TNFα, since the exact timing of its expression decides on the outcome of type 1 diabetes. It has been demonstrated that blockade of TNFα early after LCMV infection abrogates type 1 diabetes possibly by preventing the very initial steps of pancreatic inflammation (Christen et al. 2001). Interestingly this mirrors the results of a recent human trial, in which neutralization of TNFα by the TNFR75-IgG1 fusion protein etanercept was successful in treatment of children with new-onset type 1 diabetes (Mastrandrea et al. 2009). In contrast, blockade of TNFα at a time when the destruction of the islets is already ongoing has no effect on the frequency of the onset of type 1 diabetes. Moreover, β-cell-specific overexpression of TNFα at such a late phase of the immunopathogenesis resulted in the activation-induced cell death of autoaggressive T cells and a subsequent reversion of diabetic mice to normoglycemia (Christen et al. 2001). Similarly, the precise timing of intervention is also important for blocking of critical chemokines that are involved in the initial attraction of lymphocytes to the islets of Langerhans (Christen and von Herrath 2004b). The chemokine CXCL10 (IP-10, IFNγ-inducible chemokine of 10 kDa) is uniquely and massively expressed by the islets of Langerhans at days 1–4 post LCMV infection (Christen et al. 2003; Frigerio et al. 2002). However, its expression returns to pre-infection levels at days 7–10 after infection (Christen et al. 2003). Thus, it is no surprise that blockade of CXCL10 was only successful when applied during the peak of its expression but not at a later time (Christen et al. 2003). In contrast, overexpression of CXCL10 in the pancreatic β-cells results in an acceleration of type 1 diabetes in the RIP-LCMV-NP model (Rhode et al. 2005). Interestingly, both time and location of CXCL10 expression have an impact on type 1 diabetes. LCMV-specific CD8 T cells express the CXCL10 receptor (CXCR3) and seem to migrate to or remain at the location with the highest CXCL10 concentration. In analogy, overexpression of CXCL10 was found in human islets as part of the interferon signature present in human islets (Uno et al. 2010).

An interesting scenario is observed upon repeated infection of RIP-LCMV-NP mice at a time when the destruction of islets is ongoing (i.e., 4 weeks after the initial infection with LCMV strain Armstrong). At that time an LCMV variant (strain Pasteur)

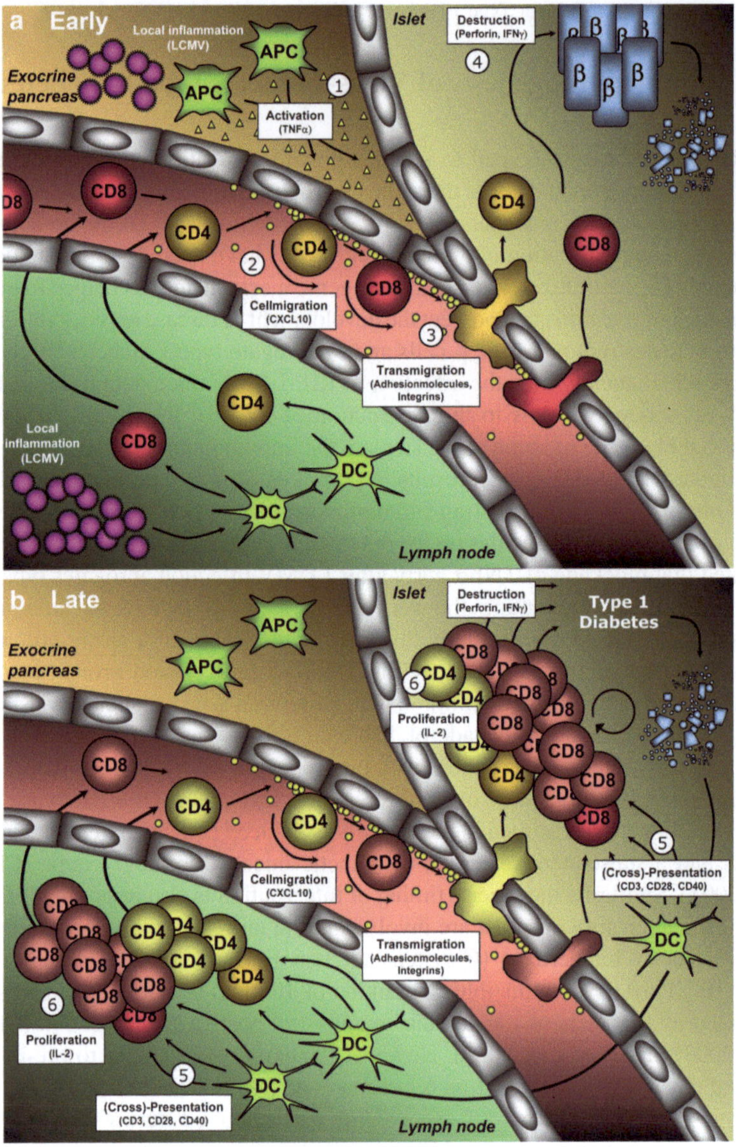

Fig. 25.1 Immunopathogenesis of virus-induced type 1 diabetes in the RIP-LCMV model. (**a**) Early: local infection of the pancreas and/or the pancreatic lymph node induces the activation of local antigen presenting cells (APC) that release inflammatory mediators, such as TNFα (*1*). The subsequent release of chemotactic cytokines (chemokines), in particular CXCL10 (*2*), by endothelial cells as well as by the β-cells themselves, results in the rolling, arrest, and transmigration of lymphocytes, including some islet antigen-specific CD4 and CD8 T cells (*3*). Some β-cells are destroyed by perforin and/or IFNγ-dependent mechanisms (*4*). However, at this early stage the mice are not yet diabetic, since the remaining intact β-cells are still able to compensate for the loss of overall β-cell mass. (**b**) Late: the destruction of some β-cells causes an augmented uptake, processing, and cross-presentation of β-cell antigens by professional APC, such as dendritic cells, in the islet and/or in the pancreatic lymph node (*5*). As a result the frequency of β-cell-specific CD4 and CD8 T cells is rapidly increasing (*6*), which leads to the destruction of most of the β-cells and ensues clinical type 1 diabetes characterized by massively elevated blood glucose levels

abrogates type 1 diabetes persistently (Christen et al. 2004). LCMV Pasteur replicates predominantly in the pancreatic lymph nodes rather than in the pancreas itself and induces a rapid and massive burst of CXCL10. Indeed, the autoaggressive CXCR3-positive T cells disappear from the islets and migrate to the lymph node, where an increased level of apoptosis can be detected (Christen et al. 2004). Thus, virus infection of a remote location and the accompanied expression of CXCL10 outside of the pancreas act as a "filter" eliminating autoaggressive T cells (Christen and von Herrath 2005). Virus infection during an ongoing autoimmune process can also abrogate disease by inducing regulatory T cells (Tregs). Infection of prediabetic NOD mice with either LCMV or CVB3 reduced the frequency of type 1 diabetes and delayed the onset of disease by increasing the number of TGF-β-producing CD4+ CD25+ Tregs and maintained long-term protection (Filippi et al. 2009). In addition, virus infection induced a transient upregulation of the programmed cell death-1 ligand 1 (PD-L1) on lymphoid cells and prevented the expansion of diabetogenic CD8 T cells expressing programmed cell death-1 (PD-1) (Filippi et al. 2009).

The induction of regulatory factors rather than the elimination of aggressive components is an alternative strategy to reestablish a neutral balance of the immune system. Virus-induced mouse models, such as the RIP-LCMV model, have helped to understand the basics of immune regulation and to evaluate therapeutic applications. A summary of some therapeutic interventions that have been successful in preventing or abrogating type 1 diabetes in the RIP-LCMV mouse model is integrated into Fig. 25.2. The induction of tolerance by oral administration of islet antigens has been tested more than a decade ago (Homann et al. 1999; von Herrath et al. 1996). Since then it has become evident that delivery of islet antigens via the oral or nasal route induces antigen-specific Tregs that suppress the function of autoaggressive T cell and thereby inhibit type 1 diabetes. Alternatively, direct transfer of either in vitro modulated Tregs or peptide-loaded tolerogenic dendritic cells (Unger et al. 2009) are considered possible therapies of type 1 diabetes. Although there has been ample discussion about the stability of Tregs in vivo (Bluestone et al. 2009; Rubtsov et al. 2010), the potential of such an antigen-specific intervention might prevent unwanted side effects such as general immunosuppression. It has also become evident from studies in the RIP-LCMV model as well as in the NOD mouse that combinational treatments might be a successful strategy. In particular, the combination of eradicating autoaggressive lymphocytes by the established anti-CD3 depletion therapy (Chatenoud and Bluestone 2007) on the one hand and enhancing suppressive functions on the other hand might influence heavily the disturbed immune balance. Indeed, treatment of RIP-LCMV mice and NOD mice with anti-CD3 monoclonal antibody combined with the administration of nasal preinsulin reduced the frequency of type 1 diabetes significantly (Bresson et al. 2006). It seems that the eradication of most of the T cells opened a window of opportunity in which the nasal administration of islet antigen induced a higher frequency of antigen-specific Tregs in the reexpanding T cell repertoire. The first steps in the direction of assessing combination therapies in patients have been initiated by the Immune Tolerance Network (ITN) and the Juvenile Diabetes Research Foundation (JDRF) in assembling a Type 1 Diabetes Combination Therapy

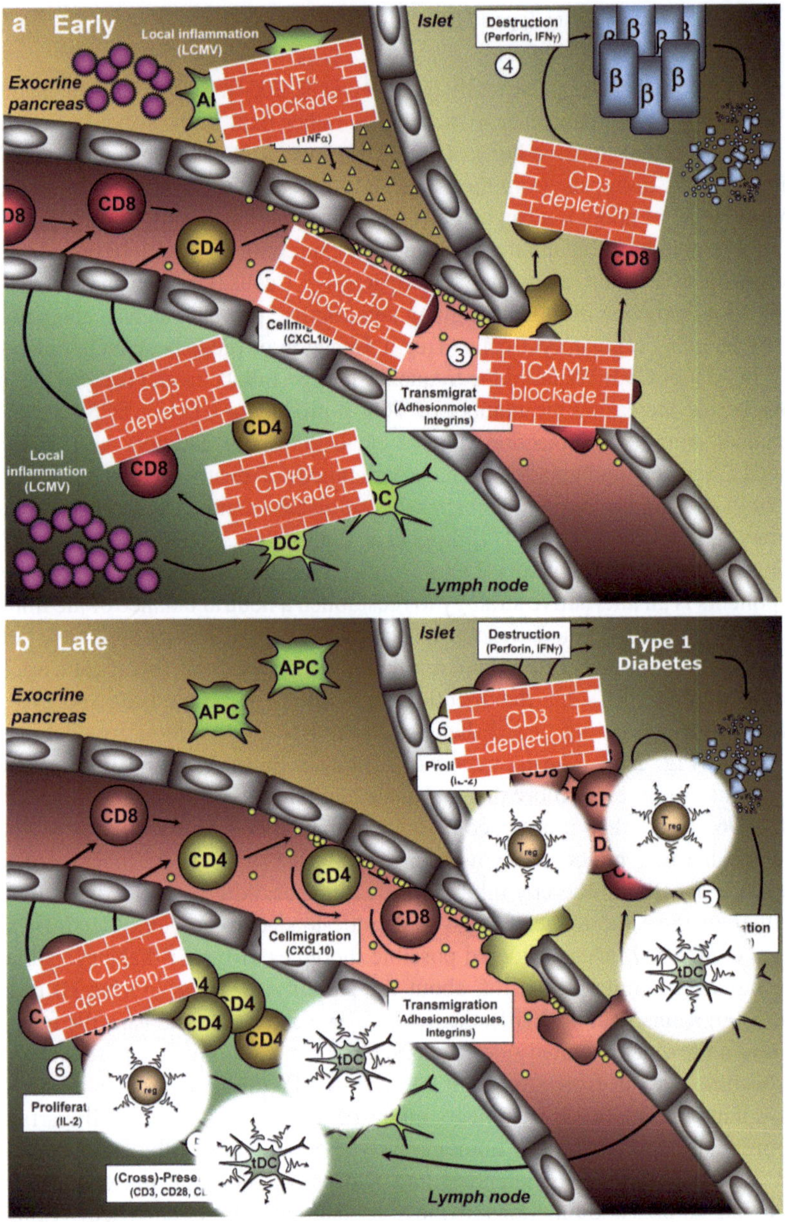

Fig. 25.2 Therapy of type 1 diabetes in the RIP-LCMV model by elimination or autoaggressive lymphocytes or by induction of regulatory cells. (**a**) The extent of inflammation directly after infection with LCMV can be reduced by blockade of the key inflammatory factors TNFα (Christen et al. 2001) and CXCL10 (Christen et al. 2003). Further, blockade of CD40L prevents proper co-stimulation and cross-presentation (Homann et al. 2002). Depletion of CD3 T cells dramatically reduces the frequency of autoaggressive lymphocytes and provides a window of opportunity for the generation of Tregs (Bresson et al. 2006). (**b**) Treatments such as CD3 T cell depletion are effective even when the autodestructive process is already ongoing (Bresson et al. 2006). Similarly, Tregs and tolerogenic DCs might have a long-lasting suppressive effect (Rubtsov et al. 2010; Unger et al. 2009)

Assessment Group (Matthews et al. 2010). The assessment group recommends that combination treatments should be limited to two agents with an independent/complementary mechanism and a safe protocol design (Matthews et al. 2010). Thus, we strongly encourage fellow researchers to consider combination therapies in their respective animal models in order to explore possible additional benefits for type 1 diabetes patients.

References

Atkinson MA, Leiter EH (1999) The NOD mouse model of type 1 diabetes: as good as it gets? Nat Med 5(6):601–604

Bluestone JA, Mackay CR, O'Shea JJ, Stockinger B (2009) The functional plasticity of T cell subsets. Nat Rev Immunol 9(11):811–816

Bresson D, Togher L, Rodrigo E, Chen Y, Bluestone JA, Herold KC, von Herrath M (2006) Anti-CD3 and nasal proinsulin combination therapy enhances remission from recent-onset autoimmune diabetes by inducing Tregs. J Clin Invest 116(5):1371–1381

Chatenoud L, Bluestone JA (2007) CD3-specific antibodies: a portal to the treatment of autoimmunity. Nat Rev Immunol 7(8):622–632

Christen U, von Herrath MG (2004a) Induction, acceleration or prevention of autoimmunity by molecular mimicry. Mol Immunol 40(14–15):1113–1120

Christen U, von Herrath MG (2004b) IP-10 and type 1 diabetes: a question of time and location. Autoimmunity 37(4):273–282

Christen U, von Herrath MG (2005) Infections and autoimmunity – good or bad? J Immunol 174(12):7481–7486

Christen U, Wolfe T, Mohrle U, Hughes AC, Rodrigo E, Green EA, Flavell RA, von Herrath MG (2001) A dual role for TNF-alpha in type 1 diabetes: islet-specific expression abrogates the ongoing autoimmune process when induced late but not early during pathogenesis. J Immunol 166(12):7023–7032

Christen U, McGavern DB, Luster AD, von Herrath MG, Oldstone MB (2003) Among CXCR3 chemokines, IFN-gamma-inducible protein of 10 kDa (CXC chemokine ligand (CXCL) 10) but not monokine induced by IFN-gamma (CXCL9) imprints a pattern for the subsequent development of autoimmune disease. J Immunol 171(12):6838–6845

Christen U, Benke D, Wolfe T, Rodrigo E, Rhode A, Hughes AC, Oldstone MB, Von Herrath MG (2004) Cure of prediabetic mice by viral infections involves lymphocyte recruitment along an IP-10 gradient. J Clin Invest 113(1):74–84

Christen U, Hintermann E, Holdener M, von Herrath MG (2010) Viral triggers for autoimmunity: is the 'glass of molecular mimicry' half full or half empty? J Autoimmun 34(1):38–44

Clough LE, Wang CJ, Schmidt EM, Booth G, Hou TZ, Ryan GA, Walker LS (2008) Release from regulatory T cell-mediated suppression during the onset of tissue-specific autoimmunity is associated with elevated IL-21. J Immunol 180(8):5393–5401

Damian RT (1964) Molecular mimicry: antigen sharing by parasite and host and its consequences. Am Nat 98(900):129–149

Driver JP, Serreze DV, Chen YG (2010) Mouse models for the study of autoimmune type 1 diabetes: a NOD to similarities and differences to human disease. Semin Immunopathol 33(1):67–87

Eizirik DL, Colli ML, Ortis F (2009) The role of inflammation in insulitis and beta-cell loss in type 1 diabetes. Nat Rev Endocrinol 5(4):219–226

Filippi CM, Estes EA, Oldham JE, von Herrath MG (2009) Immunoregulatory mechanisms triggered by viral infections protect from type 1 diabetes in mice. J Clin Invest 119(6): 1515–1523

Frigerio S, Junt T, Lu B, Gerard C, Zumsteg U, Hollander GA, Piali L (2002) Beta cells are responsible for CXCR3-mediated T-cell infiltration in insulitis. Nat Med 8(12):1414–1420
Haskins K, McDuffie M (1990) Acceleration of diabetes in young NOD mice with a CD4+ islet-specific T cell clone. Science 249(4975):1433–1436
Homann D, Dyrberg T, Petersen J, Oldstone MBA, von Herrath MG (1999) Insulin in oral immune "tolerance": a one-amino acid change in the B chain makes the difference. J Immunol 163:1833–1838
Homann D, Jahreis A, Wolfe T, Hughes A, Coon B, van Stipdonk MJ, Prilliman KR, Schoenberger SP, von Herrath MG (2002) CD40L blockade prevents autoimmune diabetes by induction of bitypic NK/DC regulatory cells. Immunity 16(3):403–415
Katz JD, Benoist C, Mathis D (1995) T helper cell subsets in insulin-dependent diabetes. Science 268(5214):1185–1188
Kurts C, Heath WR, Carbone FR, Allison J, Miller JF, Kosaka H (1996) Constitutive class I-restricted exogenous presentation of self antigens in vivo. J Exp Med 184(3):923–930
Lo D, Freedman J, Hesse S, Palmiter RD, Brinster RL, Sherman LA (1992) Peripheral tolerance to an islet cell-specific hemagglutinin transgene affects both CD4+ and CD8+ T cells. Eur J Immunol 22(4):1013–1022
Makino S, Kunimoto K, Muraoka Y, Mizushima Y, Katagiri K, Tochino Y (1980) Breeding of a non-obese, diabetic strain of mice. Jikken Dobutsu 29(1):1–13
Martinic MM, Huber C, Coppieters K, Oldham JE, Gavin AL, von Herrath MG (2010) Expression level of a pancreatic neo-antigen in beta cells determines degree of diabetes pathogenesis. J Autoimmun 35(4):404–413
Mastrandrea L, Yu J, Behrens T, Buchlis J, Albini C, Fourtner S, Quattrin T (2009) Etanercept treatment in children with new-onset type 1 diabetes: pilot randomized, placebo-controlled, double-blind study. Diabetes Care 32(7):1244–1249
Matthews JB, Staeva TP, Bernstein PL, Peakman M, von Herrath M (2010) Developing combination immunotherapies for type 1 diabetes: recommendations from the ITN-JDRF Type 1 Diabetes Combination Therapy Assessment Group. Clin Exp Immunol 160(2):176–184
Nakhooda AF, Like AA, Chappel CI, Murray FT, Marliss EB (1977) The spontaneously diabetic Wistar rat. Metabolic and morphologic studies. Diabetes 26(2):100–112
Ohashi P, Oehen S, Buerki K, Pircher H, Ohashi C, Odermatt B, Malissen B, Zinkernagel R, Hengartner H (1991) Ablation of tolerance and induction of diabetes by virus infection in viral antigen transgenic mice. Cell 65:305–317
Oldstone MBA (1989) Molecular mimicry as a mechanism for the cause and as a probe uncovering etiologic agent(s) of autoimmune disease. Curr Top Microbiol Immunol 145:127–136
Oldstone MBA, Nerenberg M, Southern P, Price J, Lewicki H (1991) Virus infection triggers insulin-dependent diabetes mellitus in a transgenic model: role of anti-self (virus) immune response. Cell 65:319–331
Redondo MJ, Rewers M, Yu L, Garg S, Pilcher CC, Elliott RB, Eisenbarth GS (1999) Genetic determination of islet cell autoimmunity in monozygotic twin, dizygotic twin, and non-twin siblings of patients with type 1 diabetes: prospective twin study. BMJ 318(7185):698–702
Rhode A, Pauza ME, Barral AM, Rodrigo E, Oldstone MB, von Herrath MG, Christen U (2005) Islet-specific expression of CXCL10 causes spontaneous islet infiltration and accelerates diabetes development. J Immunol 175(6):3516–3524
Roep BO, Atkinson M (2004) Animal models have little to teach us about type 1 diabetes: 1. In support of this proposal. Diabetologia 47(10):1650–1656
Roep BO, Atkinson M, von Herrath M (2004) Satisfaction (not) guaranteed: re-evaluating the use of animal models of type 1 diabetes. Nat Rev Immunol 4(12):989–997
Rubtsov YP, Niec RE, Josefowicz S, Li L, Darce J, Mathis D, Benoist C, Rudensky AY (2010) Stability of the regulatory T cell lineage in vivo. Science 329(5999):1667–1671
Shoda LK, Young DL, Ramanujan S, Whiting CC, Atkinson MA, Bluestone JA, Eisenbarth GS, Mathis D, Rossini AA, Campbell SE, Kahn R, Kreuwel HT (2005) A comprehensive review of interventions in the NOD mouse and implications for translation. Immunity 23(2):115–126
Todd JA (2010) Etiology of type 1 diabetes. Immunity 32(4):457–467

Tracy S, Drescher KM, Jackson JD, Kim K, Kono K (2010) Enteroviruses, type 1 diabetes and hygiene: a complex relationship. Rev Med Virol 20:106–116

Unger WW, Laban S, Kleijwegt FS, van der Slik AR, Roep BO (2009) Induction of Treg by monocyte-derived DC modulated by vitamin D3 or dexamethasone: differential role for PD-L1. Eur J Immunol 39(11):3147–3159

Uno S, Imagawa A, Saisho K, Okita K, Iwahashi H, Hanafusa T, Shimomura I (2010) Expression of chemokines, CXC chemokine ligand 10 (CXCL10) and CXCR3 in the inflamed islets of patients with recent-onset autoimmune type 1 diabetes. Endocr J 57(11):991–996

von Herrath MG, Dockter J, Oldstone MBA (1994) How virus induces a rapid or slow onset insulin-dependent diabetes mellitus in a transgenic model. Immunity 1:231–242

von Herrath MG, Dyrberg T, Oldstone MBA (1996) Oral insulin treatment suppresses virus-induced antigen-specific destruction of b-cells and prevents autoimmune diabetes in transgenic mice. J Clin Invest 98:1324–1331

von Herrath MG, Fujinami RS, Whitton JL (2003) Microorganisms and autoimmunity: making the barren field fertile. Nat Rev Microbiol 1:151–157

Yeung WC, Rawlinson WD, Craig ME (2011) Enterovirus infection and type 1 diabetes mellitus: systematic review and meta-analysis of observational molecular studies. BMJ 342:d35

Ziegler AG, Nepom GT (2010) Prediction and pathogenesis in type 1 diabetes. Immunity 32(4):468–478

Chapter 26
The Role of T Lymphocytes in the Pathogenesis of Autoimmune Type 1 Diabetes: Implications for Potential Virus-Mediated Pathways

Martin Eichmann and Mark Peakman

Abstract Evidence accumulated in the last 25 years, since the earliest characterizations of the cellular infiltrate in the pancreas in new-onset type 1 diabetes (Bottazzo et al. 1985), has led to a consensus view that type 1 diabetes (T1D) is an organ-specific autoimmune disorder in which T lymphocytes are critically involved in the process of β-cell death (Table 26.4). In the setting of a dysregulated immune system, β-cells in the pancreatic islets of Langerhans are destroyed, through a combination of targeting by autoreactive T lymphocytes and the local inflammatory milieu. The limited disease concordance in genetically identical twins (Redondo et al. 2008) implicates a complex interplay between host genetic and environmental factors in the initiation and perpetuation of islet inflammation. The possibility that viruses influence these events is the subject of this book. In this chapter the autoimmune pathways to β-cell damage are reprised, providing a knowledge-base upon which to build hypotheses and predictions in relation to a viral etiology for type 1 diabetes. The aim is to document the key immune cells involved, as well as the critical points at which pathways to disease are initiated. The potential role of viruses at these "nodal points" can then be highlighted as a means of discussing the key experiments that will enable progress in the field of the viral etiology of type 1 diabetes.

M. Eichmann, B.Sc.
Department of Immunobiology, King's College London School of Medicine, London, UK
e-mail: martin.eichmann@kcl.ac.uk

M. Peakman, M.B.B.S., B.Sc, M.S.C, Ph.D., F.R.Cpath (✉)
Department of Immunobiology, King's College London School of Medicine, London, UK

National Institute of Health Research Biomedical Research Centre at Guy's & St Thomas' NHS Foundation Trust, King's College London, London, UK
e-mail: mark.peakman@kcl.ac.uk

K.W. Taylor, H. Hyoty, A. Toniolo, A.J. Zuckerman (eds.), *Diabetes and Viruses*,
DOI 10.1007/978-1-4614-4051-2_26,

Establishment and Maintenance of T Cell Tolerance

Tolerance can be defined as controlled unresponsiveness to self and is broadly distinguished as central (thymic) and peripheral. In the thymus there is deletion of T cells bearing T cell receptors (TCR) for antigen with the potential to react against self at too high an affinity. Despite some 95–99% of T lymphocytes undergoing this process, it is likely to be incomplete (e.g., not all self-antigens available or expressed at too low levels). In the periphery, therefore, self-antigens may now become available for T cell priming. Here (i.e. in the tissues and lymph nodes), deletional and anergic mechanisms of self-tolerance induction apply. In addition, self-tolerance may operate via maintenance of a state of indifference or ignorance of the immune system to self-antigens (e.g., via lack of presentation, or lack of appropriate co-stimulation). Finally, these mechanisms are complemented by "immunological police" available in the form of regulatory T cells (Treg). These take different forms, continue to be defined, and probably have diverse induction and effector pathways (see Table 26.1). Tregs may be thymus-derived (so-called natural, or nTreg) or arise in the periphery during an adaptive response [referred to as adaptive (aTreg) or induced (iTreg)]. Tregs operate in an antigen-specific fashion—i.e. they are activated to suppress via TCR interaction with peptide–MHC (major histocompatibility complex)—but may suppress surrounding T and B lymphocytes that have different epitope and antigen specificities, as well as down-modulating the function of antigen-presenting cells (such as dendritic cells and macrophages). Through these effects, Tregs are a key factor in establishing and maintaining tissue-specific immune homeostasis in the periphery.

Loss of Tolerance and Development of Autoimmunity

The broad underlying process that promotes development of autoimmunity is the availability in the periphery of at least some T cells with self-specificity. This potential is amplified by the fact that TCRs are capable of interacting with more than 1 peptide–HLA (human leukocyte antigen) ligand (estimates are that as many as 10^6 different peptides can stimulate a single TCR) (Mason 1998). As a result there is the likelihood that the system developed to scan the ligandome of pathogenic viruses and bacteria has strong capacity for self-recognition. The fact that there are several different mechanisms through which immunological tolerance can operate leads to an obvious conclusion: several different pathological processes can break, or are required to break immunological self-tolerance and give rise to autoimmunity. The multilayered nature of self-tolerance is a failsafe mechanism: all or several control mechanisms must be breached before disease results. This would explain why autoimmune disease is typically multifactorial, often progresses much more slowly than immune reactions to pathogenic organisms (i.e., because of the countering effect of multiple parallel or serial control mechanisms), and may have a tendency to remit and relapse. Is a single self-protein responsible for the initiation

Table 26.1 Identity, properties, and characteristics of regulatory T cells in man

Property	Natural Treg (nTreg)	Induced Treg (iTreg)	Induced Treg (iTreg)—Tr1	Induced Treg (iTreg)—Th3
Development	Thymus	Periphery	Periphery	Periphery (especially in MALT)
Phenotype	$CD4^+CD25^+CTLA\text{-}4^+$ $GITR^+Foxp3^+$	$CD4^+CD25^+CTLA\text{-}4^+$ $Foxp3^+$	$CD4^+CD25^-Foxp3^-$	$CD4^+CD25^+Foxp3^-$ or $CD4^+CD25^-Foxp3^+$
Mechanism of suppression	Membrane bound TGF-β,CTLA-4, cytokines IL-10/TGF-β	Membrane bound TGF-β, CTLA-4, cytokines IL-10/TGF-β	Cell contact independent, IL-10 mediated, TGF-β	TGF-β, IL-4/IL-10
Target cells	APCs and effector T cells	APCs and effector T cells	Effector T cells	Effector T cells
In vivo role	Suppression of autoreactive T cells	Suppression of autoreactive T cells	Control of inflammatory response	Mucosal immunity, control of inflammatory response
Induction	TCR signaling, need for IL-2 or IL-15, CD28 dependent	TCR signaling, TGF-β and IL-2, RA, CTLA-4 co-stimulation	TGF-β and IL-27	Oral delivery of antigen
Specificity	Self-antigen (high avidity?)	Commensal bacteria, self (inflammation), allo-antigens	Poorly defined	Poorly defined

APC antigen-presenting cell, *CTLA-4* cytotoxic T-lymphocyte-associated antigen 4, *FoxP3* forkhead box P3, *GITR* glucocorticoid-induced TNF receptor-related protein, *IL-10* interleukin 10, *MALT* mucosa-associated lymphoid tissue, *RA* retinoic acid, *TCR* T cell receptor, *TGF-β* tumor growth factor β

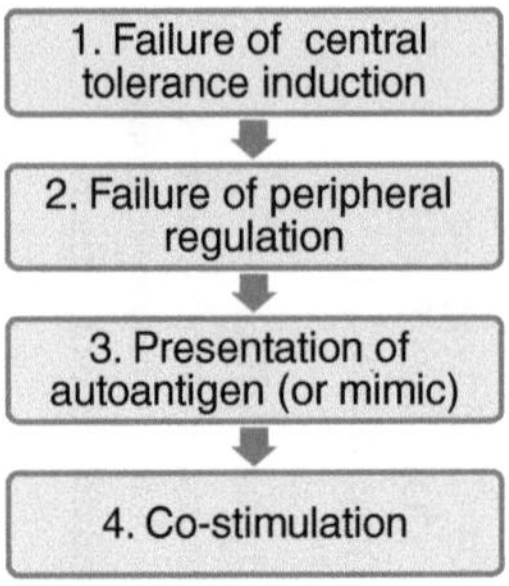

Fig. 26.1 The four major checkpoints that control progression to autoimmunity. There must be a failure at each of these to result in autoimmune disease. The sequential nature of the progression should be viewed as schematic only; clearly presentation of autoantigen may precede failure of peripheral regulation, and provision of adequate co-stimulation must happen in parallel

of type 1 diabetes? The autoantigen insulin is somewhat special, given the fact that a polymorphism in the *INS* (insulin) gene influences the risk of type 1 diabetes. Retention of T cells into islets is antigen-specific and cell-intrinsic (Lennon et al. 2009) which suggests that there can be an ongoing loss of tolerance that spreads to multiple epitopes, rapidly resulting in multiple self-antigen specificities being involved in disease progression.

Thus the key event in developing an autoimmune disease is the activation of an effector $CD4^+$ T cell that recognizes a self-peptide-HLA complex. There are four major checkpoints that must be overcome to result in this; there must be a failure of central tolerance induction, failure of peripheral regulation, presentation of the autoantigen and co-stimulation (Fig. 26.1). As discussed below, enteroviruses (EV) could impact upon each of these key steps.

At the first autoimmunity checkpoint, that of central tolerance, situations that further compromise thymic education (such as a failure to express a self-protein) would be expected to favor the development of autoimmunity. A graphic example of this scenario exists in the form of patients diagnosed with autoimmune polyglandular syndrome type 1 (APS-1) who develop multiple autoimmune disorders. APS-1 is characterized by autosomal dominant failure of expression of the *AIRE* (autoimmune regulator) gene, resulting in the absence of a transcription factor that controls thymic expression of a host of genes that encode self-antigens (Waterfield and Anderson 2010). The lack of appropriate ectopic expression of self-antigens predisposes to multiple forms of autoimmune disease. Is it conceivable that early pre- or perinatal infection events (fetal or maternal) could have an adverse effect on thymic expression of self-antigens? This is not an area that has been adequately researched to our knowledge, but may be deserving of more attention, because of its potential to profoundly affect the peripheral T cell repertoire during the period of fetal development when the thymus is, in relative terms, at its most active. Intriguingly, Coxsackievirus B4 infection of intact mice and ex vivo fetal thymus organ cultures has been reported to lead to both an impaired maturation and differentiation of lymphocytes (Brilot et al. 2008; Chatterjee et al. 1992).

Table 26.2 Factors influencing the induction of induced Tregs

Inhibited by	Facilitated by
Cytokines (IL-4, IL-6, IFN-γ) sustained Akt-mTOR activation OX40	Retinoic acid
Tim-1 co-stimulation	CTLA-4, PD-1:PD-L1 interaction

Akt-mTOR v-akt murine thymoma viral oncogene homolog—mechanistic target of rapamycin, *CTLA-4* cytotoxic T-lymphocyte-associated antigen 4, *IL-4* interleukin 4, *IFN-γ* interferon γ, *OX40* CD134, *PD-1* programmed death, *PD-L1* PD-1 ligand, *Tim-1* T cell, Ig domain, and mucin domain-1

The second checkpoint of peripheral immune regulation is easy to conceptualize. Fewer "policemen" (Tregs) results in the presence of more "criminals" (autoreactive T cells). Again, there is a graphic example in the form of mutations in the gene encoding the transcription factor, *FOXP3* (forkhead box P3). *FOXP3* deficiency results in a lack of peripheral nTregs and the disease IPEX (immune dysregulation, polyendocrinopathy, enteropathy, X-linked syndrome) (Waterfield and Anderson 2010). In terms of development of type 1 diabetes, nTreg numbers appear normal in most studies to date (Brusko et al. 2007; Lindley et al. 2005), but function may be impaired, as may be the ability of conventional effector T cells (Tconv) to be regulated and the generation of other types of Treg (e.g. interleukin-10 (IL-10)-secreting iTregs) (Arif et al. 2004; Tree et al. 2010). Additionally, type 1 diabetes-related gene polymorphisms in *IL2RA* (interleukin 2 receptor alpha chain gene) diminish IL-2 production by memory T cells and might lead to impaired T cell homeostasis (Dendrou et al. 2009). Another type 1 diabetes (T1D) susceptibility gene, *CTLA-4* (cytotoxic T lymphocyte-associated antigen 4), transcribing a protein with immunosuppressive functions, might be a core mechanism through which Tregs control APC (antigen-presenting cell) function (Wing and Sakaguchi 2010).

Against this backdrop it is easy to conceive of viral infection interfering in the generation and function of iTregs. Several factors are known to affect Treg induction/function (Table 26.2), and future studies should address the possibility that viruses impact upon such pathways.

It is self-evident that a virus infection could participate in the breaching of the third checkpoint, namely the presentation of self-antigens in the islets. Enteroviruses are known for their lytic potential against β-cells in vitro (Frisk and Diderholm 2000), and it is likely that such an event occurring in vivo will result in extensive release of intracellular antigens. Moreover, this is likely to take place in an inflammatory environment, in which activated dendritic cells can engage with autoantigens (thus breaching checkpoint 4) and, process and present them to $CD4^+$ and $CD8^+$ T cells. If an autoreactive repertoire is available (see above), then priming, with resultant breaking of tolerance, can arise. This would implicate EVs as the initiators of autoimmunity. To date the evidence for this is mixed: prospective case–control studies of pre-autoimmune high-risk subjects show EV infection contemporaneous with first signs of autoimmunity (Oikarinen et al. 2011); on the other hand, animal models suggest more of an accelerating role for EV infection (see below). Furthermore, epidemiological studies suggest that EV infection has a role in the

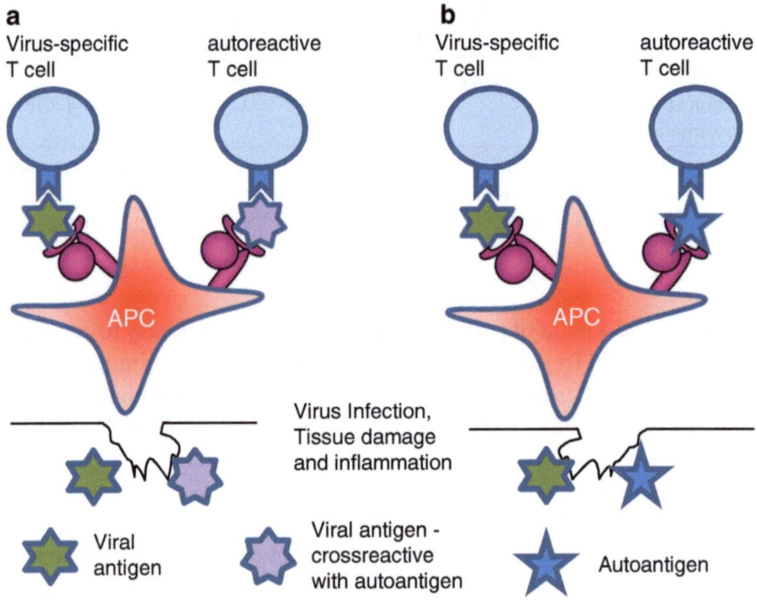

Fig. 26.2 Schematic representation of the molecular mimicry and bystander models. It is hypothesized that via one or both of these mechanisms viruses could contribute to type 1 diabetes. The image shows the two main models for the activation of autoreactive cell during the course of a virus infection. (**a**) A pathogen-derived peptide which mimics or is identical with the structure of an autoantigen can activate autoreactive T cells in a process named molecular mimicry. (**b**) Tissue damage through a virus infection leads to the presentation of autoantigen alongside virus antigen on activated APCs with accompanying inflammation. The activation of T cells not specific to the course of the infection is called bystander activation

final events of islet damage that result in overt disease but could, conceivably, not be responsible for the bulk of early islet damage (Yeung et al. 2011).

The most likely situation in which a dendritic cell is activated and becomes capable of providing robust co-stimulation (checkpoint 4) is as a result of infection. The two major models of autoimmunity that include an infectious agent are molecular mimicry and bystander activation (Fig. 26.2). Molecular mimicry assumes that a pathogen-derived peptide mimics or is identical with the structure of an autoantigen. Although such mimicry has been described, it has not usually been accompanied by robust evidence that pathogenic T cells are autoreactive and cross-react with virus. Nonetheless, evidence that there are shared sequences between the enterovirus 2C protease and glutamic-acid decarboxylase 65 (GAD65) (Vreugdenhil et al. 1998), and between the Rotavirus VP7 (viral protein 7) and, islet cell antigen 2 (IA-2) (Honeyman et al. 2010) is intriguing. The hypothesis states that under these circumstances APCs presenting the pathogen-derived peptide are able to activate autoreactive lymphocytes. Bystander activation proposes that a T cell-specific response against an autoantigen is activated during an ongoing immune response against a foreign antigen. During an infection or inflammation in the pancreas, T cells will be recruited and in the presence

Table 26.3 Predictions in relation to bystander activation and molecular mimicry models of islet autoimmunity

Model	Specific features		
	Antigen presentation	Autoreactive T cell avidity/TCR affinity	Cross-reactivity
Bystander activation	Evidence of activated antigen-presenting cells in insulitic lesion	Autoreactive T cells likely to be low avidity/have low affinity TCRs	Cross-reactivity with common (unspecified) pathogens
Molecular mimicry	Activation of antigen-presenting cells may not be islet-specific	Autoreactive T cells may be high avidity/have high-affinity TCRs	Cross-reactivity with enteroviruses

TCR T cell receptor

of concomitant β-cell damage and presentation of islet-specific antigens, activation of autoantigen-specific T cells can result in amplification of the inflammatory response. If tolerance mechanisms do not suppress these processes, multiple rounds of β-cell destruction will create a pool of autoreactive cytotoxic T lymphocytes (CTL). This will ultimately lead to chronic and ongoing β-cell destruction.

Both models are highly appealing from a theoretical standpoint, yet to date there is minimal evidence that they operate in human type 1 diabetes. A series of predictions and testable hypotheses are given in relation to each model in Table 26.3. These could form the basis for experiments designed to bring one or other of these models to the fore. Robust evidence of molecular mimicry in type 1 diabetes is lacking, but this theory cannot be completely ruled out as yet, since it can be argued that the key experiments designed to test this hypothesis in an unbiased way (e.g., examine TCR-specificities from the insulitis for both self- and antiviral reactivity) have not been done. Such studies have been used to support the mimicry model in autoimmune gastritis (Amedei et al. 2003), for example. In contrast, there is perhaps greater evidence to support the possibility of bystander activation. For example, it is known that an inflammatory process per se can contribute to autoimmunity [e.g., interferon-alfa (IFN-α) treatment for chronic viral disease results in a higher risk of autoimmune disease including type 1 diabetes (Fabris et al. 2003; Guerci et al. 1994)]. Future studies should be designed to test these models in man using as incisive a set of experimental approaches as possible.

Studies of Virus-Induced Diabetes in Animal Models

Molecular mimicry in the context of a virus infection has been studied in the RIP-LCMV ("rat insulin promoter" controls expression of "lymphocytic choriomeningitis virus" protein) mouse model where there was a perfect fit (molecular identity) between the trigger (virus) and the target (self-antigen, in this case a LCMV protein expressed in the islets) with a rapid onset of T1D (Oldstone et al. 1991). In a similar model in which LCMV was additionally expressed in the thymus to negatively select high-affinity LCMV-specific T cells, the T1D onset was CD4 dependent and occurred at a much slower rate (von Herrath et al. 1994). Furthermore a critical number of autoreactive T cells were needed (Sevilla et al. 2000). In the RIP-LCMV model a heterologous virus infection (e.g., Pichinde virus, which exhibits structural mimicry to a subdominant LCMV peptide) can increase the number of preexisting autoreactive T cells. However, in this model of molecular mimicry between a viral epitope and a self-antigen, autoimmunity can only accelerate disease, but not initiate it (Christen et al. 2004). Overall these models imply: (1) multiple events are necessary to elicit autoimmunity and type 1 diabetes and (2) molecular mimicry may have a role in accelerating an ongoing autoimmune response rather than initiating it.

Animal models have also been used to examine bystander activation. This process can be elicited by viruses which create an environment of inflammation and tissue damage and the release of sequestered islet antigens and their enhanced presentation by APCs. Coxsackievirus B4 (CVB4) infection induces diabetes in a bystander activation-style in a mouse model harboring high amounts of transgenic T cells for an islet autoantigen (Horwitz et al. 1998). Importantly, the activation of autoreactive T cells is specific and depends on the presentation of sequestered β-cell autoantigens by APCs rather than the inflammatory response itself (Horwitz et al. 2004). The onset and acceleration of the CVB4-induced type 1 diabetes is dependent on the presence of a critical mass of autoreactive T cells (Serreze et al. 2000). This critical T cell mass could be generated during multiple consecutive infections; on the other hand, this could also lead to T cell exhaustion and therefore protection from T1D.

The CVB3 strain can be either protective from, or accelerate type 1 diabetes, depending on the timing of virus infection in relation to the autoimmune process, the dose and replication capacity of the substrain (Tracy et al. 2010). Instead of accelerating type 1 diabetes, long-term protection was established after infecting young non-obese diabetic (NOD) (4 weeks of age) mice with CVB3 (Tracy et al. 2002). Infection of older, prediabetic mice (12 weeks of age) with virulent CVB3 strains led to accelerated disease onset, whereas avirulent CVB3 strains delayed the onset (Drescher et al. 2004).

Mechanistically, infection with non-β-cell-damaging viruses (CVB3 and LCMV) leads to protection in prediabetic mice in two ways (Filippi et al. 2009): (1) Increase of PD-1 signaling prevents the expansion of autoreactive $CD8^+$ T cells (2) Increase of the number of $CD4^+CD25^+$ Tregs provides long-term protection.

Infection of the BB (biobreeding) rat model with Kilham rat virus (non-β-cell-infecting) impairs Treg function (Zipris et al. 2003) and induces T1D. Additionally, this model depends on a critical stimulation of the innate immune response (via Toll-like receptors) being present at the time of virus infection (Zipris et al. 2005).

Manifestations of Autoimmunity and Inflammation in Type 1 Diabetes

The first sign of ongoing autoimmunity in type 1 diabetes is the appearance of islet-specific autoantibodies. These autoantibodies are directed against the key type 1 diabetes autoantigens: insulin, GAD65, IA-2, and zinc transporter-8 (ZnT8) (Zhang and Eisenbarth 2011). For the most part the same autoantigens have been shown to be targeted by the cellular arm of the adaptive immune system, including by $CD4^+$ and $CD8^+$ T lymphocytes.

There is substantial data suggesting that following the appearance of islet cell autoantibodies, a chronic autoimmune response ensues that underlies progression of β-cell destruction. First, the appearance of autoantibodies precedes disease onset by months/years; second, at diagnosis there is evidence of islet destruction and ongoing islet infiltration by immune cells (insulitis). This is characterized by the infiltration of the pancreatic islets by $CD4^+$ and $CD8^+$ T cells, B cells, macrophages, and dendritic cells, as well as accompanying upregulation of HLA class I molecules on β-cells (Willcox et al. 2009). Recently, a study on samples from nPOD (the Network for Pancreatic Organ Donors with Diabetes; www.jdrfnpod.org) provided evidence of asynchronous β-cell destruction leaving lobes with functional β-cells (Gianani et al. 2010) suggesting that type 1 diabetes results from a progression of the disease process in a lobule-to-lobule manner. However, to date this presumed subclinical and asymptomatic period of inflammation of the pancreatic islets has proved hard to visualize. The collection of pancreas from donors with autoantibodies for research is a logical route to follow to acquire such material and conduct studies, but results to date indicate that visualizing the prediabetic inflammatory lesion is not straightforward (In't Veld 2011).

The Role of T Cells in Type 1 Diabetes

It is proposed that the major players in β-cell destruction are T cells recognizing islet-specific autoantigens (Tree and Peakman 2004). The role for T cells in the pathology of type 1 diabetes is based on several routes of evidence:

- T cells can be found in close proximity to islets using techniques such as immunohistology (Willcox et al. 2009).
- There are several type 1 diabetes-associated susceptibility genes with putative effects on T cell function (Table 26.4).
- Both depletion of $CD4^+$ and $CD8^+$ T cells abrogates disease in mice, implying a role for both T cell subtypes in disease (Wong and Janeway 1999).
- T cell-directed immunosuppressive therapy (e.g., cyclosporine, anti-CD3 therapy) preserves β-cell mass (Herold et al. 2009).
- Several case reports indicate that type 1 diabetes may arise in nondiabetic individuals a few months or years after receiving bone marrow grafts (for hematological malignancy) from siblings with type 1 diabetes, despite the recipients

Table 26.4 Evidence that type 1 diabetes is a T cell-dependent autoimmune disease

Criteria	Reference
Presence of circulating, activated T cells directed against targets in β-cells; T cells dominate the insulitic lesion	Arif et al. (2004; 2011), Skowera et al. (2008), Velthuis et al. (2010) and Willcox et al. (2009)
Clinical response to immune suppression directed against T cells	Herold et al. (2009), Orban et al. (2011) and Sherry et al. (2011)
Passive transfer of disease using T lymphocytes (mouse model)	Wong and Janeway (1999)
Immune deficiency leads to islet autoimmunity and type 1 diabetes (IPEX syndrome and APS which affect central and peripheral tolerance mechanisms)	Husebye et al. (2009) and Wildin et al. (2002)
Genes affecting T cell function (e.g. *CD25, IL2, CTLA4*) confer risk of type 1 diabetes	Concannon et al. (2009)

APS autoimmune polyglandular syndrome type 1 which results from deficiency of the *AIRE* (autoimmune regulator) gene, a transcription factor controlling expression of autoantigens in the thymus, *CTLA-4* cytotoxic T-lymphocyte-associated antigen 4, *IL2* interleukin 2, IPEX immunodeficiency polyendocrinopathy enteropathy X-linked syndrome resulting from Treg abnormalities due to defects in *FOXP3* (forkhead box P3) gene

having undergone ablation of their own immune systems as part of the conditioning regimen. Importantly, this inadvertent "passive transfer" data implies a role for mature memory T cells. Here, type 1 diabetes can be prevented by depletion of mature T cells from the graft, strongly implying that memory T cells reactive against islet-specific autoantigens remain for many years after diabetes diagnosis and can be reactivated in the new host upon encounter with self-antigens, with diabetes as the outcome (Lampeter et al. 1993, 1998).

Specific Roles for T Cells in β-Cell Damage

A key requirement for investigating the role of autoreactive T cells in type 1 diabetes has been the identification of islet autoantigens that T cells target in the disease, and moreover their specific epitopes. Epitopes presented on disease-susceptibility HLA molecules represent targets for T cells of particular interest and potential importance (for an extensive list of β-cell-specific antigens/epitopes see Di Lorenzo et al. 2007). With the help of these epitope maps, investigators have begun to address the presence and function of autoreactive T cells. Due to limitations to the access of pancreas and draining lymph nodes, most studies are based on peripheral blood. It remains a matter of debate how representative peripheral blood can be of activities in the tissues, but there is at least some evidence to support the blood as a useful portal through which to measure disease activity (Trudeau et al. 2003; Wong et al. 2007).

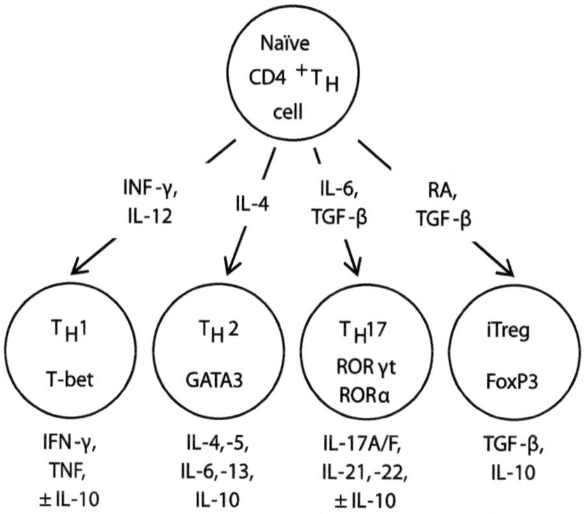

Fig. 26.3 Differentiation of CD4+ T cells. When naïve CD4+ T cells are activated by antigen there are at least four routes of functional polarization—the most frequently encountered are shown here as T_H1, T_H2, and T_H17 and at lower frequency iTreg. The cytokines present at activation dictate the route of polarization through the activation of specific transcription factors which orchestrate the differentiation process. The presence of IL-12 and IFN-γ promotes T_H1 differentiation and the synthesis of IFN-γ, TNF, and IL-10. In the presence of IL-4, T_H2 differentiation is favored, which in addition to IL-4,-5,-6, and -13, also secrete IL-10. T_H17 effectors are generated in the presence of IL-6 and TGF-β and secrete IL-17A/F, IL-21,-22, and IL-10. Differentiation TGF-β- and IL-10-producing iTreg is favored in the presence of RA (retinoic acid) and TGF-β. The specific transcription factors are: T-bet (T-box expressed in T cells) for T_H1, GATA3 (GATA-binding protein 3) for T_H2, RORγt (retinoic-acid-related orphan receptor-gamma t) and RORα (retinoic-acid-related orphan receptor-alfa) for T_H17, and FoxP3 (forkhead box P3) for iTreg

Although not strongly represented numerically in the insulitis, autoreactive CD4+ T cells have a potentially critical role in the initiation and perpetuation of the processes leading to type 1 diabetes. Thus the emergence of evidence that CD4+ T cells recognize islet epitopes presented by HLA class II molecules encoded within haplotypes that confer the highest genetic risk for type 1 diabetes is important from the standpoint of understanding disease initiators and drivers. Typically, upon activation, the surrounding microenvironment (cytokines, chemokines) dictates CD4+ T cell differentiation into one of the major lineages, mainly denoted as T helper (T_H) 1, T_H2, T_H17, and regulatory T cells (Treg) (Fig. 26.3).

The role of T_H1 cells in the pathogenesis of type 1 diabetes is two-fold. Upon stimulation, activated T_H1 cells secrete IL-2 and pro-inflammatory cytokines such as IFN-γ and tumor necrosis factor alpha (TNF-α). These pro-inflammatory cytokines have the ability to damage β-cells directly, but more importantly, they have the capacity to activate autoreactive CD8+ CTLs. Peripheral T_H1 (INF-γ secreting) cells specific for islet autoantigens, such as PPI (preproinsulin), GAD65, and IA-2 are present in subjects with new-onset type 1 diabetes, whilst T_H2 cytokine responses have not been described (Arif et al. 2004). The concept that T1D is solely a T_H1-driven disease has been challenged recently however (Arif et al. 2011), with the demonstration that there is an IL-17 signature amongst autoreactive $CD4^+$ T cells in the peripheral blood at the same stage of disease, along with preliminary evidence for the same cytokine pathway also being active in the islets. T_H17 cells secrete IL-17, a potent inducer of tissue inflammation and autoimmunity. IL-17-neutralization post insulitis induction prevents development of type 1 diabetes in mice, suggesting a role for T_H17 cells in the effector phase of the disease. This inhibition leads to a reduced infiltration of islet-specific T cells and an increased proportion of regulatory cells around the islet (Emamaullee et al. 2009). IL-17 exhibits detrimental effects on human islets, in vitro, by potentiating inflammatory and proapoptotic responses. Patients with type 1 diabetes show generally upregulated T_H17 immunity (Arif et al. 2011; Honkanen et al. 2010). In the end, it remains to be established whether T_H1 and T_H17 cells combine in the destructive process, whether there is a dominant pathway, or whether there are distinct phases of the disease at which different cytokines predominate.

If uncontrolled, the potent effector functions (cytokine production) of these T helper cells are likely to lead to or amplify autoimmune inflammation. As discussed, Tregs appear to be the primary cell population providing this control. $FoxP3^+$ Tregs act as key suppressors of immune responses, promoting tolerance via cell–cell interactions and secretion of the signature immunosuppressive cytokines IL-10 and tumor growth factor-β (TGF-β). IL-10 secreting Tregs can suppress islet-specific T_H1 cells (Tree et al. 2010).

The other major arm of the adaptive immune response is the $CD8^+$ cytotoxic T lymphocyte, which recognizes epitopes presented complexed with HLA class I molecules, of which some haplotypes also determine the risk type 1 diabetes. These $CD8^+$ CTLs are capable of "killing" infected or abnormal cells by inducing apoptosis via cytotoxins (perforin/granzyme, TNF-α, TNF-β) and Fas/Fas ligand (FasL) interaction. Animal studies suggest that the perforin/granzyme pathway plays the prime role in type 1 diabetes, but clearly other pathways contribute to β-cell death (Thomas et al. 2010). Although T_H1 cells (IFN-γ secreting) may be able to induce diabetes in mice, there is a general agreement that overall a key role exists for $CD8^+$ CTLs in the pathogenesis of type 1 diabetes. The first strand of evidence for this in man is the observation that peripheral blood $CD8^+$ T cells, recognizing islet autoantigens, such as PPI, Insulin B chain, IA-2, GAD65, and IGRP (islet-specific glucose-6-phosphatase catalytic subunit-related protein) are present in recent-onset type 1 diabetes patients (Velthuis et al. 2010). Importantly, these circulating $CD8^+$ T cells have the potential to "kill" β-cells as shown for a T cell clone restricted to HLA-A2 and specific for a naturally processed epitope from the leader sequence of PPI, kills β-cells in vitro (Skowera et al. 2008). Future work in this field aims to understand whether $CD8^+$ T cells with these specificities are also present in the islets and contribute to the killing of β-cells. There will also need to be some exploration of the nature of the priming of these cells, and whether there is any evidence for cross-reactivity with EVs.

Although not strictly considered a part of the adaptive immune response, natural killer T cells (NKT), which are nonconventional T cells that bear αβ TCRs, have a comparable potential to impact upon type 1 diabetes to that of conventional $CD4^+$ and $CD8^+$ T cells and are worthy of discussion here. These innate-like T cells are restricted to the MHC class-I-like molecule CD1d and secrete cytokines (notably IL-4 γ and IFN). Upon activation they contribute to both the innate and adaptive immune response by providing maturation signals to cells like dendritic cells and T and B cells (Diana et al. 2011). Their potential role in type 1 diabetes in man, either as an effector or regulator, has perhaps been understudied.

T Cells and the Viral Etiology of Type 1 Diabetes

Given the advances in understanding of the predominant importance of T cells in the pathogenesis of type 1 diabetes, and knowledge about the pathways through which this might operate, several key questions arise in relation to hypotheses that center upon enteroviruses:

- At which stage might an EV impact upon the process: initiation or acceleration?
- What is the nature of the impact: do EVs influence immune regulation negatively by inducing secretion of cytokines known to inhibit Tregs (e.g. IL-6); or is the effect predominantly on stoking the pro-inflammatory response?
- Is the impact a one-off or is cyclical exposure required, in which case improving or enhancing neutralizing immunity could be an important therapeutic approach?

In summary, our understanding of many aspects of the immune pathogenesis of type 1 diabetes is such that there is ample opportunity for viruses to be highly relevant to the process. Incisive experiments can now be designed and tissue, sample and population collections are being assembled that should enable these.

Acknowledgements MP acknowledges support from the Department of Health via the National Institute for Health Research (NIHR) comprehensive Biomedical Research Centre award to Guy's & St Thomas' NHS Foundation Trust in partnership with King's College London; the Juvenile Diabetes Research Foundation (7-2005-877 and 1-2007-1803 to MP) and EU FP7 Framework award for PEVNET (Persistent virus infection as a cause of pathogenic inflammation in type 1 diabetes).

References

Amedei A, Bergman MP, Appelmelk BJ, Azzurri A, Benagiano M, Tamburini C, van der Zee R, Telford JL, Vandenbroucke-Grauls CM, D'Elios MM, Del Prete G (2003) Molecular mimicry between Helicobacter pylori antigens and H+, K+-adenosine triphosphatase in human gastric autoimmunity. J Exp Med 198:1147–1156

Arif S, Tree TI, Astill TP, Tremble JM, Bishop AJ, Dayan CM, Roep BO, Peakman M (2004) Autoreactive T cell responses show proinflammatory polarization in diabetes but a regulatory phenotype in health. J Clin Invest 113:451–463

Arif S, Moore F, Marks K, Bouckenooghe T, Dayan CM, Planas R, Vives-Pi M, Powrie J, Tree T, Marchetti P, Huang GC, Gurzov EN, Pujol-Borrell R, Eizirik DL, Peakman M (2011) Peripheral and islet interleukin-17 pathway activation characterizes human autoimmune diabetes and promotes cytokine-mediated beta-cell death. Diabetes 60:2112–2119

Bottazzo GF, Dean BM, McNally JM, MacKay EH, Swift PG, Gamble DR (1985) In situ characterization of autoimmune phenomena and expression of HLA molecules in the pancreas in diabetic insulitis. N Engl J Med 313:353–360

Brilot F, Jaidane H, Geenen V, Hober D (2008) Coxsackievirus B4 infection of murine foetal thymus organ cultures. J Med Virol 80:659–666

Brusko T, Wasserfall C, McGrail K, Schatz R, Viener HL, Schatz D, Haller M, Rockell J, Gottlieb P, Clare-Salzler M, Atkinson M (2007) No alterations in the frequency of FOXP3+ regulatory T-cells in type 1 diabetes. Diabetes 56:604–612

Chatterjee NK, Hou J, Dockstader P, Charbonneau T (1992) Coxsackievirus B4 infection alters thymic, splenic, and peripheral lymphocyte repertoire preceding onset of hyperglycemia in mice. J Med Virol 38:124–131

Christen U, Edelmann KH, McGavern DB, Wolfe T, Coon B, Teague MK, Miller SD, Oldstone MB, von Herrath MG (2004) A viral epitope that mimics a self antigen can accelerate but not initiate autoimmune diabetes. J Clin Invest 114:1290–1298

Concannon P, Rich SS, Nepom GT (2009) Genetics of type 1A diabetes. N Engl J Med 360:1646–1654

Dendrou CA, Plagnol V, Fung E, Yang JH, Downes K, Cooper JD, Nutland S, Coleman G, Himsworth M, Hardy M, Burren O, Healy B, Walker NM, Koch K, Ouwehand WH, Bradley JR, Wareham NJ, Todd JA, Wicker LS (2009) Cell-specific protein phenotypes for the autoimmune locus IL2RA using a genotype-selectable human bioresource. Nat Genet 41:1011–1015

Di Lorenzo TP, Peakman M, Roep BO (2007) Translational mini-review series on type 1 diabetes: Systematic analysis of T cell epitopes in autoimmune diabetes. Clin Exp Immunol 148:1–16

Diana J, Brezar V, Beaudoin L, Dalod M, Mellor A, Tafuri A, von Herrath M, Boitard C, Mallone R, Lehuen A (2011) Viral infection prevents diabetes by inducing regulatory T cells through NKT cell-plasmacytoid dendritic cell interplay. J Exp Med 208:729–745

Drescher KM, Kono K, Bopegamage S, Carson SD, Tracy S (2004) Coxsackievirus B3 infection and type 1 diabetes development in NOD mice: insulitis determines susceptibility of pancreatic islets to virus infection. Virology 329:381–394

Emamaullee JA, Davis J, Merani S, Toso C, Elliott JF, Thiesen A, Shapiro AM (2009) Inhibition of Th17 cells regulates autoimmune diabetes in NOD mice. Diabetes 58:1302–1311

Fabris P, Floreani A, Tositti G, Vergani D, De Lalla F, Betterle C (2003) Type 1 diabetes mellitus in patients with chronic hepatitis C before and after interferon therapy. Aliment Pharmacol Ther 18:549–558

Filippi CM, Estes EA, Oldham JE, von Herrath MG (2009) Immunoregulatory mechanisms triggered by viral infections protect from type 1 diabetes in mice. J Clin Invest 119:1515–1523

Frisk G, Diderholm H (2000) Tissue culture of isolated human pancreatic islets infected with different strains of coxsackievirus B4: assessment of virus replication and effects on islet morphology and insulin release. Int J Exp Diabetes Res 1:165–175

Gianani R, Campbell-Thompson M, Sarkar SA, Wasserfall C, Pugliese A, Solis JM, Kent SC, Hering BJ, West E, Steck A, Bonner-Weir S, Atkinson MA, Coppieters K, von Herrath M, Eisenbarth GS (2010) Dimorphic histopathology of long-standing childhood-onset diabetes. Diabetologia 53:690–698

Guerci AP, Guerci B, Levy-Marchal C, Ongagna J, Ziegler O, Candiloros H, Guerci O, Drouin P (1994) Onset of insulin-dependent diabetes mellitus after interferon-alfa therapy for hairy cell leukaemia. Lancet 343:1167–1168

Herold KC, Gitelman S, Greenbaum C, Puck J, Hagopian W, Gottlieb P, Sayre P, Bianchine P, Wong E, Seyfert-Margolis V, Bourcier K, Bluestone JA (2009) Treatment of patients with new onset type 1 diabetes with a single course of anti-CD3 mAb Teplizumab preserves insulin production for up to 5 years. Clin Immunol 132:166–173

Honeyman MC, Stone NL, Falk BA, Nepom G, Harrison LC (2010) Evidence for molecular mimicry between human T cell epitopes in rotavirus and pancreatic islet autoantigens. J Immunol 184:2204–2210

Honkanen J, Nieminen JK, Gao R, Luopajarvi K, Salo HM, Ilonen J, Knip M, Otonkoski T, Vaarala O (2010) IL-17 immunity in human type 1 diabetes. J Immunol 185:1959–1967

Horwitz MS, Bradley LM, Harbertson J, Krahl T, Lee J, Sarvetnick N (1998) Diabetes induced by Coxsackie virus: initiation by bystander damage and not molecular mimicry. Nat Med 4:781–785

Horwitz MS, Ilic A, Fine C, Balasa B, Sarvetnick N (2004) Coxsackieviral-mediated diabetes: induction requires antigen-presenting cells and is accompanied by phagocytosis of beta cells. Clin Immunol 110:134–144

Husebye ES, Perheentupa J, Rautemaa R, Kampe O (2009) Clinical manifestations and management of patients with autoimmune polyendocrine syndrome type I. J Intern Med 265:514–529

In't Veld P (2011) Insulitis in the human endocrine pancreas: does a viral infection lead to inflammation and beta cell replication? Diabetologia 54:2220–2222

Lampeter EF, Homberg M, Quabeck K, Schaefer UW, Wernet P, Bertrams J, Grosse-Wilde H, Gries FA, Kolb H (1993) Transfer of insulin-dependent diabetes between HLA-identical siblings by bone marrow transplantation. Lancet 341:1243–1244

Lampeter EF, McCann SR, Kolb H (1998) Transfer of diabetes type 1 by bone-marrow transplantation. Lancet 351:568–569

Lennon GP, Bettini M, Burton AR, Vincent E, Arnold PY, Santamaria P, Vignali DA (2009) T cell islet accumulation in type 1 diabetes is a tightly regulated, cell-autonomous event. Immunity 31:643–653

Lindley S, Dayan CM, Bishop A, Roep BO, Peakman M, Tree TI (2005) Defective suppressor function in CD4(+)CD25(+) T-cells from patients with type 1 diabetes. Diabetes 54:92–99

Mason D (1998) A very high level of crossreactivity is an essential feature of the T-cell receptor. Immunol Today 19:395–404

Oikarinen S, Martiskainen M, Tauriainen S, Huhtala H, Ilonen J, Veijola R, Simell O, Knip M, Hyoty H (2011) Enterovirus RNA in blood is linked to the development of type 1 diabetes. Diabetes 60:276–279

Oldstone MB, Nerenberg M, Southern P, Price J, Lewicki H (1991) Virus infection triggers insulin-dependent diabetes mellitus in a transgenic model: role of anti-self (virus) immune response. Cell 65:319–331

Orban T, Bundy B, Becker DJ, DiMeglio LA, Gitelman SE, Goland R, Gottlieb PA, Greenbaum CJ, Marks JB, Monzavi R, Moran A, Raskin P, Rodriguez H, Russell WE, Schatz D, Wherrett D, Wilson DM, Krischer JP, Skyler JS (2011) Co-stimulation modulation with abatacept in patients with recent-onset type 1 diabetes: a randomised, double-blind, placebo-controlled trial. Lancet 378:412–419

Redondo MJ, Jeffrey J, Fain PR, Eisenbarth GS, Orban T (2008) Concordance for islet autoimmunity among monozygotic twins. N Engl J Med 359:2849–2850

Serreze DV, Ottendorfer EW, Ellis TM, Gauntt CJ, Atkinson MA (2000) Acceleration of type 1 diabetes by a coxsackievirus infection requires a preexisting critical mass of autoreactive T-cells in pancreatic islets. Diabetes 49:708–711

Sevilla N, Homann D, von Herrath M, Rodriguez F, Harkins S, Whitton JL, Oldstone MB (2000) Virus-induced diabetes in a transgenic model: role of cross-reacting viruses and quantitation of effector T cells needed to cause disease. J Virol 74:3284–3292

Sherry N, Hagopian W, Ludvigsson J, Jain SM, Wahlen J, Ferry RJ Jr, Bode B, Aronoff S, Holland C, Carlin D, King KL, Wilder RL, Pillemer S, Bonvini E, Johnson S, Stein KE, Koenig S, Herold KC, Daifotis AG (2011) Teplizumab for treatment of type 1 diabetes (Protege study): 1-year results from a randomised, placebo-controlled trial. Lancet 378:487–497

Skowera A, Ellis RJ, Varela-Calvino R, Arif S, Huang GC, Van-Krinks C, Zaremba A, Rackham C, Allen JS, Tree TI, Zhao M, Dayan CM, Sewell AK, Unger WW, Drijfhout JW, Ossendorp F,

Roep BO, Peakman M (2008) CTLs are targeted to kill beta cells in patients with type 1 diabetes through recognition of a glucose-regulated preproinsulin epitope. J Clin Invest 118:3390–3402

Thomas HE, Trapani JA, Kay TW (2010) The role of perforin and granzymes in diabetes. Cell Death Differ 17:577–585

Tracy S, Drescher KM, Chapman NM, Kim KS, Carson SD, Pirruccello S, Lane PH, Romero JR, Leser JS (2002) Toward testing the hypothesis that group B coxsackieviruses (CVB) trigger insulin-dependent diabetes: inoculating nonobese diabetic mice with CVB markedly lowers diabetes incidence. J Virol 76:12097–12111

Tracy S, Drescher KM, Jackson JD, Kim K, Kono K (2010) Enteroviruses, type 1 diabetes and hygiene: a complex relationship. Rev Med Virol 20:106–116

Tree TI, Peakman M (2004) Autoreactive T cells in human type 1 diabetes. Endocrinol Metab Clin North Am 33:113–133, ix–x

Tree TI, Lawson J, Edwards H, Skowera A, Arif S, Roep BO, Peakman M (2010) Naturally arising human CD4 T cells that recognize islet autoantigens and secrete IL-10 regulate pro-inflammatory T cell responses via linked suppression. Diabetes 59:1451–1460

Trudeau JD, Kelly-Smith C, Verchere CB, Elliott JF, Dutz JP, Finegood DT, Santamaria P, Tan R (2003) Prediction of spontaneous autoimmune diabetes in NOD mice by quantification of autoreactive T cells in peripheral blood. J Clin Invest 111:217–223

Velthuis JH, Unger WW, Abreu JR, Duinkerken G, Franken K, Peakman M, Bakker AH, Reker-Hadrup S, Keymeulen B, Drijfhout JW, Schumacher TN, Roep BO (2010) Simultaneous detection of circulating autoreactive CD8+ T-cells specific for different islet cell-associated epitopes using combinatorial MHC multimers. Diabetes 59:1721–1730

von Herrath MG, Dockter J, Oldstone MB (1994) How virus induces a rapid or slow onset insulin-dependent diabetes mellitus in a transgenic model. Immunity 1:231–242

Vreugdenhil GR, Geluk A, Ottenhoff TH, Melchers WJ, Roep BO, Galama JM (1998) Molecular mimicry in diabetes mellitus: the homologous domain in coxsackie B virus protein 2C and islet autoantigen GAD65 is highly conserved in the coxsackie B-like enteroviruses and binds to the diabetes associated HLA-DR3 molecule. Diabetologia 41:40–46

Waterfield M, Anderson MS (2010) Clues to immune tolerance: the monogenic autoimmune syndromes. Ann N Y Acad Sci 1214:138–155

Wildin RS, Smyk-Pearson S, Filipovich AH (2002) Clinical and molecular features of the immunodysregulation, polyendocrinopathy, enteropathy, X linked (IPEX) syndrome. J Med Genet 39:537–545

Willcox A, Richardson SJ, Bone AJ, Foulis AK, Morgan NG (2009) Analysis of islet inflammation in human type 1 diabetes. Clin Exp Immunol 155:173–181

Wing K, Sakaguchi S (2010) Regulatory T cells exert checks and balances on self tolerance and autoimmunity. Nat Immunol 11:7–13

Wong FS, Janeway CA Jr (1999) The role of CD4 vs. CD8 T cells in IDDM. J Autoimmun 13:290–295

Wong CP, Stevens R, Long B, Li L, Wang Y, Wallet MA, Goudy KS, Frelinger JA, Tisch R (2007) Identical beta cell-specific CD8(+) T cell clonotypes typically reside in both peripheral blood lymphocyte and pancreatic islets. J Immunol 178:1388–1395

Yeung WC, Rawlinson WD, Craig ME (2011) Enterovirus infection and type 1 diabetes mellitus: systematic review and meta-analysis of observational molecular studies. BMJ 342:d35

Zhang L, Eisenbarth GS (2011) Prediction and prevention of type 1 diabetes mellitus. J Diabetes 3:48–57

Zipris D, Hillebrands JL, Welsh RM, Rozing J, Xie JX, Mordes JP, Greiner DL, Rossini AA (2003) Infections that induce autoimmune diabetes in BBDR rats modulate CD4+CD25+ T cell populations. J Immunol 170:3592–3602

Zipris D, Lien E, Xie JX, Greiner DL, Mordes JP, Rossini AA (2005) TLR activation synergizes with Kilham rat virus infection to induce diabetes in BBDR rats. J Immunol 174:131–142

Chapter 27
Innate Immune Responses to Viruses Inducing Diabetes

Katharina Lind and Malin Flodström Tullberg

Abstract Together with physical and chemical barriers, the innate immune system is the first line of defense against infecting pathogens. By responding rapidly to conserved structures expressed by pathogens, so-called pathogen-associated molecular patterns (PAMPs), the innate immune system prevents replication and spread of pathogens, and promotes the activation of specific immune responses. Many observations have indicated that certain virus infections may contribute to the development of type 1 diabetes. Although it has been difficult to establish firmly the role of viruses in the development of human type 1 diabetes, numerous animal models have provided proof-of-concept for their involvement in the pathogenesis of the disease. It is also becoming clear that the innate immune response to the infecting pathogens may contribute to disease development. Recent genome-wide studies have identified several type 1 diabetes susceptibility loci containing genes of direct or indirect importance for the functions of the innate immune system. Increased knowledge of host–pathogen interactions as well as the functional effects of the type 1 diabetes-associated gene polymorphisms may therefore contribute to a better understanding of the potential role of viruses in type 1 diabetes.

Introduction

Studies in humans and experimental models indicate that the innate immune response plays an important role in dictating the risk of development of type 1 diabetes. In addition, many observations point to a role of viruses in the etiopathogenesis of this disease. During virus infections the innate immune system of the host acts as the first line of defense to prevent viral invasion and replication.

K. Lind, B.Sc. • M.F. Tullberg, Ph.D. (✉)
Department of Medicine HS, The Center for Infectious Medicine, Karolinska Institutet, Karolinska University Hospital Huddinge F59, 14186 Stockholm, Sweden
e-mail: malin.flodstrom-tullberg@ki.se

K.W. Taylor, H. Hyoty, A. Toniolo, A.J. Zuckerman (eds.), *Diabetes and Viruses*,
DOI 10.1007/978-1-4614-4051-2_27,

Environmental Influences Regulate the Risk for the Development of Type 1 Diabetes

Over the past few decades there has been a steep increase in the incidence of type 1 diabetes, particularly in very young children (Gale 2005; Vehik and Dabelea 2011). The same pattern has been found in many parts of the world including Europe, Australia, and the USA, although recent data from Sweden indicate a possible change (Berhan et al. 2011). The major type 1 diabetes susceptibility locus contains the HLA class II genes and has been considered as 30–50% of the genetic risk for disease development. Recent studies have demonstrated, however, that low- or moderate-risk HLA genotypes are now represented in increasing percentages among the patients (Steck et al. 2011; Vehik and Dabelea 2011). Although the reason behind this remains to be determined, it may reflect changes in the environment resulting in increased penetrance of lower-risk HLA genotypes or an increased importance of genetic loci containing non-HLA genes.

The observed increase in type 1 diabetes has been too rapid to be attributed solely to genetic factors. That environmental factors are important in the pathogenesis of the disease is supported by studies demonstrating that the concordance rates among monozygotic twins are at most 65% (Redondo et al. 2008). Further, that not all individuals carrying high-risk HLA II genotypes develop disease. People who migrate from a country or region with a low risk to a location with a high risk for type 1 diabetes face an increased risk of diabetes. This is particularly true for children moving before the teenage and for first- and second-generation immigrants. Countries with high sanitary standards have typically a higher incidence of type 1 diabetes patients than countries with poor sanitary standards. Type 1 diabetes also has a significant seasonal pattern with fewer cases diagnosed during spring and summer than in the autumn and winter. In addition, populations living in equatorial regions have a lower risk of developing diabetes than people living at higher latitudes (for recent reviews, see: Vehik and Dabelea 2011; Zipris 2009). Based upon these observations it is reasonable to suggest that environmental factors play an important role in type 1 diabetes. It is not established, however, whether environmental factors trigger disease or whether a lack of protective environmental influences contributes to the development of diabetes.

Virus Infections and Type 1 Diabetes

Over the years many suggestions have been made regarding environmental factors that may modulate the risk for the development of diabetes. Hormonal influences, dietary factors, stressful life events, and increased or lack of exposure to microbes have all been implicated in type 1 diabetes. Epidemiological studies and clinical case reports have strongly indicated viral infections as potential triggers or precipitators of a number of human autoimmune diseases such as type 1 diabetes. For example, virus infections are associated with relapses in patients with multiple sclerosis

Table 27.1 Examples of viruses that have been associated with the development of type 1 diabetes in humans and animals

Humans		
Virus	Family	Genome
Coxsackievirus A	*Picornaviridae*	ss(+)RNA
Coxsackievirus B	*Picornaviridae*	ss(+)RNA
Echovirus	*Picornaviridae*	ss(+)RNA
Rubella virus	*Togaviridae*	ss(+)RNA
Mumps virus	*Paramyxoviridae*	ss(−)RNA
Cytomegalovirus	*Herpesviridae*	dsDNA
Epstein–Barr virus	*Herpesviridae*	dsDNA
Varicella zoster	*Herpesviridae*	dsDNA
Rotavirus	*Reoviridae*	dsRNA
Endogenous retrovirus	*Retroviridae*	dsDNA

Animals			
Virus	Family	Genome	Host
Bovine viral diarrhea virus	*Flaviviridae*	ss(+)RNA	Cattle
Coxsackievirus B	*Picornaviridae*	ss(+)RNA	Mouse and nonhuman primates
Encephalomyocarditis virus	*Picornaviridae*	ss(+)RNA	Mouse
Foot-and-mouth disease virus	*Picornaviridae*	ss(+)RNA	Cattle and pig
Ljungan virus	*Picornaviridae*	ss(+)RNA	Bank vole
Mengo virus	*Picornaviridae*	ss(+)RNA	Mouse
Rubella virus	*Togaviridae*	ss(+)RNA	Rabbit and hamster
Kilham rat virus	*Parvoviridae*	ssDNA	Rat
Endogenous retrovirus	*Retroviridae*	dsDNA	Mouse

Compiled from Hyoty (2004), Jun and Yoon (2001) and Yoon and Jun (2006)

(MS) and viral myocarditis correlates with progression to chronic dilated cardiomyopathy (Flodstrom-Tullberg 2003). A significantly higher risk for the development of type 1 diabetes is also seen in individuals infected congenitally with rubella virus (Hyoty 2004). The finding of viruses and/or antibodies to certain viruses in newly diagnosed type 1 diabetes patients has provided additional support for a role of viruses in type 1 diabetes (Tauriainen et al. 2011). Certain viruses are also known to trigger diabetes in animal models such as the non-obese diabetic (NOD) mouse and the biobreeding diabetes resistant (BBDR) mouse (Jun and Yoon 2001; Yoon and Jun 2006; Zipris 2009). Examples of viruses that have been proposed to be involved in human type 1 diabetes, as well as a number of viruses associated with the development of diabetes in animals are summarized in Table 27.1.

The proposed mechanisms on how a virus infection may trigger (or prevent) type 1 diabetes are numerous and appear to be limited only by our own imagination. The models that are discussed mainly in type 1 diabetes are direct infection of target cells, molecular mimicry, and bystander activation of autoreactive T cells, and it is clear that several of these mechanisms may act simultaneously (Flodstrom-Tullberg 2003; Munz et al. 2009). Studies in animal models of virus-induced diabetes have pointed to a pivotal role for an activated host innate immune response in regulating

Table 27.2 Examples of PRRs, their cellular localization and proposed ligands

Receptor name	Cellular localization	Ligands	MyD88-dependent signaling (TLRs)
Toll-like receptors (TLRs)			
TLR1	Cell surface	Triacyl lipopeptides	Yes
TLR2	Cell surface	Peptidoglycan, hemagglutinin	Yes
TLR3	Endosome	ssRNA virus, dsRNA virus	No
TLR4	Cell surface	Lipopolysaccaride	Yes
TLR5	Cell surface	Flagellin from flagellated bacteria	Yes
TLR6	Cell surface	Diacyl lipopeptides from mycoplasma, lipoteic acid	Yes
TLR7	Endolysosome	ssRNA viruses	Yes
TLR8	Endolysosome	ssRNA from RNA viruses	Yes
TLR9	Endolysosome	CpG motifs from bacteria and viruses	Yes
RIG-I-like receptors (RLRs)			
RIG-I	Cytoplasm	Short dsRNA with triphosphate or monophosphate at 5′end, short length poly I:C, several negative-sense, ssRNA viruses	
MDA-5/IFIH1	Cytoplasm	Long length poly I:C, several positive-sense ssRNA viruses	
NOD-like receptors (NLRs)			
NOD1	Cytoplasm	gamma-D-Glutamyl-meso-diaminopimelic acid from several types of bacteria	
NOD2	Cytoplasm	Muramyl dipeptide from several types of bacteria	

Compiled from Jeong and Lee (2011) and Kumar et al. (2011)

disease development (Flodstrom et al. 2002; McCartney et al. 2011; Zipris et al. 2007). In this chapter we will discuss a group of viruses that have been strongly associated with type 1 diabetes in humans (the enteroviruses) and two viruses that cause diabetes in certain strains of mice and rats (encephalomyocarditis virus and Kilham rat virus, respectively).

The Innate Immune System

The innate immune system provides the first line of defense against microbes and stimulates the subsequent activation of the adaptive immune response. It includes barriers such as the skin and the mucosal lining of the gastrointestinal and respiratory tracts as well as immune cells such as neutrophils, mononuclear phagocytes, and natural killer (NK) cells.

Detection of invading pathogens occurs via pattern recognition receptors (PRRs) that recognize pathogen-associated molecular patterns (PAMPs) or danger-associated molecular patterns (DAMPs) (summarized in Table 27.2; Jeong and Lee

2011; Kumar et al. 2011). Virus infections activate mainly two receptor families, the toll-like receptors (TLRs) and the retinoic-acid-inducible gene-I (RIG-I)-like receptors (RLRs). Ten and 12 TLRs have been identified in humans and mice, respectively, and TLRs 1–9 are conserved in humans and mice. While most human TLRs have been studied in detail the functional role of TLR10 remains unclear. RLRs consist of RIG-I, melanoma differentiation associated factor 5 [MDA5, also denoted interferon induced helicase-1 (IFIH1)] and laboratory of genetics and physiology 2 (LGP2). RIG-I and MDA5 are helicases recognizing viral RNA (Table 27.2), and LGP2 may act to regulate RIG-I and MDA5 functions (Jeong and Lee 2011; Kumar et al. 2011).

Interactions between PRR and viral nucleic acids induce a signaling cascade, which leads to the induction of cytokines including type I interferons (type I IFNs) (Brennan and Bowie 2011; Jeong and Lee 2011; Kumar et al. 2011; Takeuchi and Akira 2011). During a virus infection, immune cells such as dendritic cells and macrophages are the main producers of type I IFNs, but parenchymal cells also express PRRs and contribute to the production of IFNs. Type I IFNs induce the expression of a set of genes, such as protein kinase R (PKR) and 2′5′-oligoadenylate synthase, to activate a so-called antiviral state in responding cells. The antiviral state aims to block viral replication and prevent infection of uninfected cells. In addition to the direct antiviral actions, IFNs modulate various cellular responses of the innate immune system including cytotoxicity of NK cells and also contribute to the shaping of the adaptive immune response (Samuel 2001; Seo and Hahm 2010).

The Innate Immune Response to Viruses Associated with Development of Diabetes

Enteroviruses

Enteroviruses are small single-stranded (ss) RNA viruses infecting the host mainly by the fecal–oral route. The majority of infections are asymptomatic or associated with mild symptoms. In rare cases, enterovirus infections result in more severe outcomes such as pancreatitis, myocarditis, and meningitis. Enteroviruses, in particular the group B Coxsackieviruses (CVBs), have also been associated with dilated cardiomyopathy and type 1 diabetes (Whitton et al. 2005).

Many epidemiological studies and clinical observations have linked enterovirus infections to the development of type 1 diabetes (Tauriainen et al. 2011; Yeung et al. 2011). Enteroviruses have been found in the pancreatic islets of patients with newly diagnosed type 1 diabetes (Dotta et al. 2007; Richardson et al. 2009; Ylipaasto et al. 2004), and in vitro studies demonstrate that pancreatic beta cells can be infected by CVBs, often with a detrimental outcome (e.g. Flodstrom et al. 2002; Roivainen et al. 2000; Szopa et al. 1986; Yoon et al. 1979). Animal models have shown that a CVB infection can precipitate diabetes in genetically susceptible hosts with an established pool of beta cell specific T cells (Horwitz et al. 1998; Serreze et al. 2000).

Much of our understanding of the host immune response to enteroviruses (mainly CVBs) is derived from studies in the mouse as the viruses show a similar, but not identical, tropism as in humans. Although an intact adaptive immune response is crucial for viral clearance, restriction of early viral replication and host survival is critically dependent on an intact innate immune response; mice lacking IFN receptors succumb early after infection with CVBs (Flodstrom et al. 2002), and the IFNs provide early protection by inducing the expression of proteins involved in antiviral defense as suggested by a high mortality following infection in mice lacking the inducible form of nitric oxide (iNOS) (Flodstrom et al. 2001; Zaragoza et al. 1999), PKR (Flodstrom-Tullberg et al. 2005) and RNaseL (Flodstrom-Tullberg et al. 2005). Even the pancreatic beta cells are highly dependent on an intact response to IFNs for their survival during CVB infection. Indeed, mice harboring pancreatic beta cells that cannot respond to IFNs develop rapidly diabetes following infection (Flodstrom et al. 2002).

Studies in the mouse and in vitro studies using human cells have provided a fairly detailed understanding for how the host sense CVBs. In vitro studies have implicated TLR4, TLR7, and TLR8 in the recognition of the virus (Triantafilou et al. 2005; Triantafilou and Triantafilou 2004). Mice deficient in the TLR-dependent signaling molecule MyD88 did not demonstrate, however, increased pathology and mortality after infection with CVB. On the contrary, *myd88–/–* mice had a better survival and less cardiac and pancreatic inflammation compared to wild-type mice (Fuse et al. 2005) (compare also MyD88-usage among the different TLRs in Table 27.2). This somewhat unexpected finding indicates that viral recognition via one or several of the MyD88-dependent TLRs contributes to pathology rather than to protection. More important roles in protecting the host seem to be played by TLR3 and MDA5/IFIH1. Both *tlr3–/–* and *mda5–/–* animals show increased mortality after infection with CVBs (Huhn et al. 2010; Negishi et al. 2008; Wang et al. 2010). This is paralleled with a reduced capacity of the host to produce IFNs, stressing further the importance for IFNs in protection from the virus. Collectively, the studies on the role of TLRs and RLRs in CVB infection highlight the complexity and possible double-edged outcome of viral recognition. On the one hand, TLR3 and MDA5 provide protection from virally instigated damage. On the other hand, MyD88-dependent TLR signals contribute to tissue damage and impaired host survival.

The identification of a role for MDA5/IFIH1 in the host immune response to CVBs is of special interest. Several non-synonymous single nucleotide polymorphisms (nsSNPs) associated with an altered risk for type 1 diabetes development have been identified in the human *ifih1* gene (now denoted IDDM19) (Nejentsev et al. 2009; Smyth et al. 2006). Although it remains to be shown whether these different polymorphisms alter the ability for human MDA5/IFIH1 to recognize and respond to enteroviruses, the observations that *mda5–/–* animals fail to restrict CVB replication (Huhn et al. 2010; Wang et al. 2010) clearly encourage further studies on the possible role of enteroviruses in human type 1 diabetes and also on how the different SNPs in the human *ifih1* gene regulate susceptibility to diabetes.

Encephalomyocarditis Virus

EMCV is a cardiovirus that, similar to the enteroviruses, belongs to the family of picornaviruses. It has a positive-sense ssRNA genome. The diabetogenic strain of EMCV (EMCV-D) infects selectively and replicates in pancreatic beta cells and induces type 1 diabetes in susceptible strains of mice. Inoculation of mice with a large dose of EMCV-D results in the development of type 1 diabetes within 4 days after infection due to extensive beta cell destruction. However, in mice infected with a small dose of virus, initial replication in the beta cells is followed by recruitment of macrophages to the pancreas. In small dose infection the macrophages are the main mediators of beta cell destruction, since activation of macrophages prior to infection increases the incidence, whereas their depletion almost completely abrogates EMCV-D-induced type 1 diabetes. The molecular mechanism behind beta cell destruction is not fully understood but cytokines secreted by the macrophage seem to play a central role. EMCV-D can infect and activate macrophages to produce cytokines such as interleukin-1 beta (IL-1β) and tumor necrosis factor alfa (TNF-α) as well as to express the inducible nitric oxide synthase (iNOS). Activation of macrophages occurs through signaling by tyrosine kinases, blockage of this signaling pathway decreases cytokine production as well as the incidence of type 1 diabetes (Jun and Yoon 2001; van der Werf et al. 2007; Yoon and Jun 2006).

The role of certain PPRs in EMCV infection has been examined. Both MDA5 and TLR3 have been shown to regulate the cytokine response and confer protection from lethal infection with EMCV (Kato et al. 2006). TLR3 is required for early IFN secretion by hematopoetic cells and for protection against EMCV-D-induced type 1 diabetes in mice normally not prone to diabetes after infection with this virus (McCartney et al. 2011). MDA5 appears to be most important for stromal cell recognition of EMCV-D, and mice deficient in *mda5* succumb early to EMCV-D infection most likely due to massive heart damage. The observation that the infected *mda5–/–* animals developed hyperglycemia suggests that MDA5 is important also in protecting the beta cell from damage (McCartney et al. 2011). Collectively, these studies demonstrate that the risk for the development of diabetes increases dramatically when host recognition of the virus is disabled.

Kilham Rat Virus

KRV is a single-stranded (ss) DNA virus belonging to the *Parvovirus* family. It replicates primarily in the nucleus of dividing cells and thus has a preference for cells in the gut, the bone marrow, and lymphoid organs. The virus induces diabetes in the so-called diabetes resistant BB rat (denoted the BBDR rat). This strain of rats was derived from the diabetes-prone BB rat (the BBDP rat), a model for type 1 diabetes with spontaneous onset of diabetes. BBDP rats suffer from lymphopenia due to a mutation in a gene called Ian4L1 and the development of diabetes is T cell dependent. In contrast to the BBDP rat, the BBDR rat has normal numbers of circulating T cells, is not lymphopenic, and does not develop spontaneous diabetes if housed under virus-free conditions. However, infection with KRV induces diabetes in 25–30% of infected BBDR rats as a result of induction of insulitis and selective beta cell destruction. The disease mechanism involves an autoimmune T cell response against the pancreatic beta cells but is independent of direct infection of beta cells. KRV can trigger diabetes in other rat strains sharing the same MHC II as the BBDR rat (e.g. LEW1.WR1 rats), clearly demonstrating that genetics regulate susceptibility. How previously silent autoreactive T cells are activated to attack the pancreatic beta cells remains unclear, although an important role for a dysregulated control by regulatory T cell subsets has been proposed (van der Werf et al. 2007; Zipris 2009).

There are observations suggesting that activation of the innate immune is crucial for diabetes to develop in KRV-infected BBDR rats. For example, the depletion of macrophages results in a nearly complete protection from diabetes (Chung et al. 1997). Moreover, pancreatic lymph nodes (PLN) from KRV-infected rats express increased levels of pro-inflammatory cytokines and genes linked to IFN production and signaling (Wolter et al. 2009). It has also been shown that a more general activation by the innate immune response by TLR ligands (e.g. poly I:C, a viral mimic triggering TLR3 activation) prior to infection with KRV increases diabetes incidence to nearly 100% in the BBRD rats (Zipris et al. 2005). In fact, such pre-activation of the innate immune response allows diabetes development with even suboptimal doses of KRV (Zipris 2009; Zipris et al. 2005).

Some observations have indicated that TLR9, a TLR known commonly as a receptor recognizing CpG DNA motifs, is important in the recognition of KRV. In addition, treatment of BBDR rats with chloroquine, an agent that prevents endosomal acidification and thereby the activation of lysosomal TLRs (see Table 27.2), provided protection from diabetes induced by KRV in conjunction with poly I:C treatment (Zipris et al. 2007). Despite the numerous examples for an involvement of the innate immune system in KRV-induced type 1 diabetes, the mechanisms by which it triggers disease still remains to be identified. Several hypotheses have been put forward including modulated macrophage and CD8+ T cell responses, increased expression of MHC I on islet cells, increased recruitment of T cells to the pancreatic islets, etc. (van der Werf et al. 2007; Zipris 2009).

Concluding Remarks

Although it is clear that environmental factors regulate type 1 diabetes and that viruses are prime candidates, it is important to stress that it has still not been possible to establish firmly a causative role for any virus in the etiology of type 1 diabetes. However, given the available clinical and epidemiological data it is still justified to hypothesize that virus infections regulate susceptibility to the disease. As reviewed above, innate immune components are keys in regulating risk for diabetes development following infection with CVB, EMCV, and KRV in rodents. Thus, a greater understanding of host–pathogen interactions will provide additional knowledge on how viruses may regulate type 1 diabetes development.

Recent genome-wide studies have identified genes related directly or indirectly to the innate immune system, which contains nsSNPs that modulate risk for human type 1 diabetes development. These include, for example, the *mda5/ifih1* gene discussed above (Nejentsev et al. 2009; Smyth et al. 2006) and genes involved in vitamin D metabolism (vitamin D is increasingly recognized for its importance in normal immune functions) (Cooper et al. 2011). In addition, a recent study identified a type 1 diabetes-associated SNP in a gene likely to operate as a "master regulator" of an IRF7-driven inflammatory gene network (Heinig et al. 2011). These studies clearly indicate that altered expression and/or functions of genes involved in innate immunity are associated with the risk of type 1 diabetes. Whether this results in a different handling of encountered viruses must be clarified. Further, whether a dysregulated innate immune response to viruses contributes to diabetes development remains to be established.

Acknowledgements Work in the M.F-T. group is supported by the European Foundation for the Study of Diabetes, the European Union, the Swedish Child Diabetes Foundation, The Swedish Diabetes Association Research Foundation, and the Swedish Research Council.

References

Berhan Y, Waernbaum I, Lind T, Mollsten A, Dahlquist G (2011) Thirty years of prospective nationwide incidence of childhood type 1 diabetes: the accelerating increase by time tends to level off in Sweden. Diabetes 60:577–581

Brennan K, Bowie AG (2011) Activation of host pattern recognition receptors by viruses. Curr Opin Microbiol 13:503–507

Chung YH, Jun HS, Kang Y, Hirasawa K, Lee BR, Van Rooijen N, Yoon JW (1997) Role of macrophages and macrophage-derived cytokines in the pathogenesis of Kilham rat virus-induced autoimmune diabetes in diabetes-resistant BioBreeding rats. J Immunol 159:466–471

Cooper JD, Smyth DJ, Walker NM, Stevens H, Burren OS, Wallace C, Greissl C, Ramos-Lopez E, Hypponen E, Dunger DB, Spector TD, Ouwehand WH, Wang TJ, Badenhoop K, Todd JA (2011) Inherited variation in vitamin d genes is associated with predisposition to autoimmune disease type 1 diabetes. Diabetes 60:1624–1631

Dotta F, Censini S, van Halteren AG, Marselli L, Masini M, Dionisi S, Mosca F, Boggi U, Muda AO, Prato SD, Elliott JF, Covacci A, Rappuoli R, Roep BO, Marchetti P (2007) Coxsackie B4 virus infection of beta-cells and NK cell insulitis in recent onset type 1 diabetic patients. Proc Natl Acad Sci USA 104:5115–5120

Flodstrom M, Horwitz MS, Maday A, Balakrishna D, Rodriguez E, Sarvetnick N (2001) A critical role for inducible nitric oxide synthase in host survival following coxsackievirus B4 infection. Virology 281:205–215

Flodstrom M, Maday A, Balakrishna D, Cleary MM, Yoshimura A, Sarvetnick N (2002) Target cell defense prevents the development of diabetes after viral infection. Nat Immunol 3:373–382

Flodstrom-Tullberg M (2003) Viral infections: their elusive role in regulating susceptibility to autoimmune disease. Microbes Infect 5:911–921

Flodstrom-Tullberg M, Hultcrantz M, Stotland A, Maday A, Tsai D, Fine C, Williams B, Silverman R, Sarvetnick N (2005) RNase L and double-stranded RNA-dependent protein kinase exert complementary roles in islet cell defense during coxsackievirus infection. J Immunol 174:1171–1177

Fuse K, Chan G, Liu Y, Gudgeon P, Husain M, Chen M, Yeh WC, Akira S, Liu PP (2005) Myeloid differentiation factor-88 plays a crucial role in the pathogenesis of Coxsackievirus B3-induced myocarditis and influences type I interferon production. Circulation 112:2276–2285

Gale EA (2005) Spring harvest? Reflections on the rise of type 1 diabetes. Diabetologia 48:2445–2450

Heinig M, Petretto E, Wallace C, Bottolo L, Rotival M, Lu H, Li Y, Sarwar R, Langley SR, Bauerfeind A, Hummel O, Lee YA, Paskas S, Rintisch C, Saar K, Cooper J, Buchan R, Gray EE, Cyster JG, Erdmann J, Hengstenberg C, Maouche S, Ouwehand WH, Rice CM, Samani NJ, Schunkert H, Goodall AH, Schulz H, Roider HG, Vingron M, Blankenberg S, Munzel T, Zeller T, Szymczak S, Ziegler A, Tiret L, Smyth DJ, Pravenec M, Aitman TJ, Cambien F, Clayton D, Todd JA, Hubner N, Cook SA (2011) A trans-acting locus regulates an anti-viral expression network and type 1 diabetes risk. Nature 467:460–464

Horwitz MS, Bradley LM, Harbertson J, Krahl T, Lee J, Sarvetnick N (1998) Diabetes induced by coxsackie virus: initiation by bystander damage and not molecular mimicry. Nat Med 4:781–785

Huhn MH, McCartney SA, Lind K, Svedin E, Colonna M, Flodstrom-Tullberg M (2010) Melanoma differentiation-associated protein-5 (MDA-5) limits early viral replication but is not essential for the induction of type 1 interferons after coxsackievirus infection. Virology 401:42–48

Hyoty H (2004) Environmental causes: viral causes. Endocrinol Metab Clin North Am 33:27–44, viii

Jeong E, Lee JY (2011) Intrinsic and extrinsic regulation of innate immune receptors. Yonsei Med J 52:379–392

Jun HS, Yoon JW (2001) The role of viruses in type I diabetes: two distinct cellular and molecular pathogenic mechanisms of virus-induced diabetes in animals. Diabetologia 44:271–285

Kato H, Takeuchi O, Sato S, Yoneyama M, Yamamoto M, Matsui K, Uematsu S, Jung A, Kawai T, Ishii KJ, Yamaguchi O, Otsu K, Tsujimura T, Koh CS, Reis e Sousa C, Matsuura Y, Fujita T, Akira S (2006) Differential roles of MDA5 and RIG-I helicases in the recognition of RNA viruses. Nature 441:101–105

Kumar H, Kawai T, Akira S (2011) Pathogen recognition by the innate immune system. Int Rev Immunol 30:16–34

McCartney SA, Vermi W, Lonardi S, Rossini C, Otero K, Calderon B, Gilfillan S, Diamond MS, Unanue ER, Colonna M (2011) RNA sensor-induced type I IFN prevents diabetes caused by a beta cell-tropic virus in mice. J Clin Invest 121:1497–1507

Munz C, Lunemann JD, Getts MT, Miller SD (2009) Antiviral immune responses: triggers of or triggered by autoimmunity? Nat Rev Immunol 9:246–258

Negishi H, Osawa T, Ogami K, Ouyang X, Sakaguchi S, Koshiba R, Yanai H, Seko Y, Shitara H, Bishop K, Yonekawa H, Tamura T, Kaisho T, Taya C, Taniguchi T, Honda K (2008) A critical link between toll-like receptor 3 and type II interferon signaling pathways in antiviral innate immunity. Proc Natl Acad Sci USA 105:20446–20451

Nejentsev S, Walker N, Riches D, Egholm M, Todd JA (2009) Rare variants of IFIH1, a gene implicated in antiviral responses, protect against type 1 diabetes. Science 324:387–389

Redondo MJ, Jeffrey J, Fain PR, Eisenbarth GS, Orban T (2008) Concordance for islet autoimmunity among monozygotic twins. N Engl J Med 359:2849–2850

Richardson SJ, Willcox A, Bone AJ, Foulis AK, Morgan NG (2009) The prevalence of enteroviral capsid protein vp1 immunostaining in pancreatic islets in human type 1 diabetes. Diabetologia 52:1143–1151

Roivainen M, Rasilainen S, Ylipaasto P, Nissinen R, Ustinov J, Bouwens L, Eizirik DL, Hovi T, Otonkoski T (2000) Mechanisms of coxsackievirus-induced damage to human pancreatic beta-cells. J Clin Endocrinol Metab 85:432–440

Samuel CE (2001) Antiviral actions of interferons. Clin Microbiol Rev 14:778–809

Seo YJ, Hahm B (2010) Type I interferon modulates the battle of host immune system against viruses. Adv Appl Microbiol 73:83–101

Serreze DV, Ottendorfer EW, Ellis TM, Gauntt CJ, Atkinson MA (2000) Acceleration of type 1 diabetes by a coxsackievirus infection requires a preexisting critical mass of autoreactive T-cells in pancreatic islets. Diabetes 49:708–711

Smyth DJ, Cooper JD, Bailey R, Field S, Burren O, Smink LJ, Guja C, Ionescu-Tirgoviste C, Widmer B, Dunger DB, Savage DA, Walker NM, Clayton DG, Todd JA (2006) A genome-wide association study of nonsynonymous SNPs identifies a type 1 diabetes locus in the interferon-induced helicase (IFIH1) region. Nat Genet 38:617–619

Steck AK, Armstrong TK, Babu SR, Eisenbarth GS (2011) Stepwise or linear decrease in penetrance of type 1 diabetes with lower-risk HLA genotypes over the past 40 years. Diabetes 60:1045–1049

Szopa TM, Gamble DR, Taylor KW (1986) Coxsackie B4 virus induces short-term changes in the metabolic functions of mouse pancreatic islets in vitro. Cell Biochem Funct 4:181–187

Takeuchi O, Akira S (2011) Pattern recognition receptors and inflammation. Cell 140:805–820

Tauriainen S, Oikarinen S, Oikarinen M, Hyoty H (2011) Enteroviruses in the pathogenesis of type 1 diabetes. Semin Immunopathol 33:45–55

Triantafilou K, Triantafilou M (2004) Coxsackievirus B4-induced cytokine production in pancreatic cells is mediated through toll-like receptor 4. J Virol 78:11313–11320

Triantafilou K, Orthopoulos G, Vakakis E, Ahmed MA, Golenbock DT, Lepper PM, Triantafilou M (2005) Human cardiac inflammatory responses triggered by coxsackie B viruses are mainly Toll-like receptor (TLR) 8-dependent. Cell Microbiol 7:1117–1126

van der Werf N, Kroese FG, Rozing J, Hillebrands JL (2007) Viral infections as potential triggers of type 1 diabetes. Diabetes Metab Res Rev 23:169–183

Vehik K, Dabelea D (2011) The changing epidemiology of type 1 diabetes: why is it going through the roof? Diabetes Metab Res Rev 27:3–13

Wang JP, Cerny A, Asher DR, Kurt-Jones EA, Bronson RT, Finberg RW (2010) MDA5 and MAVS mediate type I interferon responses to coxsackie B virus. J Virol 84:254–260

Whitton JL, Cornell CT, Feuer R (2005) Host and virus determinants of picornavirus pathogenesis and tropism. Nat Rev Microbiol 3:765–776

Wolter TR, Wong R, Sarkar SA, Zipris D (2009) DNA microarray analysis for the identification of innate immune pathways implicated in virus-induced autoimmune diabetes. Clin Immunol 132:103–115

Yeung WC, Rawlinson WD, Craig ME (2011) Enterovirus infection and type 1 diabetes mellitus: systematic review and meta-analysis of observational molecular studies. BMJ 342:d35

Ylipaasto P, Klingel K, Lindberg AM, Otonkoski T, Kandolf R, Hovi T, Roivainen M (2004) Enterovirus infection in human pancreatic islet cells, islet tropism in vivo and receptor involvement in cultured islet beta cells. Diabetologia 47:225–239

Yoon JW, Jun HS (2006) Viruses cause type 1 diabetes in animals. Ann N Y Acad Sci 1079:138–146

Yoon JW, Austin M, Onodera T, Notkins AL (1979) Isolation of a virus from the pancreas of a child with diabetic ketoacidosis. N Engl J Med 300:1173–1179

Zaragoza C, Ocampo CJ, Saura M, Bao C, Leppo M, Lafond-Walker A, Thiemann DR, Hruban R, Lowenstein CJ (1999) Inducible nitric oxide synthase protection against coxsackievirus pancreatitis. J Immunol 163:5497–5504

Zipris D (2009) Epidemiology of type 1 diabetes and what animal models teach us about the role of viruses in disease mechanisms. Clin Immunol 131:11–23

Zipris D, Lien E, Xie JX, Greiner DL, Mordes JP, Rossini AA (2005) TLR activation syjournaltitlenergizes with Kilham rat virus infection to induce diabetes in BBDR rats. J Immunol 174:131–142

Zipris D, Lien E, Nair A, Xie JX, Greiner DL, Mordes JP, Rossini AA (2007) TLR9-signaling pathways are involved in Kilham rat virus-induced autoimmune diabetes in the biobreeding diabetes-resistant rat. J Immunol 178:693–701

Chapter 28
Enterovirus Infection of Cultured Human Pancreatic Islets

Teemu Smura and Merja Roivainen

Abstract The factors involved in viral tropism for pancreatic islets and islet response to infection can be studied in an experimental model utilizing pancreatic islets isolated from organ donors and cultivated as free-floating preparations. Enteroviruses, in general, have a tropism for human pancreatic islets in vitro. Both lytic and persistent enterovirus infections have been characterized under different experimental conditions and viruses have been detected in both insulin-producing and non-insulin-producing cells. Delayed (secondary) necrosis after initial pyknosis is the major mechanism of cell death during lytic enterovirus infection in cultured human pancreatic islets, whereas apoptosis appears to play only a minor role. Pancreatic beta-cell tropism and the ability to induce beta-cell dysfunction and death probably depends both on the genetic properties of the virus and on the host cell response to the infection. Viral properties are likely to affect the phenomenon since some enterovirus strains are highly cytolytic whereas others show progeny production with no apparent islet destruction in vitro. Even for highly destructive virus strains, there is a significant delay between viral progeny production and pancreatic islet destruction in vitro, suggesting a role for secondary, virus-induced, host factors in the process.

T. Smura • M. Roivainen, Ph.D. (✉)
Intestinal Viruses Unit, Division of Health Protection, Department of Infectious Disease Surveillance and Control, National Institute for Health and Welfare (THL), P.O. Box 30, 00271 Helsinki, Finland
e-mail: merja.roivainen@thl.fi

K.W. Taylor, H. Hyoty, A. Toniolo, A.J. Zuckerman (eds.), *Diabetes and Viruses*,
DOI 10.1007/978-1-4614-4051-2_28,

Introduction

Cultured Human Pancreatic Islets: An In Vitro Model for Virus–Host Interactions

Several cell types and both viral and host factors contribute to the pathogenesis of enteroviruses. Enterovirus genome/proteins have been detected in the pancreatic islets of type 1 diabetes patients (reviewed in Roivainen and Klingel 2010) indicating that enteroviruses are capable of reaching the islets during infection of the host. The primary site of enterovirus infection, however, is the mucosal tissue of the gastrointestinal tract where virus replication can continue, often asymptomatically, for several weeks. The primary infection is followed by a brief viremic phase, during which the virus spreads through the lymphatic system and circulation and may gain access to secondary target tissues such as pancreatic islets.

Within the human pancreas enteroviruses have a tropism for islets and possibly for ductal cells but apparently not for exocrine cells (Foulis et al. 1997; Ylipaasto et al. 2005). Using electron microscopy, viral inclusions have been located specifically in the cytoplasm of pancreatic beta-cells (Dotta et al. 2007). While direct cytolysis and/or virus-targeted immunity can be a possible cause of some cases of fulminant type 1 diabetes, it is likely that autoimmune reactions triggered by viral infection are also involved in enterovirus-induced type 1 diabetes. In this scenario, a virus-induced beta-cell damage (by a viral replication and/or virally induced cytokines) would result in the exposure of autoantigens in a local inflammatory milieu, leading to upregulation of MHC molecules and uptake and presentation of autoantigens by activated antigen-presenting cells, which in turn would promote further beta-cell damage by activating autoreactive T-cells (Fig. 28.1) (reviewed in Filippi and von Herrath 2005).

The key factors of enterovirus-induced type 1 diabetes would therefore be virus-induced beta-cell damage and inflammation. In addition to damaging beta-cells directly, enteroviruses may affect the fate of beta-cells and immune reactions within islets by infecting endothelial and/or ductal cells, which may provide virus a route to infect pancreatic islets, amplify the infection and/or modify genetic properties of the virus population and local inflammatory milieu within pancreatic islets (Fig. 28.1) (Smura et al. 2010).

The factors involved in viral tropism for pancreatic islets and islet response to infection can be studied in an experimental model utilizing pancreatic islets isolated from organ donors and cultivated as free-floating preparations. The general patterns of enterovirus tropism, enterovirus-induced cell destruction, and virus-induced host responses in human pancreatic islets in vitro are reviewed and the viral and host factors involved in these processes are discussed.

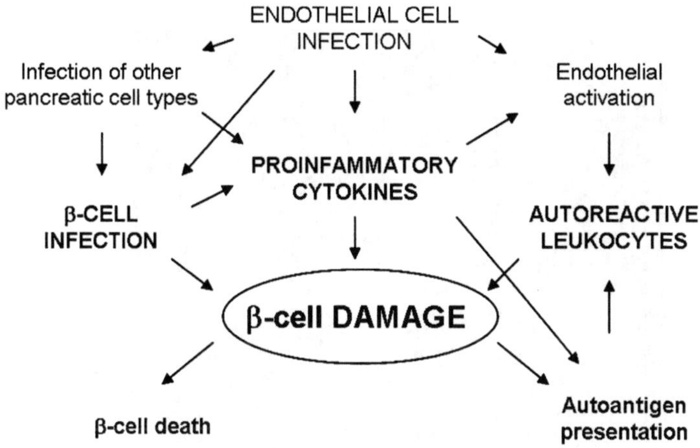

Fig. 28.1 Schematic representation on hypothetical factors involved in virally induced type 1 diabetes. During viremia a virus may gain access to pancreatic islets by infecting the endothelial cell lining that forms a barrier between vascular space and tissue parenchyma. Alternatively, the virus can be transported through the endothelial barrier by trancytosis or by infected leukocytes migrating to the target tissues. The infection of various pancreatic cell types may lead to production of proinflammatory cytokines, inducing endothelial cell activation, leukocyte migration and activation of antigen-presenting cells. Moreover, together with direct adverse effects of beta-cell infection, they may induce beta-cell damage and death, possibly leading to autoantigen exposure. Presentation of autoantigens by activated antigen-presenting cells in a local inflammatory milieu would promote further beta-cell damage by activating autoreactive T-cells

Characteristics of Experimental Pancreatic Islet Infection In Vitro

Enteroviruses, in general, have a tropism for human pancreatic islets in vitro (Chehadeh et al. 2000; Elshebani et al. 2007; Paananen et al. 2003; Roivainen et al. 2000; Roivainen et al. 2002; Smura et al. 2010; Vuorinen et al. 1992; Yin et al. 2002; Yoon et al. 1978). Both lytic and persistent enterovirus infections have been characterized under different experimental conditions, and viruses have been detected in both insulin-producing and non-insulin-producing cells, suggesting infection of alpha-, delta-, PP- or islet-associated ductal epithelial cells in addition to beta-cells (Chehadeh et al. 2000; Roivainen et al. 2000, 2002; Smura et al. 2010; Vuorinen et al. 1992).

In a well-defined in vitro model for enterovirus infection in pancreatic islets (Roivainen et al. 2000), viruses typically show exponential progeny production by 24 h post infection (Fig. 28.2b). Thereafter, the amount of infective virus either remains the same or decreases steadily, possibly due to heat inactivation and/or by cellular proteases. Intriguingly, despite confirmed beta-cell infection the viabilities

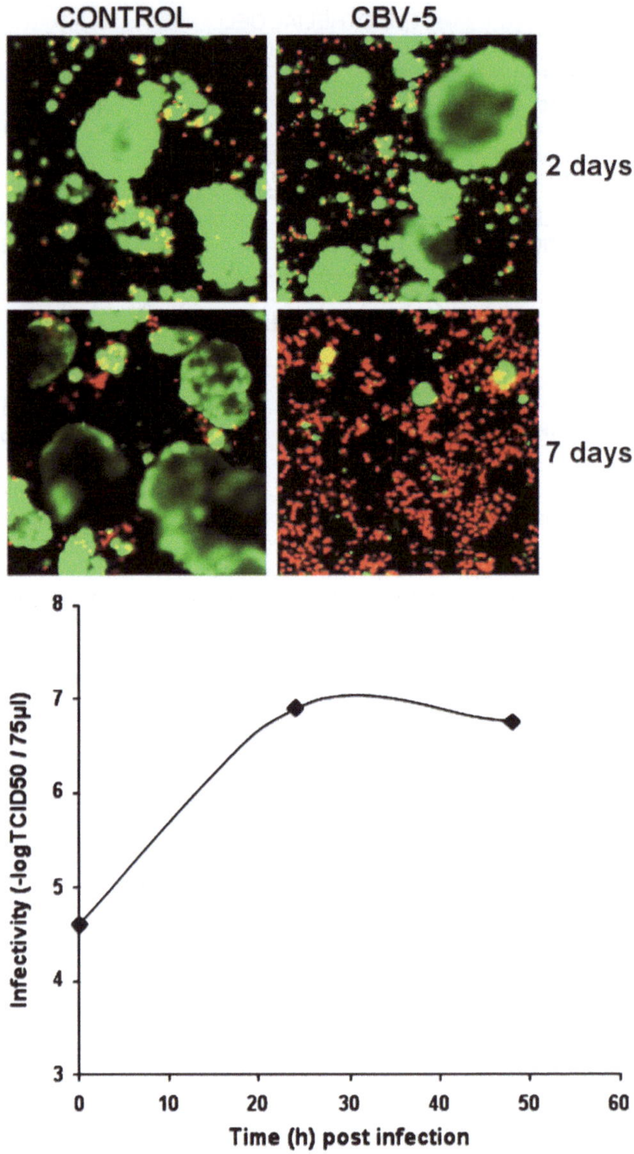

Fig. 28.2 Lytic enterovirus infection viral progeny production in human pancreatic islets in vitro. The islets were infected with CBV-5. Virus multiplication was detected with end-point titration method (**b**) and the viabilities of the islets were assessed using Live/Dead assay (Molecular Probes) and a confocal microscopy (**a**). Because of their esterase activity, live cells are stained *green* by calcein, while nuclei of dead cells are stained *red* by ethidium homodimer-1. The virus shows strong progeny production by 24 h post infection. By this time the viabilities of virus-infected and mock-infected control islets were essentially identical. The virus-induced secondary necrosis is visible by 7 days post infection

of both infected and mock-infected control islets typically remain high for several days after infection (Fig. 28.2a). CBV-5 infection of human pancreatic islets initially (by 2 days post infection) induces morphological changes characteristic of pyknosis, i.e. highly distorted nuclei with condensed but intact chromatin (Roivainen et al. 2000). Both mitochondria and plasma membrane were shown to be intact in these cells. With CBV-5 (and several other enterovirus strains), the initial pyknosis is followed by secondary necrosis usually by 1 week post infection. Apoptosis, on the contrary, appears to play a minor role during CBV infection in beta-cells in vitro (Rasilainen et al. 2004b; Roivainen et al. 2000). The significant delay between viral progeny production and pancreatic islet destruction suggests a role for secondary virus-induced host factors in the process. In addition, some enterovirus strains (such as the prototype strains of CVA-9 and EV-68 as well as several E-9 and E-30 strains), when studied with the same experimental setup, show replication comparable to lytic strains without apparent islet destruction, suggesting that enteroviral replication and islet destruction overlap but are (at least partially) separate phenomena.

In addition to secondary necrosis described above, persistent infection with no apparent decrease in islet viability has also been characterized with CBV-3 and CBV-4 (Chehadeh et al. 2000; Yin et al. 2002). In these experiments the viral titers in cell culture supernatants peaked at 3 days after infection, after which virus production decreased, but continued for 30 days.

The different outcomes of experimental islet infections may be due to details of experimental conditions such as multiplicity of infection and/or serum concentration in the culture medium. The persistently infected islets showed IFN-alpha expression that protected islets from virus-induced cytolysis and reduced viral progeny production (Chehadeh et al. 2000). Treatment with IFN-alpha and (to a lesser extent) IFN-gamma has been shown to induce antiviral state and repress viral replication in human pancreatic islets (Hultcrantz et al. 2007). Production of the viral progeny also appears to be somewhat slower in the experimental models leading to persistent infection (Chehadeh et al. 2000; Yin et al. 2002) suggesting that they may represent models with restricted virus replication, whereas the experimental settings leading to necrosis would represent models with unrestricted viral replication.

The distinction between restricted and unrestricted viral replication may have important consequences, since, at least in some continuous cell lines, the mode of cellular death (i.e. apoptosis vs. necrosis) appears to differ between these (Agol et al. 2000; Rasilainen et al. 2004a; Tolskaya et al. 1995). The type of cell death has importance in vivo, since necrosis represents more severe form of death where the cell content leaks into the surrounding tissue possibly causing damage to adjoining cells and inducing strong inflammatory response, whereas apoptosis represents more controlled cell death where excessive inflammatory response is avoided by rapid clearance of apoptotic cells from the tissues (Taylor et al. 2008; Virag and Szabo 2002). This suggests that, in addition to possible autoantigen exposure by enterovirus-induced beta-cell damage, the mechanism of cellular death after enterovirus infection is likely to have a profound effect on the local inflammatory milieu at the site of infection and consequently on the induction of autoimmunity.

All these results suggest that enteroviruses are capable of inducing both lytic and non-lytic infections in pancreatic islets. Further, both viral genetic determinants and host response to infection (which is probably affected by host genetics) are likely to affect the outcome of the infection. In the following, we will briefly outline how these factors may affect the fate of cultured pancreatic islets after infection with enterovirus.

Viral Genetic Determinants of Pancreatic Islet Tropism and Destruction

Species and Serotype Specificity

Human enteroviruses are classified into four species; *Human enterovirus A–D* (HEV-A–D), which together with three human rhinovirus species and three enterovirus species infecting nonhuman primates, pigs and bovines (SEV-A, PEV-B and BEV) comprise Enterovirus genus. Each species is divided further into serotypes based on sequence variation in the capsid protein VP1-coding region. Among human enteroviruses (referred in this chapter as enteroviruses) over 100 enterovirus serotypes have been characterized so far (Knowles et al. 2012), and new virus types are discovered continuously (reviewed in Smura et al. 2011).

All of the enterovirus strains isolated from patients with diabetes or patients with autoantibodies have to date been members of HEV-B species (reviewed in Tracy et al. 2010). However, the in vitro studies with isolated human pancreatic islets indicate that also HEV-C and HEV-D species contain strains (e.g. prototype strains of CAV-13, PV-1, and EV-94) that are capable of infecting, replicating, and inducing damage in the islets in vitro (Roivainen et al. 2002; Smura et al. 2010), suggesting that (assuming that these viruses are able to gain access to pancreas in vivo) pancreatic islet and beta-cell tropism is not restricted to HEV-B species.

As discussed above, in vitro studies suggest that several enterovirus strains are capable of inducing severe functional impairment of pancreatic beta-cells culminating in the destruction of islets, whereas some strains replicate in beta-cells without apparent adverse effects to the islets (Roivainen et al. 2000, 2002; Smura et al. 2010). In addition, some serotypes, such as E-9 and E-30, contain both strains with destructive and strains with nondestructive phenotypes. The laboratory strains E-9-Barty and E-9-Hill replicated in the islets but induced only minor cellular death and no functional impairment whereas E-9-DM strain, isolated from a 6-week-old child with acute type 1 diabetes (Vreugdenhil et al. 2000), induced drastic functional impairment and destruction of the islets (Paananen et al. 2003; Roivainen et al. 2002). Similarly, despite efficient progeny production, E-30 strains formed a continuum of beta-cell destructive phenotypes ranging from near benign to severe impairment as measured by insulin content of the islet preparations (Roivainen et al. 2002). For neither E-9 nor E-30 strains, could the phylogenetic position based on capsid-coding sequence be associated with the viral phenotype in pancreatic islets (Paananen et al. 2003, 2007).

These results suggest that genetic properties defining the serotype (i.e. capsid-coding region) are, at least to some extent, different to those defining the phenotype of a virus in pancreatic islets. While for many serotypes all of the strains studied so far have been highly destructive, for some serotypes the phenotypic properties of virus strains seem to form a continuum ranging from highly destructive to nearly benign. However, the possible strain/serotype-specific differences such as cell tropism within islet, viral replication kinetics, and cellular responses to infection have not been characterized in detail.

The incomplete correlation between destructive phenotype and serotype of a virus is perhaps not surprising, since the evolution of capsid-coding region (and thus serotype designation) is probably dominated by the selection pressure imposed by host immune system, resulting in antigenic change, a phenomenon likely to be largely independent from pancreatic islet tropism. However, capsid-coding region also determines the receptor specificity, which is an integral cell tropism determinant for a virus. Intriguingly, in the case of many viruses there is overlap between receptor binding sites and antigenic sites, suggesting that antibody—directed selection might affect indirectly the evolution of receptor tropism (Baranowski et al. 2001).

Cell surface molecules utilized by enteroviruses as receptors in beta-cells include coxsackie-adenovirus receptor (CAR), αv intergrins and poliovirus receptor (PVR) (Ylipaasto et al. 2004, 2010). However, although integral for infection, the receptor specificity of a virus cannot be a sole determinant for viral phenotype (e.g. virus-induced islet destruction) since both destructive and nondestructive strains of E-9 use the same $\alpha v\beta 3$ receptor in pancreatic islets (Paananen et al. 2003). Likewise, CBV-2, which uses CAR as a receptor, did not induce islet destruction, whereas other Coxsackie B viruses with apparently similar receptor tropism did (Roivainen et al. 2002). In addition to receptor specificity, other factors, including host cell factors participating in translation and replication initiation (reviewed by Whitton et al. 2005) and the local cytokine (alpha/beta interferon) milieu at the site of infection (Flodstrom et al. 2002; Ida-Hosonuma et al. 2005), are likely to be involved in the cell/tissue tropism of a virus.

The Roles of Recombination, Point Mutations, and Quasispecies Biology

Genetic recombination is likely to add another level of complexity in the relationship of a virus serotype and pancreatic islet tropism. Recombination between serotypes of a given enterovirus species is very common (Lukashev 2005). Sites of frequent recombination have been located in the 5′NCR and in the nonstructural part of the genome. As a consequence of the semi-independent evolution of different genome regions, any circulating enterovirus strain has probably undergone several recombination events during its evolutionary history. Therefore, the members of an enterovirus serotype usually have a congruent phylogeny only in the capsid-coding region of the genome, and thus the serotype designation provides information about this region only. It is known that the genome regions outside the capsid-coding region also contain determinants affecting, for example, neurotropism of polioviruses and cardiovirulence of coxsackieviruses (Dunn et al. 2000; Evans et al. 1985) and reviewed in Pallansch and Roos (2001) suggesting that the 5′NCR or nonstructural protein coding region might harbor determinants affecting pancreatic islet tropism also.

On the other hand, experimental studies suggest that very small changes in the viral capsid proteins may change viral cell type specificity, tissue tropism, and host species and affect cytopathogenicity, host cell response to infection and virulence and immune response in mouse models (Al-Hello et al. 2009; Caggana et al. 1993; Kim and Racaniello 2007; Knowlton et al. 1996; Ramsingh et al. 1997) reviewed in Kim et al. (2006). Accordingly, the extent of pancreatic islet destruction by different E-9 clones was shown to be affected by few amino acids in the capsid-coding region (Paananen et al., unpublished).

Although the genetic determinants of a distinct genome can affect the pathogenicity of a virus, the sequence complexity of an enterovirus population within a host may provide an additional virulence determinant (Domingo et al. 2008). Enterovirus RNA-dependent RNA polymerase lacks mechanisms for proof-reading/repair and post-replicative error correction. Due to their error prone replication machinery, circulating enterovirus populations are genetically extremely variable.

Due to their extremely high mutation frequency, any enterovirus population can be considered to consist of interconnected mutant genomes (or microvariants) termed quasispecies (reviewed by Domingo et al. 2008). The individual genomes of quasispecies differ from each other by one or more nucleotides. At the population level high mutation rate affords the viral quasispecies a greater probability to evolve and adapt to changing environments and selection pressures during infection. Therefore, the high mutation frequency opens up a possibility for an emergence of a quasispecies minority population with pancreatic islet tropism out of an enterovirus population replicating in the mucosa of the intestine (Fig. 28.3). In analogy, in the case of poliovirus, it has been shown that a virus strain with increased replication fidelity (thus producing virus population with less genetic diversity) was less successful at accessing the secondary target tissue (i.e. the central nervous system) (Pfeiffer and Kirkegaard 2005; Vignuzzi et al. 2006). Also, according to these experiments, in vivo the quasispecies diversity itself, rather than selection of individual adaptive mutations, correlated with enhanced pathogenesis, suggesting that the subpopulations within the diverse quasispecies cooperate during the course of systemic infection (Vignuzzi et al. 2006). According to this hypothesis, different viral subpopulations might ensure replication in the intestine, induction of viremia, migration through the endothelial cell lining, and replication in the secondary target tissue. The quasispecies nature of enterovirus populations therefore further obscures the genetic determinants of islet tropism, since while replication in islets might be determined by a distinct set of microvariants, much wider virus population diversity might be needed for islet tropism in vivo.

In conclusion, the incomplete association between serotype identity and islet tropism, frequent intertypic recombination and quasispecies biology hamper the

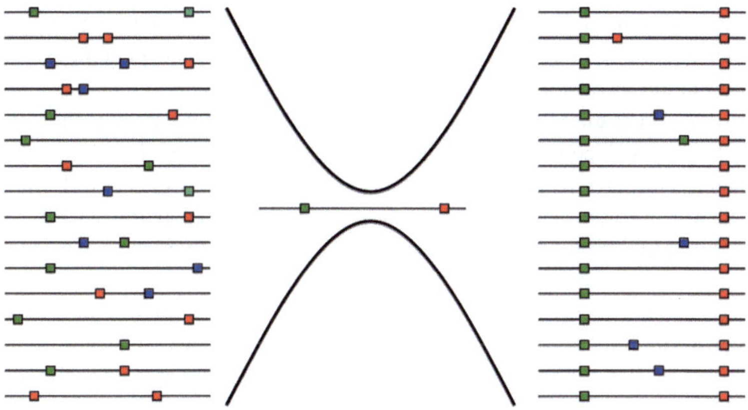

Fig. 28.3 A simplified schematic representation of viral quasispecies diversity and hypothetical changes in quasispecies population structure during colonization of a secondary target tissue. Due to high mutation rate, a virus population within host (or within organ) is composed of variant genomes that differ from each other by one or more nucleotides. A quasispecies subpopulation may contain properties needed for adaptation to a new niche (such as colonization of another organ, cell type). An adaptation to new niche is likely to represent a genetic bottleneck for a virus population. After successful adaptation the quasispecies diversity is likely to increase again due to accumulation of new mutations. If a certain mutation is needed for replication in a new niche, it would become fixed in the viral quasispecies population. During systemic infection in vivo a virus is likely to go through several such bottlenecks probably leading to changes in quasispecies population structure

attempts to detect "diabetogenic" virus strains out of the circulating enteroviruses, since viral genetic determinants behind islet tropism and viral phenotype may be scattered around the genome, be reassorted frequently and ultimately be derived from dynamic distribution of quasispecies subpopulations. Therefore, none of the enterovirus serotypes can be determined unequivocally as "diabetogenic" or "non-diabetogenic" by serotypic determinants only, rather there seems to be a continuum of "diabetogenicity" among enteroviruses.

Islet Response to Enterovirus Infection

In addition to viral genetic factors, also the host cell response to infection affects both beta-cell damage and inflammatory milieu needed for induction of type 1 diabetes. The cells of pancreatic islets respond to enterovirus infection by changing the expression profiles of hundreds of genes (Ylipaasto et al. 2005). The temporal effects of enterovirus infection on islet gene expression are largely unknown at the moment. Moreover, due to lack of established protocol to isolate beta-cells from human pancreatic islets and the lack of human beta-cell lines, the gene expression patterns monitored so far reflect combined effects of virus and virus-induced factors on all of the cell types present in the islet preparation. On the other hand, this may mimic situation in vivo, where the fate of a beta-cell is likely to be affected also by neighboring non-beta cells. Despite these limitations, some general patterns of enterovirus-induced islet responses can be deduced.

A prominent feature of the islet response to enterovirus infection is the upregulation of genes involved in the innate immune reactions. The cellular response against pathogens is initiated by pattern recognition receptors (PRR) such as toll-like receptors (TLRs), retinoic acid-inducible gene (RIG)-I-like receptors (RLRs) (RIG-I, MDA-5, and LGP2) and nod-like receptors (NLR). The activation of these receptors (by pathogen associated molecular patterns) is mediated by various signal transduction cascades ultimately leading to upregulation of innate immunity molecules.

Enterovirus infection induces a strong interferon response. Chronic IFN-alpha synthesis was detected in the beta-cells of islets persistently infected with CBV-3 or CBV-4 (Chehadeh et al. 2000), whereas strong IFN-beta and IFN-lambda induction has been observed during CBV-5 infection in pancreatic islets (Ylipaasto et al. 2005). The role of IFN-gamma in enterovirus-induced gene expression modifications of cultured islets is unclear, since upregulation for IFN-gamma gene was not detected 48 h after infection with CBV-5 (Ylipaasto et al. 2005). However, expression of IFN-gamma has been detected in a rat insulinoma cell line (INS-1) after CBV infection (Nair et al. 2010), suggesting a role for all types I–III interferons in the islet response to enterovirus infection.

The interferons utilize STAT-mediated signaling pathway for the regulation of interferon-stimulated genes (ISGs). These include several molecules with direct antiviral properties as well as PRRs. Accordingly, several PRRs involved in the recognition of and response to intracellular (viral) dsRNA and ssRNA are upregulated after CBV-5 infection. These include toll-like receptor TLR-3, RIG-I-like receptors DDX58 (RIG-I) and IFIH5 (MDA-5) as well as many of the components of respective signaling pathways and the downstream effectors of these pathways (Ylipaasto et al. 2005).

The expression of PRRs is upregulated by interferons thus forming a positive feedback loop that has to be tightly regulated. The lack of decent negative regulation of PRRs would lead to overt immune reactions and possibly to autoimmunity, whereas lack of decent positive regulation would lead to impaired antimicrobial response. Hypothetically enterovirus infections might drive this positive feedback loop towards overt immune response if an enterovirus infection induced sustained upregulation of interferons and PRRs. The strong upregulation of IFN-beta, IFN-lambda, and dsRNA receptors after CBV-5 infection suggests that, at least in vitro, the overt innate immune response may have a role in islet destruction.

As discussed above, innate immunity and inflammatory mediators have a major role in the development of type 1 diabetes (reviewed by Eizirik et al. 2009) and several cell types are involved in the process (reviewed by Lehuen et al. 2010). A strong upregulation of cytokine and chemokine gene expression has been detected after enterovirus infection (Berg et al. 2006; Ylipaasto et al. 2005). The chemokines induced by CBV-5 infection (Ylipaasto et al. 2005) bind to a wide variety of receptors (i.e. CCR1, -2, -3, -4, -5, -6 and -10, CXCR-1, -2, -3A, -3B, -6 and CX3CR) suggesting a wide scale recruitment of different leukocyte types (including neutrophils, monocytes/macrophages, dendritic cells, NK cells, and T-cells) to islets. Concurrently, class I MHC molecules and the immunoproteasome complex responsible for antigen processing for MHC I are upregulated by CBV-5 infection, probably leading to enhanced antigen presentation (Ylipaasto et al. 2005).

Enterovirus infection of cultured pancreatic islets also induces strong expression of proinflammatory cytokines (Ylipaasto et al. 2005). Proinflammatory cytokines, such as IL-1, TNF-alpha, and IFN-gamma, are directly harmful for beta-cells and probably have a major role in the development of type 1 diabetes (reviewed by Eizirik et al. 2009). These results are in accordance with the hypothesis that enterovirus-induced immune reaction may contribute the damage induced by virus infection. Evidently neither innate immune response nor virus replication alone can explain the necrosis observed after enterovirus infection, since a prolonged cytokine treatment (IL-1, TNF-alpha, and IFN-gamma) of human pancreatic islets induces apoptosis rather than necrosis (Delaney et al. 1997; Ylipaasto et al. 2005), and some enterovirus strains are capable of replicating in pancreatic islet with no or mild destruction.

Together these results suggest that enterovirus infection in pancreatic islets is capable of inducing strong innate immune response and metabolic changes, which might affect the recruitment of leukocytes to the site of infection and subsequent (possible) induction or proliferation of autoimmune reactions as well as the fate of islet in vitro (i.e. necrosis vs. replication with no or mild damage).

Conclusion

In conclusion, there is a significant degree of heterogeneity in the infectivity, virulence, cell types affected, host response and outcome of infection in the experimental in vitro models of enterovirus infection in pancreatic islets. The phenotypic variation depends both on viral and host factors and underlines the complexity of enterovirus infections in vivo and the difficulty of establishing causal evidence on enterovirus-induced type 1 diabetes.

Acknowledgements This work is supported by grants from the European Union (EP7-HEALTH-2007, DIAPREPP N202013), the Juvenile Diabetes Research Foundation (USA), The Academy of Finland, and Päivikki and Sakari Sohlberg Foundation.

References

Agol VI, Belov GA, Bienz K, Egger D, Kolesnikova MS, Romanova LI, Sladkova LV, Tolskaya EA (2000) Competing death programs in poliovirus-infected cells: commitment switch in the middle of the infectious cycle. J Virol 74:5534–5541

Al-Hello H, Ylipaasto P, Smura T, Rieder E, Hovi T, Roivainen M (2009) Amino acids of Coxsackie B5 virus are critical for infection of the murine insulinoma cell line, MIN-6. J Med Virol 81:296–304

Baranowski E, Ruiz-Jarabo CM, Domingo E (2001) Evolution of cell recognition by viruses. Science 292:1102–1105

Berg Anna-Karin , Korsgren O, Frisk G (2006) Induction of the chemokine interferon-gamma-inducible protein-10 in human pancreatic islets during enterovirus infection. Diabetologia 49:2697–2703

Caggana M, Chan P, Ramsingh A (1993) Identification of a single amino acid residue in the capsid protein VP1 of coxsackievirus B4 that determines the virulent phenotype. J Virol 67: 4797–4803

Chehadeh W, Kerr-Conte J, Pattou F, Alm G, Lefebvre J, Wattre P, Hober D (2000) Persistent infection of human pancreatic islets by coxsackievirus B is associated with alpha interferon synthesis in beta cells. J Virol 74:10153–10164

Delaney CA, Pavlovic D, Hoorens A, Pipeleers DG, Eizirik DL (1997) Cytokines induce deoxyribonucleic acid strand breaks and apoptosis in human pancreatic islet cells. Endocrinology 138:2610–2614

Domingo E, Martin V, Perales C, Escarmis C (2008) Coxsackieviruses and quasispecies theory: evolution of enteroviruses. Curr Top Microbiol Immunol 323:3–32

Dotta F, Censini S, van Halteren AG, Marselli L, Masini M, Dionisi S, Mosca F, Boggi U, Muda AO, Prato SD, Elliott JF, Covacci A, Rappuoli R, Roep BO, Marchetti P (2007) Coxsackie B4 virus infection of beta cells and natural killer cell insulitis in recent-onset type 1 diabetic patients. Proc Natl Acad Sci USA 104:5115–5120

Dunn JJ, Chapman NM, Tracy S, Romero JR (2000) Genomic determinants of cardiovirulence in coxsackievirus B3 clinical isolates: localization to the 5′ nontranslated region. J Virol 74:4787–4794

Eizirik DL, Colli ML, Ortis F (2009) The role of inflammation in insulitis and beta-cell loss in type 1 diabetes. Nat Rev Endocrinol 5:219–226

Elshebani A, Olsson A, Westman J, Tuverno T, Korsgren O, Frisk G (2007) Effects on isolated human pancreatic islet cells after infection with strains of enterovirus isolated at clinical presentation of type 1 diabetes. Virus Res 124:193–203

Evans DMA, Dunn G, Minor PD, Schild GC, Cann AJ, Stanway G, Almond JW, Currey K, Maizel JV (1985) Increased neurovirulence associated with a single nucleotide change in a noncoding region of the sabin type-3 poliovaccine genome. Nature 314:548–550

Filippi C, von Herrath M (2005) How viral infections affect the autoimmune process leading to type I diabetes. Cell Immunol 233:125–132

Flodstrom M, Maday A, Balakrishna D, Cleary MM, Yoshimura A, Sarvetnick N (2002) Target cell defense prevents the development of diabetes after viral infection. Nat Immunol 3:373–382

Foulis AK, McGill M, Farquharson MA, Hilton DA (1997) A search for evidence of viral infection in pancreases of newly diagnosed patients with IDDM. Diabetologia 40:53–61

Hultcrantz M, Huhn MH, Wolf M, Olsson A, Jacobson S, Williams BR, Korsgren O, Flodstrom-Tullberg M (2007) Interferons induce an antiviral state in human pancreatic islet cells. Virology 367:92–101

Ida-Hosonuma M, Iwasaki T, Yoshikawa T, Nagata N, Sato Y, Sata T, Yoneyama M, Fujita T, Taya C, Yonekawa H, Koike S (2005) The alpha/beta interferon response controls tissue tropism and pathogenicity of poliovirus. J Virol 79:4460–4469

Kim MS, Racaniello VR (2007) Enterovirus 70 receptor utilization is controlled by capsid residues that also regulate host range and cytopathogenicity. J Virol 81:8648–8655

Kim K, Kanno T, Chapman NM, Tracy S (2006) Genetic determinants of virulence in the group B coxsackieviruses. Future Virol 1:597–604

Knowles NJ, Hovi T, Hyypiä T, King AMQ, Lindberg M, Pallansch MA, Palmenberg AC, Simmonds P, Skern T, Stanway G, Yamashita T, Zell R. (2012). Picornaviridae. In Virus Taxonomy: Classification and Nomenclature of Viruses: Ninth Report of the International Committee on Taxonomy of Viruses pp. 855–880. Edited by King AMQ, Adams MJ, Carstens EB. & Lefkowitz EJ. San Diego: Elsevier

Knowlton KU, Jeon ES, Berkley N, Wessely R, Huber S (1996) A mutation in the puff region of VP2 attenuates the myocarditic phenotype of an infectious cDNA of the Woodruff variant of coxsackievirus B3. J Virol 70:7811–7818

Lehuen A, Diana J, Zaccone P, Cooke A (2010) Immune cell crosstalk in type 1 diabetes. Nat Rev Immunol 10:501–513

Lukashev AN (2005) Role of recombination in evolution of enteroviruses. Rev Med Virol 15:157–167

Nair S, Leung K, Rawlinson WD, Naing Z, Craig ME (2010) Enterovirus infection induces cytokine and chemokine expression in insulin-producing cells. J Med Virol 82:1950–1957

Paananen A, Ylipaasto P, Rieder E, Hovi T, Galama J, Roivainen M (2003) Molecular and biological analysis of echovirus 9 strain isolated from a diabetic child. J Med Virol 69:529–537

Paananen A, Savolainen-Kopra C, Kaijalainen S, Vaarala O, Hovi T, Roivainen M (2007) Genetic and phenotypic diversity of echovirus 30 strains and pathogenesis of type 1 diabetes. J Med Virol 79:945–955

Pallansch MA, Roos RP (2001) Enteroviruses: polioviruses, coxsackieviruses, echoviruses and newer enteroviruses. Lippincot, Williams and Wilkins, Philadelphia

Pfeiffer JK, Kirkegaard K (2005) Increased fidelity reduces poliovirus fitness and virulence under selective pressure in mice. PLoS Pathog 1:102–110

Ramsingh AI, Lee WT, Collins DN, Armstrong LE (1997) Differential recruitment of B and T cells in coxsackievirus B4-induced pancreatitis is influenced by a capsid protein. J Virol 71:8690–8697

Rasilainen S, Ylipaasto P, Roivainen M, Bouwens L, Lapatto R, Hovi T, Otonkoski T (2004a) Mechanisms of beta cell death during restricted and unrestricted enterovirus infection. J Med Virol 72:451–461

Rasilainen S, Ylipaasto P, Roivainen M, Lapatto R, Hovi T, Otonkoski T (2004b) Mechanisms of coxsackievirus B5 mediated beta-cell death depend on the multiplicity of infection. J Med Virol 72:586–596

Roivainen M, Klingel K (2010) Virus infections and type 1 diabetes risk. Curr Diab Rep 10:350–356

Roivainen M, Rasilainen S, Ylipaasto P, Nissinen R, Ustinov J, Bouwens L, Eizirik DL, Hovi T, Otonkoski T (2000) Mechanisms of coxsackievirus-induced damage to human pancreatic beta-cells. J Clin Endocrinol Metab 85:432–440

Roivainen M, Ylipaasto P, Savolainen C, Galama J, Hovi T, Otonkoski T (2002) Functional impairment and killing of human beta cells by enteroviruses: the capacity is shared by a wide range of serotypes, but the extent is a characteristic of individual virus strains. Diabetologia 45:693–702

Smura T, Ylipaasto P, Klemola P, Kaijalainen S, Kyllonen L, Sordi V, Piemonti L, Roivainen M (2010) Cellular tropism of human enterovirus D species serotypes EV-94, EV-70, and EV-68 in vitro: implications for pathogenesis. J Med Virol 82:1940–1949

Smura T, Savolainen-Kopra C, Roivainen M (2011) Evolution of newly described enteroviruses. Future Virol 6:109–131

Taylor RC, Cullen SP, Martin SJ (2008) Apoptosis: controlled demolition at the cellular level. Nat Rev Mol Cell Biol 9:231–241

Tolskaya EA, Romanova LI, Kolesnikova MS, Ivannikova TA, Smirnova EA, Raikhlin NT, Agol VI (1995) Apoptosis-inducing and apoptosis-preventing functions of poliovirus. J Virol 69:1181–1189

Tracy S, Drescher KM, Jackson JD, Kim K, Kono K (2010) Enteroviruses, type 1 diabetes and hygiene: a complex relationship. Rev Med Virol 20:106–116

Vignuzzi M, Stone JK, Arnold JJ, Cameron CE, Andino R (2006) Quasispecies diversity determines pathogenesis through cooperative interactions in a viral population. Nature 439:344–348

Virag L, Szabo C (2002) The therapeutic potential of poly(ADP-ribose) polymerase inhibitors. Pharmacol Rev 54:375–429

Vreugdenhil GR, Schloot NC, Hoorens A, Rongen C, Pipeleers DG, Melchers WJG, Roep BO, Galama JMD (2000) Acute onset of type I diabetes mellitus after severe echovirus 9 infection: putative pathogenic pathways. Clin Infect Dis 31:1025–1031

Vuorinen T, Nikolakaros G, Simell O, Hyypia T, Vainionpaa R (1992) Mumps and coxsackie-B3 virus-infection of human fetal pancreatic islet-like cell clusters. Pancreas 7:460–464

Whitton JL, Cornell CT, Feuer R (2005) Host and virus determinants of picornavirus pathogenesis and tropism. Nat Rev Microbiol 3:765–776

Yin H, Berg AK, Westman J, Hellerstrom C, Frisk G (2002) Complete nucleotide sequence of a coxsackievirus B-4 strain capable of establishing persistent infection in human pancreatic islet cells: effects on insulin release, proinsulin synthesis, and cell morphology. J Med Virol 68:544–557

Ylipaasto P, Klingel K, Lindberg AM, Otonkoski T, Kandolf R, Hovi T, Roivainen M (2004) Enterovirus infection in human pancreatic islet cells, islet tropism in vivo and receptor involvement in cultured islet beta cells. Diabetologia 47:225–239

Ylipaasto P, Kutlu B, Rasilainen S, Rasschaert J, Salmela K, Teerijoki H, Korsgren O, Lahesmaa R, Hovi T, Eizirik DL, Otonkoski T, Roivainen M (2005) Global profiling of coxsackievirus- and cytokine-induced gene expression in human pancreatic islets. Diabetologia 48:1510–1522

Ylipaasto P, Eskelinen M, Salmela K, Hovi T, Roivainen M (2010) Vitronectin receptors, alpha v integrins, are recognized by several non-RGD-containing echoviruses in a continuous laboratory cell line and also in primary human Langerhans' islets and endothelial cells. J Gen Virol 91:155–165

Yoon JW, Onodera T, Jenson AB, Notkins AL (1978) Virus-induced diabetes-mellitus.11. Replication of coxsackie B3 virus in human pancreatic beta cell-cultures. Diabetes 27: 778–781

Chapter 29
Innate Immunity of Human Pancreatic Islets Infected with Different Enterovirus Types

Gun Frisk

Abstract Human enterovirus (HEV) infections are believed to be an environmental factor in the pathogenesis of type 1 diabetes pathogenesis, but the exact mechanism behind beta-cell death still remains unclear. Accumulating evidence suggests that viral induction of cytokines and chemokines promoting insulitis could be the link between virus infection and type 1 diabetes. When isolated human pancreatic islets are infected with HEV, IP-10 and MCP-1 are secreted from the islets. HEV infection of human islets induces many genes involved in the innate immune response or sensing viral dsRNA such as IL-6, IL-8, RANTES and INF-β, TLR3 and MDA5. If these proteins were expressed in the pancreas, they would promote β-cell death, directly or indirectly by attracting immune cells. Enterovirus-positive pancreatic sections from recent-onset type 1 diabetes cases also expressed the IP10 chemokine. T cells infiltrating the same areas expressed CXCR3, the IP10 receptor. In pancreatic sections from type 1 diabetes patients at onset stained positive for HEV protein 1 IP-10 was detected and, in addition, CCXCR3 was expressed on islet infiltrating was T-cells. These findings support the idea of HEV infection as a trigger of the immune-mediated beta-cell destruction and also suggest a possible mechanism for HEV-induced type 1 diabetes. The induction and secretion of the chemokine IP-10 with a prominent role in the induction of insulitis might be one of the key targets for immune intervention in this group of patients.

Type 1 diabetes is a multifactorial disease characterized by inflammation of the pancreatic islets and immune-mediated destruction of the islet β-cells (Foulis et al. 1986). The causes of this immune reaction are not shown, but in addition to genetic susceptibility factors, the etiology may involve single or multiple infections with

G. Frisk, Ph.D. (✉)
Department of Women's and Children's Health, Uppsala University,
Akademiska Hospital, 75185 Uppsala, Sweden
e-mail: gun.frisk@kbh.uu.se

K.W. Taylor, H. Hyoty, A. Toniolo, A.J. Zuckerman (eds.), *Diabetes and Viruses*,
DOI 10.1007/978-1-4614-4051-2_29,

beta-cell tropic viruses that could trigger a localized inflammatory response (Horwitz et al. 1998; Christen et al. 2003; Ylipaasto et al. 2004). Despite several years of research, it has remained difficult to establish a causal link between specific viruses and the induction of autoimmunity in humans, leading to type 1 diabetes. This suggests that the majority of persons infected with a particular virus will not develop type 1 diabetes and thereby making the determination of a single causative agent difficult. Furthermore, establishment of type 1 diabetes will often occur long after viral clearance making it difficult to identify the causative agent(s). Despite all these difficulties in linking type 1 diabetes to a virus infection, the main candidate is at present the genera enterovirus. Several viruses in the family *Picornaviridae*, in particular the Coxsackie B viruses (CBV) belonging to the human enterovirus B (HEVB) species, have long been implicated in the etiology of type 1 diabetes (Frisk et al. 1985; Gamble et al. 1969; King et al. 1983; Yin et al. 2002; Yoon et al. 1979).

Viruses do not benefit from inflicting harm on their host, the main reason why viruses evolved the ability to damage infected cells is their need to disarm the cellular defensive machinery. The first line of host defense against viruses is the innate immune response. Viral components have been shown to trigger the production of type I interferons (IFN alpha/beta) and other pro-inflammatory cytokines through recognition by several pattern recognition receptors (PRRs), including toll-like receptors and cytoplasmic helicases (Akira et al. 2006). Double-stranded RNA (dsRNA), making-up the genome of the virus or being formed as an intermediate during single-stranded RNA (ssRNA) virus replication, is recognized by toll-like receptor 3 (TLR-3) in endosomal compartments (Alexopoulou et al. 2001) and by the cytoplasmic helicases retinoic acid-inducible gene I (RIG-I) (Yoneyama et al. 2004) and melanoma-differentiation-associated gene 5 (MDA5) (Kato et al. 2006). The relative importance of these different components of the innate immune system for virus recognition is not clarified fully and varies among different types of viruses. Detection and induction of pro-inflammatory cytokines by members of the family *Picornaviridae* have been shown to be mediated by MDA5. Most interestingly, a recent genetic study found four rare variants of MDA5 that confer protection against type 1 diabetes (Nejentsev et al. 2009), suggesting a role for MDA5 in type 1 diabetes development, and also providing a possible link between type 1 diabetes and EV infection. HEV can infect human islets in vitro (Frisk and Diderholm 2000; Roivainen et al. 2000) and has been found in vivo in islets of infants who died of systemic HEV infections (Ylipaasto et al. 2004) as well as in islets of recent onset type 1 diabetes patients (Dotta et al. 2007; Richardson et al. 2009). Thus, understanding the innate immune response triggered by HEV in human pancreatic islets is essential since it provides an immediate reply to a viral challenge. Also controlling the later antigen-specific adaptive immune response, possibly contributing to the beta-cell destruction in type 1 diabetes. Fully understanding the innate immune response triggered by beta-cell tropic viruses might make it possible to prevent or reduce the massive beta-cell death seen in type 1 diabetes.

The chemokine CXCL10, also known as interferon-γ-inducible protein (IP)-10 in particular, has been identified as a major contributor to the type 1 cellular infiltration of the islets in mouse models of type 1 diabetes (Christen et al. 2003;

Rhode et al. 2005). It is also believed that this is the major initiator of the T-cell infiltration seen in human islet at onset of type 1 diabetes. CXCL10 acts via a single receptor, CXCR3, which is primarily expressed on activated type 1 T-cells (Baggiolini et al. 1997). This chemokine has also been implicated in several other immune inflammatory and autoimmune diseases (Wenzel et al. 2008).

Islet Studies

When human islets were infected with two different strains of HEV in vitro genes encoding for proteins with powerful biological activities, i.e. IL-6, IL-8, MIP-1, MIP-2, MIP, RANTES, MCP-1, IP-10, TNF-α, and INF-β were induced. If they were expressed also in vivo they would much likely induce rapidly an inflammation of the islets (Olsson et al. 2005). Most strains of HEV used to infect isolated human islets in vitro are beta-cell tropic, which was shown by double staining for insulin and HEV. Such stainings show that only beta-cells are stained for HEV (Smura et al. 2010) suggesting that these viruses not only beta-cell tropic but it seems like the insulin-producing cells are the only type of cell infected. Also HEV particles could only be found in beta-cells by electron microscopy (Elshebani et al. 2007).

It has been shown that IP-10 is not expressed or secreted from isolated human islets but it is induced strongly after infection with certain HEV strains (Fig. 29.1) (Berg et al. 2006; Moell et al. 2009; Skog et al. 2011). It is known that HEV infections shut down almost all cap-dependent translation of proteins and also, to some extent, block the intracellular transport of proteins(Cornell et al. 2006), despite the fact that IP 10 is synthesized and secreted from islet cells infected with CBV. The cytokine monocyte cell chemotactic protein-1 (MCP-1) is also induced in human islets infected in vitro with CBV (Moell et al. 2009), although MCP-1 is also secreted from noninfected control islets, although to a significantly lower extent. A possible explanation for the secretion of MCP-1 from control islets is most probably the stress they experience during the isolation procedure, since it is not likely that MCP-1 is secreted from healthy islets in vivo. Supplementation of human islet cultures with nicotinamide (a vitamin B analog) reduces the secretion of MCP-1 both in CBV-infected and in noninfected control islets. CBV-induced synthesis and secretion of IP-10 are totally blocked when the culture medium is supplemented with nicotinamide (Fig. 29.2) (Moell et al. 2009). This suggests that there are two separate pathways that lead to the induction and secretion of MCP-1 in isolated islets, since the secretion is of MCP-1 is not totally blocked by nicotinamide in contrast to the virus-induced secretion of IP-10 or that other endocrine cells contribute to the MCP-1 secretion. The HEV-induced secretion of IP-10 and MCP-1 from human islets supports the idea of an immune-mediated destruction of beta-cells seen in human type 1 diabetes.

It has also been shown that the female sex hormone, 17-beta-estradiol can modulate induction of the chemokine IP-10 in isolated human islets infected with a CVB5

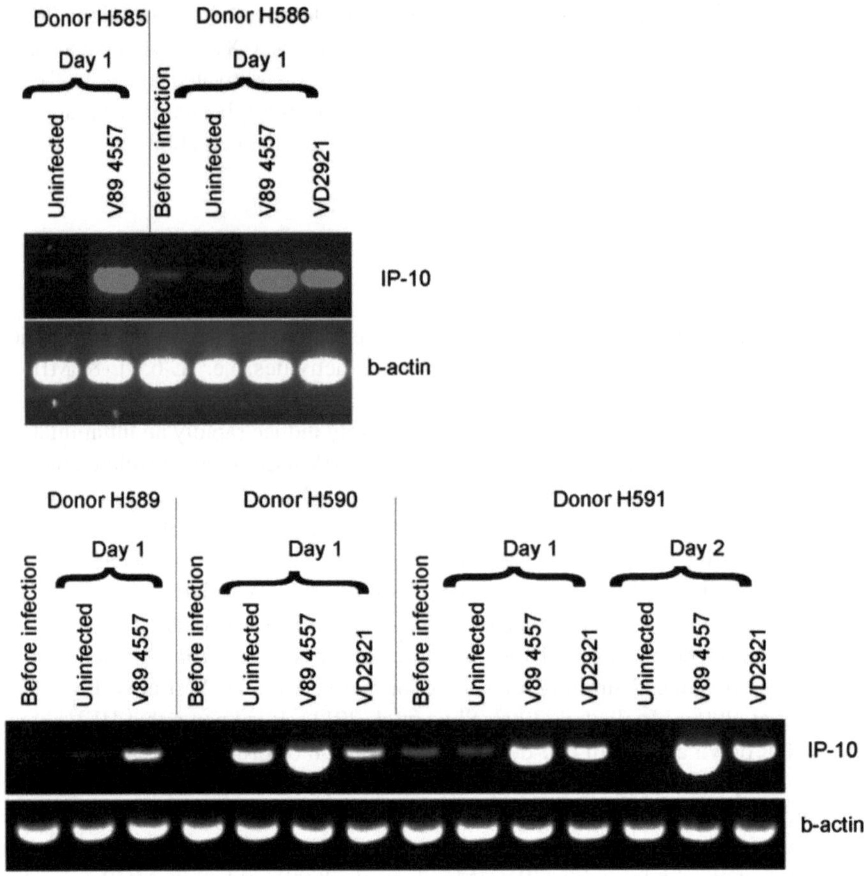

Fig. 29.1 Expression of IP-10 and beta-actin mRNA before and after CBV-4 infection of human islets. Isolated islets from five organ donors were infected with two strains of CBV-4 (VD2921 and V89 4557) or left uninfected. RT-PCR testing was performed before infection (day 0), 24 h after infection (day 1) and 48 h after infection (day 2)

strain isolated from a patient at the onset of type 1 diabetes (Fig. 29.3) (Skog et al. 2011). The secretion of IP-10 from infected islet cultured, with the addition of 17β-estradiol, is reduced significantly compared to untreated HEV-infected controls. Nicotinamide inhibits poly (ADP-ribose) polymerases (PARP-1), and it is believed that high doses of nicotinamide affect ADP-ribosylation reactions in beta-cells. As a consequence, cell death pathways and gene expression patterns are modified, leading to improved beta-cell survival and an altered immunoregulatory balance. 17β-Estradiol binds to the estrogen receptor and the main function of that receptor is as a DNA-binding transcription factor that regulates gene expression. Treatment of isolated human islets with 17β-estradiol results in lower NF-κB nuclear translocation, cytochrome *c* release, and caspase 9 activation

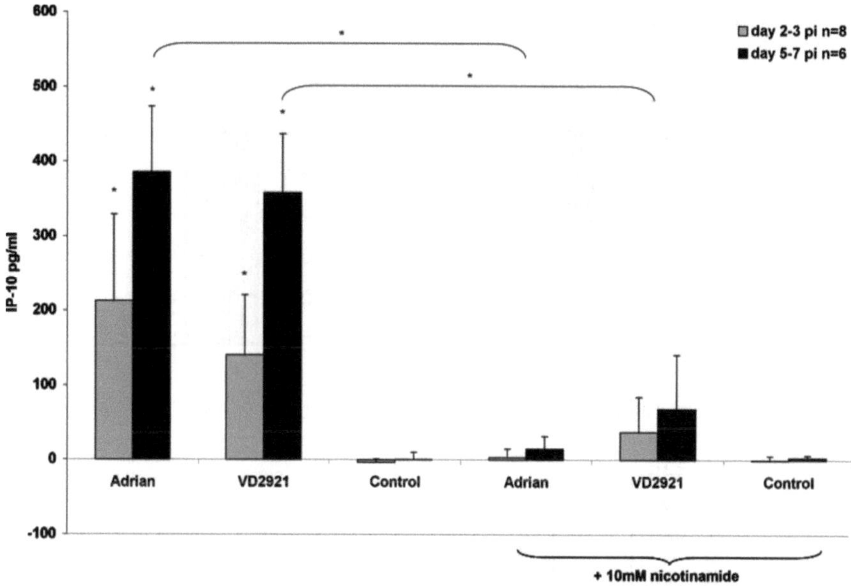

Fig. 29.2 IP-10 secretion from HEV-infected and uninfected human islets cultured with or without nicotinamide. Human islets were infected with one of two HEV strains, Adrian or VD2921, and cultured 5–7 days, with or without pretreatment with 10 mM nicotinamide. Mock-infected islets were treated identically and served as controls. The accumulation of IP-10 in the culture medium, from day 0 to days 2–3 and days 5–7 is presented as mean ± SEM. $^{*}P<0.05$ Wilcoxon signed-rank test. The *n*-values refer to the number of individual islet donors examined

(Contreras et al. 2002). Both compounds affect gene expression although by different pathways. The different pathways affected by nicotinamide and 17β-estradiol most likely explain the different outcome of treatment of HEV-infected islets, e.g. nicotinamide treatment totally blocks the virus-induced secretion of IP-10, whereas 17β-estradiol only reduces it.

The effect of 17β-estradiol on HEV-induced IP-10 secretion introduces one possible contributor to the high male-to-female incidence ratio seen in this disease after puberty.

In the same study, it was also shown that in human islets, different signaling cascades are involved in the induction of IP-10 by HEV or synthetic dsRNA (poly I:C). Poly I:C treatment of islets also induces IP-10 but addition of 17β-estradiol to the culture medium has no effect on IP-10 secretion (Fig. 29.3). This highlights the importance of care when interpreting results of studies performed in different cell systems and with poly I:C as a substitute for viruses.

Type I interferon has been shown to inhibit picornavirus replication in human islets (Hultcrantz et al. 2007), so in order to replicate, the virus needs to counteract the induction of an antiviral state. Both infection of human islets with CBV5 and treatment with poly I:C induced the genes encoding the INFβ and the genes encoding the sensors for virus RNA, MDA5, and TLR3. The induction of the genes by addition

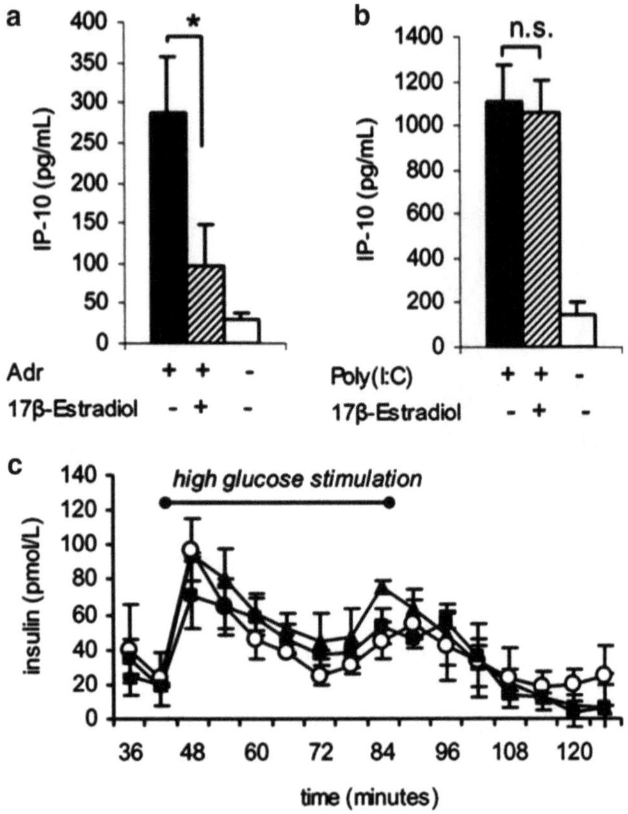

Fig. 29.3 17β-Estradiol reduces HEV but not poly(I:C)-induced IP-10/CXCL10 secretion from human islets. Fifty handpicked human islets were cultured for 24 h, with or without the addition of 17β-estradiol (1 mM) to the culture medium, before infection with HEV (Adr) (**a**) or exposure to poly(I:C) (**b**). Accumulation of IP-10/CXCL10 in the culture media was analyzed by ELISA at 3 days post-exposure to 1 mg/ml poly(I:C) or at 5–6 dpi with HEV. Exposure of islets to 17β-estradiol did not affect their ability to respond to high glucose, as tested in a dynamic perifusion system (**c**). Isolated islets were cultured for 5 days either untreated (&), or in the presence of 1 μM (*) or 10 μM (~) 17β-estradiol. The figure shows insulin release after perifusion with glucose (1.67, 16.7, and 1.67 mmol/L)

of poly I:C to the islet cultures was more rapid and the increase in gene expression higher than the induction by the CBV5 infection. In the former the genes peaked after 24 h and in the latter system the genes peaked 3 days post-infection. Interestingly, the basic gene expression levels of MDA5 varied from very low to high between the different organ donors, although the polymorphism of this gene was not studied. Treating human islets with poly I:C 24 h before or at the time of infection with CBV5 induced an antiviral state in the islets resulting in total inhibition of viral replication (islets from 5/8 donors) or significantly reduced replication (islets from 3/8 donors) suggesting that the virus has means of disarming the defense machinery during infection (Skog et al. 2011).

It has been shown that HEV can infect dendritic cells directly. This seems in agreement with the above scenario where no induction of IL-6 IL-8, RANTES or type 1 INFs was detected (Kramer et al. 2007).

In contrast, induction of innate immunity in human dendritic cells after phagocytosis of human islets infected with HEV has been shown by Schulte et al. (2010). They showed that two of the sensors for dsRNA were induced, RIG-I and MDA5, but only after phagocytosis of HEV-infected islets cells.

These results show that the effect on the innate immune induction in primary human cells after infection with HEV differs among different cell types types of cells (in this case endocrine cells and immune cells), and a dysfunction of any of these viral RNA sensing proteins in human islets might lead to persistent HEV infection of beta-cells.

Studies in Mice

Since the initiating event for human type 1 diabetes is still unclear, several rodent models have been developed. One of these is the RIP-LCMC model for diabetes where breaking of tolerance to a defined target autoantigen is the main goal. The RIP-LCMV diabetes is initiated by i.p. infection of mice with LCMV that only results in a modest infection of the pancreas (von Herrath and Holz 1997). CXCR3 chemokines are among the first chemoattractants factors expressed in the pancreas after such an infection (Christen et al. 2003). IP-10 is highly expressed already after 24 h in the pancreas of these animals. In addition it was shown by IHC also that islets infected with LCMV expressed IP-10. A co-localization of IP-10 and insulin was found in islets two days post-LCMV infection (Christen et al. 2004). Most prominently, more than 90% of the LCMV-specific CD8 lymphocytes expressed CXCR3 (Christen et al. 2003) suggesting that IP-10 was the key factor in the process of imprinting a pattern for the subsequent killing of the beta-cell.

Clinical Studies

Even though mice models and human in vitro studies have suggested that a virus-induced innate immune response could be the cause of the T-cell-mediated beta-cell death in type 1 diabetes, studies in humans are of major importance. Aida et al. (2011) have shown that in a proportion of cases of fulminant onset of type 1 diabetes three sensors of virus RNA were expressed in the pancreatic islets, RIG-I was strongly expressed in beta-cell, MDA-5 was expressed in islet cells and TLR3 was expressed in mononuclear cells that infiltrated the islets found positive for the HEV structural protein 1 (VP1). This suggests that in vivo a virus infection also triggered the innate immune response and that the activation of this immune response likely attracted the immune cells. Richardson et al. (2009) stained pancreatic sections from children with recent onset type 1 diabetes with an antibody against HEV-VP1

Table 29.1 Serum CXCL10 concentrations in subjects diagnosed with acute virus infection

Subject groups	*n*	CXCL10[a] (pg/ml)
Acute HEV infection	25	172 (0–585)
Infection with other viruses (herpes simplex virus, adenovirus, cytomegalovirus, mumps virus)	13	419* (34–611)

*A statistically significant difference $P<0.05$ compared to the HEV-positive group
[a]Median (range)

protein and found that sections from 61% ($n=72$) of the cases were positive, and the positivity was found in insulin-positive cells. Among the controls the corresponding figure was 4%. In the same islet cells protein kinase R (PKR), a protein activated by virus RNA or by INF in the latent state of virus infection, was also stained clearly suggesting an ongoing or recent virus infection.

As mentioned above Roep et al. (2010) showed that IP-10 was detected in sections from pancreas from type 1 diabetes cases close to onset. It was also shown that there are CD8 positive T-cells expression CXCR3, the receptor for IP-10 within the islets. Expression of IP10 was seen also in sections stained negative for HEV VP1.

Studies of serum IP-10 levels in patients with type 1 diabetes have been contradictory, some studies have shown elevated levels (Shimada et al. 2001), but others did not find any difference between cases and controls (Berg et al. 2010; Nicoletti et al. 2002; Rotondi et al. 2003). In a study where IP-10 levels in serum were compared to levels in serum from patients with virus infection the CXCL10 serum levels were not elevated in children with type 1 diabetes at the onset (Berg et al. 2010). There was a considerable overlap between the groups with 99 (8–498) pg/ml in the serum in type 1 diabetes, 120 (17–538) pg/ml in controls and 117 (7–448) pg/ml in siblings. The CXCL10 serum levels in patients with confirmed HEV infection were somewhat increased compared to the other groups, 172 (0–585) pg/ml, but there was no statistically significant difference. In contrast, CXCL10 serum levels in patients with other confirmed virus infection were clearly elevated 418 (34–611) pg/ml, Table 29.1. Despite the fact that elevated serum CXCL10 levels have been shown in some groups of patients with type 1 diabetes, the mean CXCL10 serum levels were not elevated in type 1 diabetes patients or in patients with confirmed HEV infection in this study. In contrast in patients with other virus infection the CXCL10 levels were elevated, presumably reflecting the severity of the infection or which organ that was infected. This suggests that local production of CXCL10 in the affected organ cannot be measured reproducibly in serum and its potential use in clinical practice is limited.

Conclusions

IP-10 secretion from human pancreatic islets seems to be associated with the onset of type 1 diabetes, and as shown by Roep et al. (2010). It also seems to be secreted only from beta-cells. The beta-cell specificity of this secretion also explains why these cells are targeted by the cytotoxic T-cells and not the other endocrine cells that contain most of the autoantigens associated with type 1 diabetes such as GAD65 and IA-2.

The question is then, what triggers this synthesis of IP-10 in these cells. Both INF-gamma and virus infections are the main causes of induction of this chemokine.

It has also been clearly shown that human pancreatic islets do not secret IP-10 unless they are infected with a virus. The most studied virus in this respect is HEV, and during such infections, virus particles and viral proteins are found mainly in the beta-cell. Taken together, a beta-cell tropic virus that triggers IP-10 secretion from the infected cell would result in a T-cell-mediated killing of only insulin-producing cells.

Substances with anti-inflammatory properties such as nicotinamide and 17β-estradiol that blocked and/or reduced the secretion of IP-10 from HEV/CBV-infected human islet suggest that some kind of anti-inflammatory therapy might prevent or reduce the death of beta-cells.

References

Aida K, Nishida Y, Tanaka S, Maruyama T, Shimada A, Awata T, Suzuki M, Shimura H, Takizawa S, Ichijo M, Akiyama D, Furuya F, Kawaguchi A, Kaneshige M, Itakura J, Fujii H, Endo T, Kobayashi T (2011) RIG-I- and MDA5-initiated innate immunity linked with adaptive immunity accelerates beta-cell death in fulminant type 1 diabetes. Diabetes 60:884–889

Akira S, Uematsu S, Takeuchi O (2006) Pathogen recognition and innate immunity. Cell 124:783–801

Alexopoulou L, Holt AC, Medzhitov R, Flavell RA (2001) Recognition of double-stranded RNA and activation of NF-kappaB by Toll-like receptor 3. Nature 413:732–738

Baggiolini M, Moser B (1997) Blocking chemokine receptors. J Exp Med 186:1189–1191

Berg AK, Korsgren O, Frisk G (2006) Induction of the chemokine interferon-gamma-inducible protein-10 in human pancreatic islets during enterovirus infection. Diabetologia 49:2697–2703

Berg AK, Tuvemo T, Frisk G (2010) Enterovirus markers and serum CXCL10 in children with type 1 diabetes. J Med Virol 82:1594–1599

Christen U, McGavern DB, Luster AD, von Herrath MG, Oldstone MB (2003) Among CXCR3 chemokines, IFN-gamma-inducible protein of 10 kDa (CXC chemokine ligand (CXCL) 10) but not monokine induced by IFN-gamma (CXCL9) imprints a pattern for the subsequent development of autoimmune disease. J Immunol 171:6838–6845

Christen U, Benke D, Wolfe T, Rodrigo E, Rhode A, Hughes AC, Oldstone MB, von Herrath MG (2004) Cure of prediabetic mice by viral infections involves lymphocyte recruitment along an IP-10 gradient. J Clin Invest 113:74–84

Contreras JL, Smyth CA, Bilbao G, Young CJ, Thompson JA, Eckhoff DE (2002) 17beta-Estradiol protects isolated human pancreatic islets against proinflammatory cytokine-induced cell death: molecular mechanisms and islet functionality. Transplantation 74:1252–1259

Cornell CT, Kiosses WB, Harkins S, Whitton JL (2006) Inhibition of protein trafficking by coxsackievirus b3: multiple viral proteins target a single organelle. J Virol 80:6637–6647

Dotta F, Censini S, van Halteren AG, Marselli L, Masini M, Dionisi S, Mosca F, Boggi U, Muda AO, Prato SD, Elliott JF, Covacci A, Rappuoli R, Roep BO, Marchetti P (2007) Coxsackie B4 virus infection of beta cells and natural killer cell insulitis in recent-onset type 1 diabetic patients. Proc Natl Acad Sci U S A 104:5115–5120

Elshebani A, Olsson A, Westman J, Tuvemo T, Korsgren O, Frisk G (2007) Effects on isolated human pancreatic islet cells after infection with strains of enterovirus isolated at clinical presentation of type 1 diabetes. Virus Res 124:193–203

Foulis AK, Liddle CN, Farquharson MA, Richmond JA, Weir RS (1986) The histopathology of the pancreas in type 1 (insulin-dependent) diabetes mellitus: a 25-year review of deaths in patients under 20 years of age in the United Kingdom. Diabetologia 29:267–274
Frisk G, Diderholm H (2000) Tissue culture of isolated human pancreatic islets infected with different strains of coxsackievirus B4: assessment of virus replication and effects on islet morphology and insulin release. Int J Exp Diabetes Res 1:165–175
Frisk G, Fohlman J, Kobbah M, Ewald U, Tuvemo T, Diderholm H, Friman G (1985) High frequency of Coxsackie-B-virus-specific IgM in children developing type I diabetes during a period of high diabetes morbidity. J Med Virol 17:219–227
Gamble DR, Kinsley ML, FitzGerald MG, Bolton R, Taylor KW (1969) Viral antibodies in diabetes mellitus. Br Med J 3:627–630
Horwitz MS, Bradley LM, Harbertson J, Krahl T, Lee J, Sarvetnick N (1998) Diabetes induced by Coxsackie virus: initiation by bystander damage and not molecular mimicry. Nat Med 4:781–785
Hultcrantz M, Huhn MH, Wolf M, Olsson A, Jacobson S, Williams BR, Korsgren O, Flodstrom-Tullberg M (2007) Interferons induce an antiviral state in human pancreatic islet cells. Virology 367:92–101
Kato H, Takeuchi O, Sato S, Yoneyama M, Yamamoto M, Matsui K, Uematsu S, Jung A, Kawai T, Ishii KJ, Yamaguchi O, Otsu K, Tsujimura T, Koh C-S, Reis E, Sousa C, Matsuura Y, Fujita T, Akira S (2006) Differential roles of MDA5 and RIG-I helicases in the recognition of RNA viruses. Nature 441:101–105
King ML, Shaikh A, Bidwell D, Voller A, Banatvala JE (1983) Coxsackie-B-virus-specific IgM responses in children with insulin-dependent (juvenile-onset; type I) diabetes mellitus. Lancet 1:1397–1399
Kramer M, Schulte BM, Toonen LW, de Bruijni MA, Galama JM, Adema GJ, van Kuppeveld FJ (2007) Echovirus infection causes rapid loss-of-function and cell death in human dendritic cells. Cell Microbiol 9:1507–1518
Moell A, Skog O, Ahlin E, Korsgren O, Frisk G (2009) Antiviral effect of nicotinamide on enterovirus-infected human islets in vitro: effect on virus replication and chemokine secretion. J Med Virol 81:1082–1087
Nejentsev S, Walker N, Riches D, Egholm M, Todd JA (2009) Rare variants of IFIH1, a gene implicated in antiviral responses, protect against type 1 diabetes. Science 324:387–389
Nicoletti F, Conget I, Di Mauro M, Di Marco R, Mazzarino MC, Bendtzen K, Messina A, Gomis R (2002) Serum concentrations of the interferon-gamma-inducible chemokine IP-10/CXCL10 are augmented in both newly diagnosed type I diabetes mellitus patients and subjects at risk of developing the disease. Diabetologia 45:1107–1110
Olsson A, Johansson U, Korsgren O, Frisk G (2005) Inflammatory gene expression in Coxsackievirus B-4-infected human islets of Langerhans. Biochem Biophys Res Commun 330:571–576
Richardson SJ, Willcox A, Bone AJ, Foulis AK, Morgan NG (2009) The prevalence of enteroviral capsid protein vp1 immunostaining in pancreatic islets in human type 1 diabetes. Diabetologia 52:1143–1151
Rhode A, Pauza ME, Barral AM, Rodrigo E, Oldstone MB, von Herrath MG, Christen U (2005) Islet-specific expression of CXCL10 causes spontaneous islet infiltration and accelerates diabetes development. J Immunol 175:3516–3524
Roep BO, Kleijwegt FS, van Halteren AG, Bonato V, Boggi U, Vendrame F, Marchetti P, Dotta F (2010) Islet inflammation and CXCL10 in recent-onset type 1 diabetes. Clin Exp Immunol 159:338–343
Roivainen M, Rasilainen S, Ylipaasto P, Nissinen R, Ustinov J, Bouwens L, Eizirik DL, Hovi T, Otonkoski T (2000) Mechanisms of coxsackievirus-induced damage to human pancreatic beta-cells. J Clin Endocrinol Metab 85:432–440
Rotondi M, Romagnani P, Brozzetti A, Santeusanio F, Serio M, Falorni A (2003) to: Nicoletti F, Conget L, Di Mauro M et al. (2002) Serum concentrations of the interferon-alpha-inducible chemokine IP-10/CXCL10 are augmented in both newly-diagnosed Type I diabetes mellitus

patients and subjects at risk of developing the disease. Diabetologia 45:1107–1110. Diabetologia 46:1020–1021

Schulte BM, Kramer M, Ansems M, Lanke KH, van Doremalen N, Piganelli JD, Bottino R, Trucco M, Galama JM, Adema GJ, Van Kuppeveld FJM (2010) Phagocytosis of enterovirus-infected pancreatic beta-cells triggers innate immune responses in human dendritic cells. Diabetes 59:1182–1191

Shimada A, Morimoto J, Kodama K, Suzuki R, Oikawa Y, Funae O, Kasuga A, Saruta T, Narumi S (2001) Elevated serum IP-10 levels observed in type 1 diabetes. Diabetes Care 24:510–515

Skog O, Korsgren O, Frisk G (2011) Modulation of innate immunity in human pancreatic islets infected with enterovirus in vitro. J Med Virol 83:658–664

Smura T, Ylipaasto P, Klemola P, Kaijalainen S, Kyllonen L, Sordi V, Piemonti L, Roivainen M (2010) Cellular tropism of human enterovirus D species serotypes EV-94, EV-70, and EV-68 in vitro: implications for pathogenesis. J Med Virol 82:1940–1949

von Herrath M, Holz A (1997) Pathological changes in the islet milieu precede infiltration of islets and destruction of beta-cells by autoreactive lymphocytes in a transgenic model of virus-induced IDDM. J Autoimmun 10:231–238

Wenzel J, Lucas S, Zahn S, Mikus S, Metze D, Ständer S, von Stebut E, Hillen U, Bieber T, Tüting T (2008) CXCR3-ligand-mediated skin inflammation in cutaneous lichenoid graft-versus-host disease. J Am Acad Dermatol 58:437–442

Yin H, Berg AK, Tuvemo T, Frisk G (2002) Enterovirus RNA is found in peripheral blood mononuclear cells in a majority of type 1 diabetic children at onset. Diabetes 51:1964–1971

Ylipaasto P, Klingel K, Lindberg AM, Otonkoski T, Kandolf R, Hovi T, Roivainen M (2004) Enterovirus infection in human pancreatic islet cells, islet tropism in vivo and receptor involvement in cultured islet beta cells. Diabetologia 47:225–239

Yoneyama M, Kikuchi M, Natsukawa T, Shinobu N, Imaizumi T, Miyagishi M, Taira K, Akira S, Fujita T (2004) The RNA helicase RIG-I has an essential function in double-stranded RNA-induced innate antiviral responses. Nat Immunol 5:730–737

Yoon JW, Austin M, Onodera T, Notkins AL (1979) Isolation of a virus from the pancreas of a child with diabetic ketoacidosis. N Engl J Med 300:1173–1179

Chapter 30
Antibody-Dependent Enhancement of Coxsackievirus-B Infection: Role in the Pathogenesis of Type 1 Diabetes

Didier Hober, Famara Sane, Karena Riedweg, Rachel Desailloud, and Anne Goffard

Abstract Antibodies can prevent viruses from infecting target cells, but antibodies against viruses can also enhance the infection of target cells. This phenomenon is called antibody-dependent enhancement (ADE) of infection. The mechanisms of ADE of infection and the results of this phenomenon in vivo are discussed. The ADE of CVB infection has been observed in animal models. In the human system it has been shown in vitro that CVB4, in combination with non-neutralizing antibodies, can infect human monocytes and stimulate the production of IFN-α by these cells through interactions with a specific viral receptor (CAR) and receptors for the Fc portion of IgG (FcγRII and FcγRIII). It cannot be excluded that the ADE of CVB infection in humans can increase the infection of peripheral blood mononuclear cells by these viruses, can cause viral escape from the immune response, and may contribute to the spread of CVBs in the host. Therefore, antibodies enhancing CVB infection may play a role in the pathogenesis of type 1 diabetes induced by or associated with these viruses.

Introduction

Viruses have been associated with type 1 diabetes and those belonging to the Enterovirus genus of the *Picornaviridae* family are major candidates. A relationship between these viruses and type 1 diabetes, especially coxsackie B viruses (CVB) belonging to the human enterovirus B species, has been demonstrated (Jaidane and Hober 2008; Jaidane et al. 2010; Yeung et al. 2011). The hypothesis of the role of CVB in type 1 diabetes is strengthened by the results of experimental investigations in vitro and in animal models (Jaidane et al. 2009). This link probably involves an

---

D. Hober, Ph.D. (✉) • F. Sane • K. Riedweg • A. Goffard
University Lille 2 CHRU, Laboratory of Virology/EA3610, Loos-Lez-Lille, France

R. Desailloud
UPJV CHU, Service d'Endocrinologie-Diabétologie-Nutrition, 80054 Amiens, France

K.W. Taylor, H. Hyoty, A. Toniolo, A.J. Zuckerman (eds.), *Diabetes and Viruses*,
DOI 10.1007/978-1-4614-4051-2_30, © Springer Science+Business Media New York 2013

interplay between enteroviruses, pancreatic beta cells, the innate and adaptive immune systems, and host genes that regulate the immune response to virus infections (Hober and Sane 2010, 2011; Hober and Sauter 2010).

Following a viral infection the host immune response may develop into a pathogenic process and lead to autoimmune disease (Tauriainen et al. 2011). CVBs may induce or aggravate the development of type 1 diabetes through several mechanisms which are not mutually exclusive such as repeated and/or persistent infection of beta cells resulting in inflammation, bystander activation of T cells directed towards beta cell antigens, auto-reactive response to islet self-antigens by molecular mimicry, and thymus infection. Another possible mechanism relies on antibodies (Sane et al. 2011). Antibodies can prevent viruses from infecting target cells, but at the opposite, antibodies against viruses can increase the infection of target cells. This phenomenon is designed as antibody-dependent enhancement (ADE) of infection (Sauter and Hober 2009).

The mechanisms of ADE of virus infection and its consequences in vivo are discussed. Emphasis placed on ADE of CVB infection, first in animal models, then in the human system. The possible consequences of ADE of CVB infections in humans are reviewed especially in the context of the viral pathogenesis of type 1 diabetes.

Antibodies Can Enhance Viral Infections

In the case of immune system cells, complexes constituted by virus-bound antibodies recognize one or several receptors for the Fc portion of antibodies, alternatively antibodies bound to Fc receptor(s) can recognize viruses. Furthermore, virus-bound antibodies can interact with C1q which activates complement and results in the binding of C3d protein to the virus, as described in the case of flavivirus infection of macrophages (Cardosa et al. 1983).

In the model of macrophage infection by the Ross River virus (Alphavirus genus of the *Togaviridae* family), a virus can interact, on the one hand, with a specific viral receptor and, on the other hand, with an Fcγ receptor through virus-bound antibodies. In that model a higher amount of virus can infect cells, and, through the FcγR-activated cellular signaling, and the synthesis of antiviral protein is suppressed, which may increase the viral multiplication (Lidbury and Mahalingam 2000).

In the case of human immunodeficiency virus that binds the CD4 receptor through the gp120 viral protein, antibodies can enhance the infection through binding to this protein which thereafter adopts a conformation enabling the attachment to a co-receptor (CCR5) (Sullivan et al. 1998).

In cells that do not belong to the immune system, the ADE of viral infection has also been described, and in that case a complement-derived protein plays a role. In the Ebola virus model, the expression of C1qR onto target cells enables the attachment of C1q which is bound to antibodies involved in immune complexes (Takada et al. 2003).

Whatever the target cell type, the concentration of antibodies can play a role in ADE of viral infections. When small amounts of neutralizing antibodies are present, most of the sites onto the viral capsid involved in binding to specific receptors are

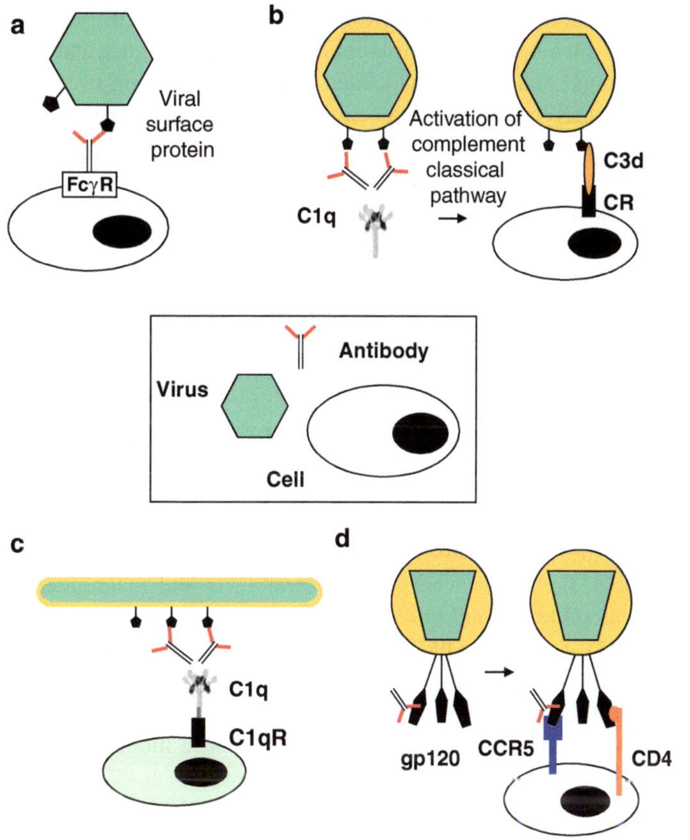

Fig. 30.1 Mechanisms of antibody-dependent enhancement of virus infection. (**a**) Fc part of virus-bound antibody recognized by Fcγ receptor (FcγR), as observed in the case of foot-and-mouth disease virus infection. (**b**) Flavivirus-bound antibody interaction with C1q activates complement which results in the binding of the C3d protein to the virus and to complement receptor (CR) at the surface of macrophages. (**c**) Ebola virus-bound antibody interacts with C1q which binds to the C1q receptor (C1qR) at the surface of cells not belonging to the immune system. (**d**) In the case of HIV, the binding of antibodies to the viral surface glycoprotein gp120 induces a conformational change enabling its attachment to the co-receptor CCR5

available. In this context, the antibodies can bind C1q or one FcγR which facilitates infection with the virus (Takada and Kawaoka 2003).

Viruses use specific receptors at least for binding host cells, whereas the next steps for infecting can rely onto coreceptors, the role of which can be played by FcγR (Montgomery et al. 1996). Indeed it has been reported that antibodies and FcγR, irrespective of the presence of specific viral receptor, can be the sole prerequisite for the infection with enveloped viruses. Furthermore this phenomenon has been observed also with foot-and-mouth disease virus (FMDV), a non-enveloped virus belonging to the Aphtovirus genus of the *Picornaviridae* family (Mason et al. 1994).

The role of enhancing antibodies in the infection of cells with viruses has been displayed in vitro (see Fig. 30.1 for an overview of mechanisms). In addition, there are examples in animal and human virology showing that ADE plays a role in the pathogenesis of enveloped RNA virus-induced diseases. Higher and longer lasting viral titers in blood and organs associated with the worsening of symptoms were obtained in pigs carrying maternal specific antibodies that were inoculated intranasally with porcine reproductive and respiratory syndrome virus (PRRSV), a virus belonging to the *Arteriviridae* family (Shibata et al. 1998; Brown et al. 2006; Chen et al. 2001; Halstead 1979; Halstead 1981). In humans, dengue virus infection is a well-documented example of ADE. There are four serotypes of dengue virus that belong to the *Flaviviridae* family. Preexisting antibodies to one dengue virus serotype expose to a higher risk of severe disease when the individual is infected with a homologous or heterologous serotype of the virus (Dejnirattisai et al. 2010).

Antibody-Dependent Enhancement of Coxsackievirus-B Infection in Animals

Antibodies can increase in vitro the infection of animal cells with FMDV. In this particular model, host cells were mononuclear cells and macrophages extracted from pig peripheral blood and the mouse macrophage-like cell line P388D1 (Baxt and Mason 1995). A restricted FMDV replication was observed in pig cells, whereas both structural and nonstructural proteins were detected in the mouse cell line.

Besides FMDV, antibodies can enhance infection in animals with viruses belonging to the *Picornaviridae* family. Indeed, in the mouse model, the ADE of CVB infection has been investigated both in vitro and in vivo.

Murine heterologous antibodies directed against CVB2 have been shown to increase the infection with CVB3 of mouse peritoneal macrophages and the mouse macrophage-like cell line P388D1 (Girn et al. 2002). In vivo, murine anti-CVB2 IgGs inoculated together with CVB3 enhanced the level of virus in blood and different organs (heart, pancreas, spleen) as well as tissue damage of exocrine pancreas and heart (Girn et al. 2002). It has been shown that mice (C3H/He, male) with low levels of anti-CVB3 following a first inoculation with a nonmyocarditic CVB3 had a severe (and sometimes lethal) myocarditis with an extended inflammatory infiltrate in the myocardium as well as an increased level of chemokine macrophage inflammatory protein-2 (MIP-2) in blood after inoculation with a myocardiotropic strain of CVB3 (Kishimoto et al. 2002). In contrast, mice with high levels of anti-CVB3 antibodies or without anti-CVB3 antibodies had no heart damage or moderate lesions and inflammatory reaction together with low viral titers in the heart (Kishimoto et al. 2002).

Antibody-Dependent Enhancement of CVB Infection in Human Cells

Poliovirus with its three serotypes belongs to the Enterovirus genus. In the human system, the enhancing activity of IgG towards poliovirus has been observed in vitro. Human polyvalent IgG enhanced the poliovirus 3-induced production of IFN-α by

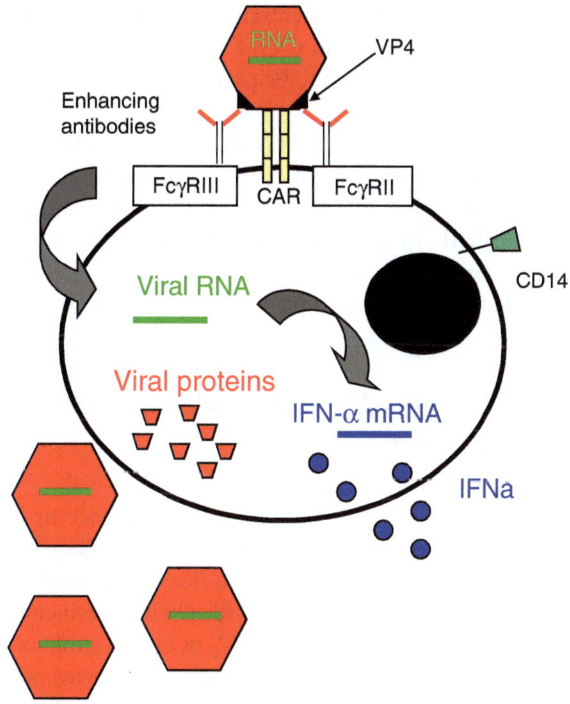

Fig. 30.2 Antibody-dependent enhancement of coxsackievirus B4 infection. CVB4 binds enhancing antibodies (IgG), then, at the surface of monocytes/macrophages, the virus interacts with the coxsackievirus and adenovirus receptor (CAR) and the bound antibodies interact with the Fc IgG receptors (FcγRII and FcγRIII). Afterwards virus-antibody complexes are internalized and viral RNA is uncoated. Entry of viral RNA induces IFN-α synthesis. Viral RNA and proteins are produced, assembled, and new viral particles are released. The target of enhancing antibodies is a region of VP4 (amino acids 11–30) represented as dark triangles at the vertex of virus

PBMCs, which resulted from the entry of viral RNA in cells but was not dependent on viral replication (Palmer et al. 2000). More recently, studies regarding ADE of enterovirus infection have been dealing with CVBs; therefore, the rest of this chapter will focus on these viruses (Fig. 30.2).

Antibody-Dependent Enhancement of CVB4-Induced Production of IFN-α by PBMCs

We have been investigating the relationship between virus and type 1 diabetes for several years. A marker of viral infection, interferon-alpha (IFN-α), was found in plasma samples of patients at various stages of diseases associated with CVB infections, but was not detected in controls. Interferon-α mRNA was detected in PBMCs of these patients, suggesting that CVB were involved in the activation of the IFN-α system in these cells. These observations prompted us to study the mechanisms of CVB-induced production of IFN-α by PBMCs. Thus, it was shown that the in vitro inoculation of PBMCs with CVB4 that had been pre-incubated with diluted plasma containing anti-CVB4 antibodies provoked the production of high concentrations of IFN-α by these cells. In contrast, inoculation of CVB4 alone resulted in a very weak production of IFN-α (Chehadeh et al. 2001). The ADE of IFN-α production has been attributed to non-neutralizing anti-CVB4 IgG (probably IgG1) that form immune complexes with CVB4. The phenomenon depended on the CVB receptor CAR and receptors for the Fc portion of IgG (FcγRII and FcγRIII) (Chehadeh et al. 2001).

In contrast with PBMC of healthy subjects, PBMC of patients with type 1 diabetes produced high amounts of IFN-α in presence of CV-B4. The response of PBMCs from patients was due to anti-CVB4 IgG bound to the cell surface through FcγRI and FcγRIII. In addition, experiments in which CVB4 was pre-incubated with plasma before infection, the enhancing activity of plasma from patients was higher than that associated with plasma from healthy controls (Hober et al. 2002).

Identification of IFN-α-Producing Cells in Response to CVB4

The cell type producing IFN-α in the presence of diluted plasma and CVB4 has been identified. Monocyte-enriched and monocyte-depleted PBMCs populations were obtained by separation in a density-gradient medium. Separated cells were stimulated, on the one hand, with CVB4 pre-incubated with plasma and, on the other hand, with Sendai virus (SV) or herpes simplex virus 1 (HSV-1). Sendai virus stimulated the production of IFN-α by monocytes, whereas HSV1 stimulated circulating dendritic cells. The results obtained with CVB4 compared with SV and HSV-1 showed that the cells producing IFN-α in response to CVB4/IgG complexes were within the monocyte-enriched PBMCs population. In additional experiments, monocytes have been isolated from PBMCs using magnetic beads coated with anti-CD14 antibodies. Detection of IFN-α-positive cells by immunofluorescence staining of monocyte-enriched (but not in monocyte-depleted) PBMCs cultures together with the production of IFN-α by CD14+ cells, suggested CD14+ monocytes as the major interferon-producing cell population in response to CVB4/IgG complexes (Chehadeh et al. 2001).

Antibody-Dependent Enhancement of CVB4 Infection

The relationship between the production of IFN-α by PBMCs inoculated with CVB4 and IgG and the infection of these cells has been investigated. Infectivity of CVB4 for circulating monocytes was facilitated in vitro by non-neutralizing anti-CVB4 IgG, as suggested by the double indirect immunofluorescence staining of PBMCs with CD14 antibodies and VP1 viral capsid antibodies (Hober et al. 2001). The replication of CVB4 in monocytes isolated through density-gradient medium was confirmed by detection of plus- and minus-sense CVB RNA strand by RT-PCR, only when the virus was pre-incubated with plasma. In cultures of isolated monocytes infected with CVB4 pre-incubated with plasma, the release of infectious particles occurred when anti IFN-α neutralizing antibodies were added to the medium before inoculation of permissive cells (Hep-2 cell line) with CVB4.

The ADE of CVB4 infection of monocytes was inhibited by pre-treating cells with anti-human CAR, FcγRII and FcγRIII antibodies, offering evidence that these receptors were needed for infection. Infection of monocytes with CVB4, due to anti-CVB4 antibodies, resulted in the synthesis of IFN-α. Transfection experiments with RNA from infectious CVB4 preparations or UV-inactivated CVB4 showed that virus-induced IFN-α synthesis by PBMCs was dependent on viral RNA entry into the cells, irrespective of its ability to replicate (Hober et al. 2001).

Together, these data showed that CVB4 in combination with non-neutralizing antibodies could infect monocytes and stimulate the production of IFN-α by these cells through interactions with a specific viral receptor and receptors for the Fc portion of IgG (FcγRII and FcγRIII).

The Viral Target of CVB4 Enhancing Antibodies

The mechanism of ADE of CVB4-induced production of IFN-α by PBMCs has been investigated. The plasma-dependent enhancement of CVB4-induced production of IFN-α by PBMCs was inhibited by pre-incubating plasma with the viral protein VP4 that had been obtained from heat-inactivated CVB4 (Chehadeh et al. 2005). In that study, the plasma-dependent enhancement of CVB3-induced production of IFN-α by PBMCs was also reported. It was shown that there were no cross reaction between the VP4 protein of CVB4-E2 and that of CVB3 using competition experiments aimed at inhibiting the plasma-dependent enhancement of CVB4-E2 or CVB3-induced production of IFN-α in PBMCs. An enzyme-linked immunosorbent assay (ELISA) based on CVB4 VP4-coated plates allowed the detection of antibodies more frequently and at higher rates in patients with type 1 diabetes than in control individuals (Chehadeh et al. 2005).

It was then shown that the VP4 capsid protein and anti-VP4 antibodies isolated from plasma (through elution from VP4-coated plates) were both involved in the ADE of CVB4 infection of PBMCs (Sauter et al. 2007).

The target amino acid sequence of antibodies increasing the CVB4-E2- or CVB3-induced IFN-α production by PBMCs was localized between amino acids 11 and 30 of the VP4 protein (69 amino acids in total) (Sauter et al. 2008). The titer and prevalence of antibodies to the VP4 11–30 peptide were higher in patients with type 1 diabetes than in non-type 1 diabetes subjects (Sauter et al. 2008).

The binding of antibodies to CVB through VP4 is intriguing. The capsid of CVBs is composed of four structural proteins. Whereas VP1, VP2, and VP3 are exposed at the virion surface, VP4 is buried in the capsid according to X-ray crystallography studies performed with viral particles at −196°C (Muckelbauer et al. 1995). However, at physiological temperatures, the viral conformation could be different since antibodies in plasma can bind CVBs through VP4 at 37°C. Whether a part of VP4 recognized by enhancing antibodies is continuously exhibited on the virion surface or becomes exposed following discontinuous conformational changes remains to be determined.

Concluding Comments: ADE of CVB Infection and Pathogenesis of Type 1 Diabetes

In humans, antibodies enhancing CVB4 and CVB3 have been detected in vitro. These antibodies can be involved in the production of IFN-α by PBMCs associated with the presence of CVB that is observed in the blood of patients with type 1 diabetes. It has been shown in vitro that in diabetic patients, the CVB4-induced production of IFN-α depends on enhancing anti-CVB4 IgG contained in plasma or bound at the surface of their mononuclear cells (see "Antibody-dependent enhancement of CV-B4-induced production of IFN-α by PBMC" section).

In 50% of patients with IFN-α mRNA in their peripheral blood cells, combined or not with detectable IFN-α levels in plasma, enteroviral RNA sequences presenting homologies with CVB2, CVB3, or CVB4 were detected (Hober et al. 2001). It cannot be discarded that this high infection rate of PBMCs with these viruses may be the result of an antibody-dependent mechanism.

Repeated infections with CVB in the presence of facilitating antibodies can induce an iterative production of IFN-α, which may play a part in the autoimmune circuit directed against β-cells in genetically predisposed individuals. In fact, abnormal activation of IFN-α is associated with the development of autoimmune reactions (Chehadeh et al. 2000a, b).

Several investigators have shown that enteroviruses may target β-cells, which is in agreement with the hypothesis that they can play a role in the initial disturbance of β-cells through the induction of an inflammatory response. How enteroviruses can reach these cells? The spread of CVB to β-cells may be promoted by monocytes, the productive infection of which can be enhanced by antibodies as demonstrated in vitro.

In vivo, the ADE of infection can cause a viral escape from the immune response and the spreading of viruses in the host. It can thus be responsible for an increased production of viral particles. It is noteworthy that the viral load may represent a critical parameter in the CVB-induced or aggravated pathogenesis of type 1 diabetes as suggested by studies in animal models (Kanno et al. 2006).

Experimental models of CVB3 infection and consequent heart tissue lesions in mice brought information with regard to the pathogenic role of enhancing antibodies in CVB-induced diseases. Preexisting anti-CVB antibodies are known to exacerbate CVB infection and to enhance the damage of target organs (Girn et al. 2002; Kishimoto et al. 2002). Similarly, it cannot be excluded that antibodies play a role in CVB infections resulting in increased damage of pancreatic beta cells, leading to type 1 diabetes.

A relationship between enhancing antibodies and the persistence of CVB infections cannot be ruled out. On the one hand, enhancing antibodies could be involved in the persistence of CVB, and on the other hand a persistent CVB infection could result in a maintained stimulus for the production of enhancing antibodies playing a role in subsequent CVB infections (see Fig. 30.3 for an overview of ADE of CVB infection in type 1 diabetes).

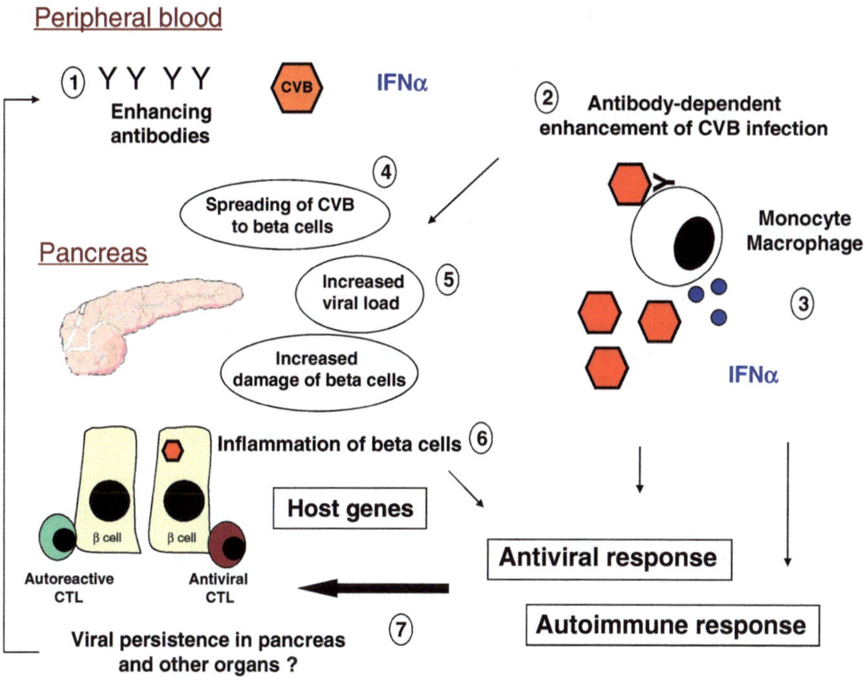

Fig. 30.3 Antibody-dependent enhancement of coxsackievirus-B infection and pathogenesis of type 1 diabetes. (*1*) CVB RNA and IFN-α have been detected in PBMCs of patients with type 1 diabetes. Anti-CVB enhancing antibodies have been found in their plasma. Enhancing antibodies can be associated with the CVB-induced production of IFN-α by PBMCs. (*2*) These antibodies can enhance CVB infection of PBMCs, which may account for the presence of CVB RNA in peripheral blood cells. (*3*) Persistent or repeated infections with CVB can be responsible for prolonged or iterative production of IFN-α that may be associated with the development of autoimmunity. (*4*) Monocytes infected with CVB through an antibody-dependent mechanism may act as a Trojan horse for immune escape and for spreading of viruses to pancreatic beta cells. (*5*) The ADE of CVB infection results in increased viral load and damage to beta cells. (*6*) CVB infection of beta cells activates inflammation and innate immune response that is influenced by the genetic background. An antiviral response occurs to clear virus-infected cells through antiviral cytotoxic T lymphocytes (CTL) which disrupt infected cells contributing to the release of beta cell-specific antigens. Viruses, host genes, innate and adaptive immune responses interact to participate in the development of an autoimmune attack to beta cells through auto-reactive CTL. (*7*) Enhancing antibodies can play a role in the persistence of CVB in the host (pancreas and/or other organs) which, in turn, may further stimulate IFN-α and the production of enhancing antibodies

Whether ADE of CVB infection does participate in diabetogenicity of CVBs deserves further investigation. These studies, however, incite to consider the potential deleterious effect of ADE of infection in the design of safe vaccines against these viruses.

Acknowledgements The authors thank Delphine Caloone for technical assistance and all their collaborators. The studies performed by the authors or in progress have been or are supported by EU FP6 Integrated Project EURO-THYMAIDE (Contract LSHB-CT-2003-503410), EU FP7 PEVNET

Project (FP7-HEALTH-2010-single-stage No. 261441), a grant from *Nord-Pas-de-Calais Région* (ArCir convention 2004/018), CHRU Lille, the *Ministère de l'Education Nationale de la Recherche et de la Technologie, Université de Lille 2*, France, and the *Comité mixte de Coopération Universitaire Franco-Tunisien* (CMCU 2008 N808/G0808). Didier Hober was awarded the 2008 prize of the Fondation pour la Recherche Médicale.

References

Baxt B, Mason PW (1995) Foot-and-mouth disease virus undergoes restricted replication in macrophage cell cultures following Fc receptor-mediated adsorption. Virology 207:503–509

Brown MG, King CA, Sherren C, Marshall JS, Anderson R (2006) A dominant role for FcgammaRII in antibody-enhanced dengue virus infection of human mast cells and associated CCL5 release. J Leukoc Biol 80:1242–1250

Cardosa MJ, Porterfield JS, Gordon S (1983) Complement receptor mediates enhanced flavivirus replication in macrophages. J Exp Med 158:258–263

Chehadeh W, Kerr-Conte J, Pattou F, Alm G, Lefebvre J, Wattre P, Hober D (2000a) Persistent infection of human pancreatic islets by coxsackievirus B is associated with alpha interferon synthesis in beta cells. J Virol 74:10153–10164

Chehadeh W, Weill J, Vantyghem MC, Alm G, Lefebvre J, Wattre P, Hober D (2000b) Increased level of interferon-alpha in blood of patients with insulin-dependent diabetes mellitus: relationship with coxsackievirus B infection. J Infect Dis 181:1929–1939

Chehadeh W, Bouzidi A, Alm G, Wattre P, Hober D (2001) Human antibodies isolated from plasma by affinity chromatography increase the coxsackievirus B4-induced synthesis of interferon-alpha by human peripheral blood mononuclear cells in vitro. J Gen Virol 82:1899–1907

Chehadeh W, Lobert PE, Sauter P, Goffard A, Lucas B, Weill J, Vantyghem MC, Alm G, Pigny P, Hober D (2005) Viral protein VP4 is a target of human antibodies enhancing coxsackievirus B4- and B3-induced synthesis of alpha interferon. J Virol 79:13882–13891

Chen RF, Yeh WT, Yang MY, Yang KD (2001) A model of the real-time correlation of viral titers with immune reactions in antibody-dependent enhancement of dengue-2 infections. FEMS Immunol Med Microbiol 30:1–7

Dejnirattisai W, Jumnainsong A, Onsirisakul N, Fitton P, Vasanawathana S, Limpitikul W, Puttikhunt C, Edwards C, Duangchinda T, Supasa S, Chawansuntati K, Malasit P, Mongkolsapaya J, Screaton G (2010) Cross-reacting antibodies enhance dengue virus infection in humans. Science 328:745–748

Girn J, Kavoosi M, Chantler J (2002) Enhancement of coxsackievirus B3 infection by antibody to a different coxsackievirus strain. J Gen Virol 83:351–358

Halstead SB (1979) In vivo enhancement of dengue virus infection in rhesus monkeys by passively transferred antibody. J Infect Dis 140:527–533

Halstead SB (1981) The Alexander D. Langmuir Lecture. The pathogenesis of dengue. Molecular epidemiology in infectious disease. Am J Epidemiol 114:632–648

Hober D, Sane F (2010) Enteroviral pathogenesis of type 1 diabetes. Discov Med 10:151–160

Hober D, Sane F (2011) Enteroviruses and type 1 diabetes. BMJ 342:c7072. doi:10.1136/bmj.c7072

Hober D, Sauter P (2010) Pathogenesis of type 1 diabetes mellitus: interplay between enterovirus and host. Nat Rev Endocrinol 6:279–289

Hober D, Chehadeh W, Bouzidi A, Wattre P (2001) Antibody-dependent enhancement of coxsackievirus B4 infectivity of human peripheral blood mononuclear cells results in increased interferon-alpha synthesis. J Infect Dis 184:1098–1108

Hober D, Chehadeh W, Weill J, Hober C, Vantyghem MC, Gronnier P, Wattre P (2002) Circulating and cell-bound antibodies increase coxsackievirus B4-induced production of IFN-alpha by peripheral blood mononuclear cells from patients with type 1 diabetes. J Gen Virol 83:2169–2176

Jaidane H, Hober D (2008) Role of coxsackievirus B4 in the pathogenesis of type 1 diabetes. Diabetes Metab 34:537–548
Jaidane H, Sane F, Gharbi J, Aouni M, Romond MB, Hober D (2009) Coxsackievirus B4 and type 1 diabetes pathogenesis: contribution of animal models. Diabetes Metab Res Rev 25:591–603
Jaidane H, Sauter P, Sane F, Goffard A, Gharbi J, Hober D (2010) Enteroviruses and type 1 diabetes: towards a better understanding of the relationship. Rev Med Virol 20:265–280
Kanno T, Kim K, Kono K, Drescher KM, Chapman NM, Tracy S (2006) Group B coxsackievirus diabetogenic phenotype correlates with replication efficiency. J Virol 80:5637–5643
Kishimoto C, Kurokawa M, Ochiai H (2002) Antibody-mediated immune enhancement in coxsackievirus B3 myocarditis. J Mol Cell Cardiol 34:1227–1238
Lidbury BA, Mahalingam S (2000) Specific ablation of antiviral gene expression in macrophages by antibody-dependent enhancement of Ross River virus infection. J Virol 74:8376–8381
Mason PW, Rieder E, Baxt B (1994) RGD sequence of foot-and-mouth disease virus is essential for infecting cells via the natural receptor but can be bypassed by an antibody-dependent enhancement pathway. Proc Natl Acad Sci U S A 91:1932–1936
Montgomery RI, Warner MS, Lum BJ, Spear PG (1996) Herpes simplex virus-1 entry into cells mediated by a novel member of the TNF/NGF receptor family. Cell 87:427–436
Muckelbauer JK, Kremer M, Minor I, Diana G, Dutko FJ, Groarke J, Pevear DC, Rossmann MG (1995) The structure of coxsackievirus B3 at 3.5 Å resolution. Structure 3:653–667
Palmer P, Charley B, Rombaut B, Daeron M, Lebon P (2000) Antibody-dependent induction of type I interferons by poliovirus in human mononuclear blood cells requires the type II fcgamma receptor (CD32). Virology 278:86–94
Sane F, Moumna I, Hober D (2011) Group B coxsackieviruses and autoimmunity: focus on Type 1 diabetes. Expert Rev Clin Immunol 7:357–366
Sauter P, Hober D (2009) Mechanisms and results of the antibody-dependent enhancement of viral infections and role in the pathogenesis of coxsackievirus B-induced diseases. Microbes Infect 11:443–451
Sauter P, Lobert PE, Lucas B, Varela-Calvino R, Alm G, Wattre P, Hober D (2007) Role of the capsid protein VP4 in the plasma-dependent enhancement of the Coxsackievirus B4E2-infection of human peripheral blood cells. Virus Res 125:183–190
Sauter P, Chehadeh W, Lobert PE, Lazrek M, Goffard A, Soumillon M, Caloone D, Vantyghem MC, Weill J, Fajardy I, Alm G, Lucas B, Hober D (2008) A part of the VP4 capsid protein exhibited by coxsackievirus B4 E2 is the target of antibodies contained in plasma from patients with type 1 diabetes. J Med Virol 80:866–878
Shibata I, Okada M, Urono K, Samegai Y, Ono M, Sakano T, Sato S (1998) Experimental dual infection of cesarean-derived, colostrum-deprived pigs with Mycoplasma hyopneumoniae and pseudorabies virus. J Vet Med Sci 60:295–300
Sullivan N, Sun Y, Binley J, Lee J, Barbas CF 3rd, Parren PW, Burton DR, Sodroski J (1998) Determinants of human immunodeficiency virus type 1 envelope glycoprotein activation by soluble CD4 and monoclonal antibodies. J Virol 72:6332–6338
Takada A, Kawaoka Y (2003) Antibody-dependent enhancement of viral infection: molecular mechanisms and in vivo implications. Rev Med Virol 13:387–398
Takada A, Feldmann H, Ksiazek TG, Kawaoka Y (2003) Antibody-dependent enhancement of Ebola virus infection. J Virol 77:7539–7544
Tauriainen S, Oikarinen S, Oikarinen M, Hyoty H (2011) Enteroviruses in the pathogenesis of type 1 diabetes. Semin Immunopathol 33:45–55
Yeung WC, Rawlinson WD, Craig ME (2011) Enterovirus infection and type 1 diabetes mellitus: systematic review and meta-analysis of observational molecular studies. BMJ 342:d35. doi:10.1136/bmj.d35

Jaïdane H, Hober D (2008) Role of coxsackievirus B4 in the pathogenesis of type 1 diabetes. [illegible]

[illegible]

Part VI
Perspectives

Chapter 31
Speculation on Prevention of Type 1 Diabetes

Richard Insel

Abstract The increasing incidence, decreasing age of onset, and the lowered threshold for developing type 1 diabetes make its prevention even more imperative and suggest changes over time in environmental etiologies contributing to the disease. A testable hypothesis is that these epidemiological changes are arising from defective development of or alteration in intestinal microbiome-induced immunoregulation in infancy and early childhood. Viruses contribute to the intestinal microbiome and the intestinal microbial flora may be altered by enteric infections. In addition, viruses may be contributing to the pathogenesis of type 1 diabetes by activation of innate immunity and/or infection of pancreatic beta cells or islets and the ensuing inflammatory response, which may lead to beta cell stress with modification of beta cell protein expression to generate beta cell protein neoepitopes that lead to breaking of immune tolerance to native beta cell antigens. Both primary and secondary prevention of childhood-onset type 1 diabetes should be pursued on a childhood population-wide basis. Primary prevention should focus on developing diabetes vaccines, including enteroviral vaccines if a restricted number of enteroviral serotypes account for a significant proportion of type 1 diabetes in different geographical regions and over time. Vaccine-based approaches to augment, accelerate, or induce robust microbiome-induced immunoregulation in childhood should be investigated.

Prevention of type 1 diabetes represents a cure for those at-risk of developing the disease, is the only affordable means toward a cure for some countries, and represents the most logical and cost-effective strategy for controlling the rising incidence and prevalence of the disease.

R. Insel, M.D. (✉)
JDRF, 26 Broadway, Floor 14, New York, NY 10004, USA

K.W. Taylor, H. Hyoty, A. Toniolo, A.J. Zuckerman (eds.), *Diabetes and Viruses*,
DOI 10.1007/978-1-4614-4051-2_31,

The incidence of type 1 diabetes has increased markedly over the last three to four decades, particularly in Europe, Canada, Australia, and the USA. Most strikingly, this increased incidence is occurring at an earlier age (Patterson et al. 2009; Harjutsalo et al. 2008), with the disease occurring in populations that had once been considered at low to moderate genetic risk (Steck et al. 2011), indicating a lowered threshold for its development. Among children 1–5 years of age, the incidence in Europe is increasing at a rate of 5.4% annually, a rate that is considerably higher than other childhood age groups in that geographic region, where the overall rate of incidence is 3.9% under the age of 15 years. If that trend continues, it will lead to a doubling of the number of new cases in that age group within Europe in a time span of 15 years, from 2005 to the end of this decade (Patterson et al. 2009). In fact, projections suggest that 1- to 5-year-old children will soon represent the peak age incidence in several European countries (Harjutsalo et al. 2008).

This alteration in the epidemiology of type 1 diabetes must be arising from changes in the environment because the "at-risk" genetic pool cannot change in such a short period of time. Unfortunately, we do not know what has changed specifically in the environment. The Environmental Determinants of Diabetes in the Young (TEDDY) Study (TEDDY Study Group 2007) and the multiple pilot studies that preceded TEDDY will hopefully provide definitive insights into specific environmental etiologies.

When considering environmental etiologies, including viruses, it is critical to distinguish those that may: (1) alter immunoregulation; (2) precipitate insulitis; (3) precipitate breaking of immune tolerance and the generation of beta cell-specific autoimmunity; (4) promote progression of the autoimmune process and overcome specific checkpoints; and/or (5) precipitate loss of functional beta cell mass prior to, and at the onset of, insulin dependence or overt diabetes. Some of the controversy surrounding the role of environmental etiologies has also resulted from a failure to consider fully the heterogeneity of the disease and the possible requirement for multiple repetitive environmental exposures. Exposure to potentiating etiologies coupled with failure to be exposed to protective factors may also play critical roles in the pathogenesis of human type 1 diabetes, which disappointingly has not been fully elucidated. Therefore, it is important to identify environmental etiologies that affect each stage in the development of the disease and may be involved in circumventing factors that control progression.

Defective Immunoregulation and the Role of the Intestinal Microbial Flora (Microbiome)

In addition to the increase in childhood-onset type 1 diabetes, there has also been a striking increase over the past several decades in childhood allergic disorders, such as peanut and other food allergies (Branum and Lukacs 2009) as well as asthma, raising the question of whether the set point or "tone" for immunoregulation in early childhood is being set differently than in the past and leading to increased susceptibility to various immune diseases. What these immune diseases all have in common is defective immunoregulation. It is also worth noting that not only type 1 diabetes is increasing in incidence in childhood, but the diagnosis of type 1 diabetes is often accompanied by evidence of other autoimmune disease, including celiac disease, thyroid disease, and autoantibodies (Triolo et al. 2011).

This increasing burden of autoimmune and allergic diseases in childhood may reflect incomplete, dysfunctional, or delayed immunoregulation and education of the immune system in early infancy and childhood. Immunoregulation is conferred early in life through maturation and education of the immune system by the microbial flora or microbiome, which reside in the gastrointestinal tract and live symbiotically with the human host (Kaplan et al. 2011; Mazmanian et al. 2005; Round and Mazmanian 2010; Cerf-Bensussan and Gaboriau-Routhiau 2010). It is conceivable that the combination of increasing incidence, younger age, and lower threshold of childhood-onset type 1 diabetes is due to defective development or alteration of the gastrointestinal microbial flora that leads to less robust or delayed maturation of immunoregulation in early childhood.

It is known that the autoimmune process associated with childhood-onset type 1 diabetes often begins in the first few years of life regardless of whether the onset of insulin dependence occurs during the first 5 years of life or more than a decade later (Ziegler and Nepom 2010; Simell et al. 2010; Parikka et al. 2012; Ziegler and Bonafacio 2012). Therefore, it is conceivable that defective development of the intestinal microbiome or changes in the infant's intestinal microbiome, known as dysbiosis, may be conferring susceptibility to type 1 diabetes throughout the childhood years.

It will be critical to determine whether and how an altered microbiome contributes to the pathogenesis of human type 1 diabetes. The serial sampling of the intestinal microbial flora from birth to early childhood has demonstrated an evolving and dynamic pattern of microbial colonization such that the composition of intestinal bacterial species is altered with various life events. These include the introduction of different foods, illnesses including viral infections, and the use of antibiotics (Koenig et al. 2011; Kau et al. 2011). Infection with viruses, which may influence critically the composition and dynamics of the gastrointestinal microbiome (Reyes et al. 2010), may also be contributing to defective establishment or maintenance of the intestinal microbial flora. It would be surprising if systemic or enteric viral infections, such as enteroviral infections, which are ubiquitous in early childhood, failed to alter the evolving intestinal microbiome in childhood (Koenig et al. 2011). It can be hypothesized that viral infections alone or with the concomitant administration of antibiotics, which are often prescribed inadvertently for viral infections, lead to changes in the intestinal microbial flora resulting in altered immunoregulation and increased susceptibility to autoimmune disease. It is worth pointing out that childhood viral infections are often accompanied by the presence of lymphopenia. Thymic repopulation of the T lymphocyte pool to correct viral-induced lymphopenia could therefore be occurring in the face of dysbiosis and defective immunoregulation, which may increase further the susceptibility to loss of immune tolerance and the development of type 1 diabetes. Although speculative, these concepts should be explored.

To investigate whether defective microbiome-induced immunoregulation accounts for the rising incidence of type 1 diabetes and explains, in part, the so-called "Hygiene Hypothesis" (Okada et al. 2010), biosamples from TEDDY and the pilot projects preceding TEDDY can be exploited. Recent pilot data has, in fact, demonstrated that a decrease in overall bacterial diversity and an increase or decrease in specific bacterial species correlate with the development of type 1 diabetes-associated autoantibodies (Giongo et al. 2011; Brown et al. 2011). In addition to understanding the pathogenesis of type 1 diabetes, investigating the role of the intestinal microbiome may provide insights into developing safe approaches to augment, accelerate, or induce robust microbiome-induced immunoregulation in childhood. If successful, this approach may have the potential to decrease the incidence of childhood-onset type 1 diabetes to levels observed three to four decades ago. The development of therapeutic approaches will, however, require better insights into the mechanisms of microbiome-induced immunoregulation and identification of specific microbe–host interactions. This will likely prove to be a fertile area of research.

Other Roles for Viral Contributions to Type 1 Diabetes

Environmental causes are likely contributing to multiple stages in the pathogenesis of type 1 diabetes and may have the potential to be targeted for either primary (prevention of type 1 diabetes-associated autoimmunity) or secondary (prevention of the onset of insulin dependence after the development of type 1 diabetes-associated autoimmunity) prevention of the disease.

Viruses have been implicated both in the activation of innate immunity early in the pathogenesis of type 1 diabetes (Nejentsev et al. 2009; Heinig et al. 2010; Hober and Sauter 2010) and in the precipitation of autoimmunity (Hober and Sauter 2010; Oikarinen et al. 2011). Inflammation in or around pancreatic islets may prove to be a necessary precursor to the development of type 1 diabetes. Some viruses have been shown to traffic specifically to islets and beta cells, whereas others have been shown to replicate in the beta cell and induce an inflammatory response by activating the production of cytokines and chemokines (Hober and Sauter 2010).

A viral infection in or around the islets could lead acutely to a number of effects, including release of beta cell autoantigens, beta cell dysfunction, activation of the unfolded protein response following beta cell endoplasmic reticulum stress or oxidative stress, and/or generation of beta cell cytokines and chemokines that act in an autocrine or paracrine manner to exacerbate further beta cell stress. The beta cell is then forced to resolve this stress or will die by apoptosis (Oslowski and Urano 2010). Beta cell stress may be associated with beta cell protein modifications, which may include novel post-translational modifications (Engelhard et al. 2006), altered protein expression from alternative splicing of mRNA with generation of novel exon junctions (Ortis et al. 2009), translational infidelity through misacylation of tRNA that leads to specific amino acid changes (Netzer et al. 2009), or protein misfolding that causes altered antigen processing and recognition (Todd et al. 2008). All of these processes could generate neoepitopes that are immunogenic because they were never expressed in the thymus to induce central immune tolerance. Generation of immune responses to neoepitopes could then lead to breaking of immune tolerance for unmodified self-antigens and precipitate autoimmune disease. Also, protein modifications may alter the binding affinity of the TCR–MHC–peptide trimolecular complex and lead directly to the breaking of peripheral immune tolerance. The overall role of beta cell antigen modifications in the pathogenesis of type 1 diabetes and a role for viral infections in precipitating these modifications have not been elucidated, but if beta cell protein modifications occurred and were conserved, they could represent targets for novel diagnostic biomarkers and therapeutics.

Although there is no conclusive data suggesting a role for viral persistence or latency within beta cells or islets in the pathogenesis of type 1 diabetes, it is conceivable that a viral infection in the beta cell could "damage" the cell and leave behind, in a "hit-and-run" manner, a cell that is more susceptible to autoimmune destruction. Possibly a subsequent systemic infection with a virus that shares cross-reactive antigens may lead to recrudescence of autoimmune-mediated beta cell destruction, even if this infection failed to target beta cells or islets directly. It is conceivable that a footprint of such an event is the expression of IFN-α or the upregulation of MHC class I proteins on beta cells, which are detected prominently during the incubation period of type 1 diabetes (Foulis 2008). Characterization of beta cells isolated from at-risk and new-onset subjects that are being collected by JDRF's Network for Pancreatic Organ Donors (http://www.jdrfnpod.org/) may provide critical insights into these concepts.

The occurrence of geographically isolated "outbreaks" of type 1 diabetes (McNally et al. 2006), evidence of recent viral infection near the time of onset of overt diabetes (Stene et al. 2010), and detection and isolation of viruses in autopsy material from new-onset type 1 diabetes (Dotta et al. 2007) all suggest that acute viral infections may play a role in precipitating the onset of insulin dependence and overt type 1 diabetes. Whether viral infections are causing acute beta cell dysfunction or death, insulin resistance with increased insulin demand, and/or acting through other mechanisms to precipitate type 1 diabetes, as suggested above, is not known. The concept of acute viral infection precipitating disease raises the question of whether antiviral drugs may have therapeutic potential in new-onset type 1 diabetes, a concept that could be explored.

To understand the potential role for viruses in the etiology of type 1 diabetes, systems biology-based approaches should be exploited. One of the approaches to address the challenges of developing viral vaccines has been to exploit large-scale gene expression patterns in the blood after viral immunization (Pulendran et al. 2010; Rappuoli and Aderem 2011). Similar approaches could be applied to look for molecular signature patterns that may serve as a roadmap for identifying specific viral etiologies contributing to the pathogenesis of type 1 diabetes.

Preventive Approaches

Secondary prevention clinical studies in first- and second-degree relatives of individuals affected by type 1 diabetes should be supported in order to better understand the pathogenesis of type 1 diabetes and to evaluate preventive interventions (http://www.diabetestrialnet.org/). Additionally, population-based primary and secondary prevention approaches should also be pursued with a focus on childhood-onset type 1 diabetes. Only 10–15% of newly diagnosed individuals with type 1 diabetes have an affected first- or second-degree relative (Eisenbarth 2010) and thus, the major burden of the disease is not being addressed by only screening relatives of affected individuals. Furthermore, a targeted approach to prevention of childhood-onset type 1 diabetes is urgently required to address the rising incidence of type 1 diabetes in childhood. Developing a population-based approach to prevention will require insights into the natural history of type 1 diabetes, which currently only exists for childhood-onset and not adult-onset type 1 diabetes. In general, population-based prevention has been shown to be proven easier to deliver to children compared to adults.

Although primary prevention of childhood-onset type 1 diabetes is the preferred goal because it will likely prove to be more cost-effective, both primary and secondary prevention should be pursued in parallel because secondary prevention appears to be closer at hand and has the potential to accelerate primary prevention strategies.

One approach to primary prevention should be the development of diabetes vaccines for universal infant and childhood immunization. At least three different types of diabetes vaccines should be investigated: diabetes-related viral vaccines, beta cell antigen-specific immunoregulatory vaccines, and vaccines that augment or accelerate intestinal microbiome-induced immunoregulation.

Development of viral-based vaccines will likely be predicated on demonstrating a restricted viral etiology conserved in different geographies over time. Although enteroviruses have been implicated in both the onset of the beta cell-specific autoimmune response and the onset of insulin dependence (Oikarinen et al. 2011; Stene et al. 2010; Yeung et al. 2011), there is a lack of robust epidemiological data demonstrating a limited number of enteroviral serotypes associated with type 1 diabetes occurring in multiple locations and at different periods of time. Based on the current vaccine technology, this is a critical prerequisite for a viable enteroviral vaccine strategy.

Multiple different vaccines that induce beta cell antigen-specific immune tolerance are being developed (Tian and Kaufman 2009; Peakman and von Herrath 2010; Tsai et al. 2010), and if they are shown to be highly safe and efficacious at prevention, they would represent an alternative vaccine approach for primary prevention. Vaccines that safely augment, accelerate, or induce robust microbiome-induced immunoregulation in childhood, as described above, represent an alternative approach for primary prevention.

Until primary prevention approaches are at hand, secondary prevention strategies should be pursued. Secondary prevention will require not only cost-effective, childhood population-based approaches for screening high risk for developing type 1 diabetes, but also precise staging of disease progression, with validated biomarkers to allow tailored, stage-specific interventions to prevent the onset of overt diabetes with insulin dependence. Although several natural history studies have identified at-risk children by neonatal screening for genetic susceptibility (TEDDY Study Group 2007), the changing patterns of genetic susceptibility to type 1 diabetes may make this approach problematic (Steck et al. 2011). Detection of autoantibodies as a risk factor for type 1 diabetes has been validated as an approach to screen for risk, but current technologies are not cost effective and are not highly predictive when only one autoantibody is detected. It is possible, however, that there may be alternative molecular biomarkers (metabolites, proteomics, gene expression patterns, other autoantibodies, etc.) that may prove to be expressed earlier, more highly predictive, and/or more cost-effective for detecting risk of type 1 diabetes compared to diabetes-associated autoantibodies or may serve as an adjunct to autoantibodies. Biomarkers that detect activation of innate immunity or inflammation, or beta cell stress, dysfunction, or damage may be demonstrated to serve this role.

To tailor interventions for secondary prevention will require more precise staging of the progression from the onset of autoimmunity to the onset of insulin dependence with prognostic biomarkers (innate immunity/autoimmune/inflammatory biomarkers, beta cell stress and death biomarkers, biomarkers of impaired glucose and metabolic regulation) and imaging for islet inflammation (Gaglia et al. 2011). Secondary prevention interventions may be used alone or in combination to target islet inflammation, beta cell-specific autoimmunity, beta cell stress and survival, and/or dysglycemia and insulin resistance. The clinical development of preventive interventions will be accelerated by identifying and validating both prognostic biomarkers, in order to aid in tailoring therapy to responders, and predictive biomarkers, in order to serve as surrogates of efficacy of delay of the onset of insulin dependence.

The development and implementation of public health-based approaches for the prevention of childhood-onset type 1 diabetes will require broad partnering and leveraging of both financial and nonfinancial resources with roles for academia, industry, government organizations, payers, and providers. Organizations such as JDRF (www.jdrf.org) will be needed to provide leadership and advocacy to help champion and organize such an effort to increase prospects for success.

References

Branum AM, Lukacs SL (2009) Food allergy among children in the United States. Pediatrics 124:1549–1555

Brown CT, Davis-Richardson AG, Giongo A, Gano KA, Crabb DB, Mukherjee N, Casella G, Drew JC, Ilonen J, Knip M, Hyöty H, Veijola R, Simell T, Simell O, Neu J, Wasserfall CH, Schatz D, Atkinson MA, Triplett EW (2011) Gut microbiome metagenomics analysis suggests a functional model for the development of autoimmunity for type 1 diabetes. PLoS One 6(10):e25792

Cerf-Bensussan N, Gaboriau-Routhiau V (2010) The immune system and the gut microbiota: friends or foes? Nat Rev Immunol 10:735–744

Dotta F, Censini S, van Halteren AG, Marselli L, Masini M, Dionisi S, Mosca F, Boggi U, Muda AO, Prato SD, Elliott JF, Covacci A, Rappuoli R, Roep BO, Marchetti P (2007) Coxsackie B4 virus infection of beta cells and natural killer cell insulitis in recent-onset type 1 diabetic patients. Proc Natl Acad Sci U S A 104:5115–5120

Eisenbarth GS (2010) Banting Lecture 2009: An unfinished journey: molecular pathogenesis to prevention of type 1A diabetes. Diabetes 59:759–774

Engelhard VH, Altrich-Vanlith M, Ostankovitch M, Zarling AL (2006) Post-translational modifications of naturally processed MHC-binding epitopes. Curr Opin Immunol 18:92–97

Foulis AK (2008) Pancreatic pathology in type 1 diabetes in human. Novartis Found Symp 292:2–13

Gaglia JL, Guimaraes AR, Harisinghani M, Turvey SE, Jackson R, Benoist C, Mathis D, Weissleder R (2011) Noninvasive imaging of pancreatic islet inflammation in type 1A diabetes patients. J Clin Invest 121:442–445

Giongo A, Gano KA, Crabb DB, Mukherjee N, Novelo LL, Casella G, Drew JC, Ilonen J, Knip M, Hyöty H, Veijola R, Simell T, Simell O, Neu J, Wasserfall CH, Schatz D, Atkinson MA, Triplett EW (2011) Toward defining the autoimmune microbiome for type 1 diabetes. ISME J 5:82–91

Harjutsalo V, Sjöberg L, Tuomilehto J (2008) Time trends in the incidence of type 1 diabetes in Finnish children: a cohort study. Lancet 371:1777–1782

Heinig M, Petretto E, Wallace C, Bottolo L, Rotival M, Lu H, Li Y, Sarwar R, Langley SR, Bauerfeind A, Hummel O, Lee YA, Paskas S, Rintisch C, Saar K, Cooper J, Buchan R, Gray EE, Cyster JG, Cardiogenics Consortium, Erdmann J, Hengstenberg C, Maouche S, Ouwehand WH, Rice CM, Samani NJ, Schunkert H, Goodall AH, Schulz H, Roider HG, Vingron M, Blankenberg S, Münzel T, Zeller T, Szymczak S, Ziegler A, Tiret L, Smyth DJ, Pravenec M, Aitman TJ, Cambien F, Clayton D, Todd JA, Hubner N, Cook SA (2010) A trans-acting locus regulates an anti-viral expression network and type 1 diabetes risk. Nature 467:460–464

Hober D, Sauter P (2010) Pathogenesis of type 1 diabetes mellitus: interplay between enterovirus and host. Nat Rev Endocrinol 5:279–289

Kaplan JL, Shi HN, Walker WA (2011) The role of microbes in developmental immunologic programming. Pediatr Res 69:465–472

Kau AL, Ahern PP, Griffin NW, Goodman AL, Gordon JI (2011) Human nutrition, the gut microbiome and the immune system. Nature 474:327–336

Koenig JE, Spor A, Scalfone N, Fricker AD, Stombaugh J, Knight R, Angenent LT, Ley RE (2011) Succession of microbial consortia in the developing infant gut microbiome. Proc Natl Acad Sci U S A 108:4578–4585

Mazmanian SK, Liu CH, Tzianabos AO, Kasper DL (2005) An immunomodulatory molecule of symbiotic bacteria directs maturation of the host immune system. Cell 122:107–118

McNally RJ, Feltbower RG, Parker L, Bodansky HJ, Campbell F, McKinney PA (2006) Space-time clustering analyses of type 1 diabetes among 0- to 29-year-olds in Yorkshire, UK. Diabetologia 49:900–904

Nejentsev S, Walker N, Riches D, Egholm M, Todd JA (2009) Rare variants of IFIH1, a gene implicated in antiviral responses, protect against type 1 diabetes. Science 324:387–389

Netzer N, Goodenbour JM, David A, Dittmar KA, Jones RB, Schneider JR, Boone D, Eves EM, Rosner MR, Gibbs JS, Embry A, Dolan B, Das S, Hickman HD, Berglund P, Bennink JR, Yewdell JW, Pan T (2009) Innate immune and chemically triggered oxidative stress modifies translational fidelity. Nature 462:522–526

Oikarinen S, Martiskainen M, Tauriainen S, Huhtala H, Ilonen J, Veijola R, Simell O, Knip M, Hyöty H (2011) Enterovirus RNA in blood is linked to the development of type 1 diabetes. Diabetes 60:276–279

Okada H, Kuhn C, Feillet H, Bach JF (2010) The 'hygiene hypothesis' for autoimmune and allergic diseases: an update. Clin Exp Immunol 160:1–9

Ortis F, Naamane N, Flamez D, Ladrière L, Moore F, Cunha DA, Colli ML, Thykjaer T, Thorsen K, Orntoft TF, Eizirik DL (2009) Cytokines interleukin-1beta and tumor necrosis factor-alpha regulate different transcriptional and alternative splicing networks in primary beta-cells. Diabetes 59:358–374

Oslowski CM, Urano F (2010) The binary switch between life and death of endoplasmic reticulum-stressed beta cells. Curr Opin Endocrinol Diabetes Obes 17:107–112

Parikka V, Näntö-Salonen K, Saarinen M, Simell T, Ilonen J, Hyöty H, Veijola R, Knip M, Simell O (2012) Early seroconversion and rapidly increasing autoantibody concentrations predict prepubertal manifestation of type 1 diabetes in children at genetic risk. Diabetologia. [Epub ahead of print]

Patterson CC, Dahlquist GG, Gyürüs E, Green A, Soltész G, EURODIAB Study Group (2009) Incidence trends for childhood type 1 diabetes in Europe during 1989–2003 and predicted new cases 2005–20: a multicentre prospective registration study. Lancet 373:2027–2033

Peakman M, von Herrath M (2010) Antigen-specific immunotherapy for type 1 diabetes: maximizing the potential. Diabetes 59:2087–2093

Pulendran B, Li S, Nakaya HI (2010) Systems vaccinology. Immunity 33:516–529

Rappuoli R, Aderem A (2011) A 2020 vision for vaccines against HIV, tuberculosis and malaria. Nature 473:463–469

Reyes A, Haynes M, Hanson N, Angly FE, Heath AC, Rohwer F, Gordon JI (2010) Viruses in the faecal microbiota of monozygotic twins and their mothers. Nature 466:334–338

Round JL, Mazmanian SK (2010) The gut microbiota shapes intestinal immune responses during health and disease. Nat Rev Immunol 9:313–323

Simell S, Hoppu S, Simell T, Ståhlberg MR, Viander M, Routi T, Simell V, Veijola R, Ilonen J, Hyöty H, Knip M, Simell O (2010) Age at development of type 1 diabetes- and celiac disease-associated antibodies and clinical disease in genetically susceptible children observed from birth. Diabetes Care 33:774–779

Steck AK, Armstrong TK, Babu SR, Eisenbarth GS, Type 1 Diabetes Genetics Consortium (2011) Stepwise or linear decrease in penetrance of type 1 diabetes with lower-risk HLA genotypes over the past 40 years. Diabetes 60:1045–1049

Stene LC, Oikarinen S, Hyöty H, Barriga KJ, Norris JM, Klingensmith G, Hutton JC, Erlich HA, Eisenbarth GS, Rewers M (2010) Enterovirus infection and progression from islet autoimmunity to type 1 diabetes: the Diabetes and Autoimmunity Study in the Young (DAISY). Diabetes 59:3174–3180

TEDDY Study Group (2007) The Environmental Determinants of Diabetes in the Young (TEDDY) study: study design. Pediatr Diabetes 8:286–298

Tian J, Kaufman DL (2009) Antigen-based therapy for the treatment of type 1 diabetes. Diabetes 58:1939–1946

Todd DJ, Lee AH, Glimcher LH (2008) The endoplasmic reticulum stress response in immunity and autoimmunity. Nat Rev Immunol 8:663–674

Triolo TM, Armstrong TK, McFann K, Yu L, Rewers MJ, Klingensmith GJ, Eisenbarth GS, Barker JM (2011) Additional autoimmune disease found in 33% of patients at type 1 diabetes onset. Diabetes Care 34:1211–1213

Tsai S, Shameli A, Yamanouchi J, Clemente-Casares X, Wang J, Serra P, Yang Y, Medarova Z, Moore A, Santamaria P (2010) Reversal of autoimmunity by boosting memory-like autoregulatory T cells. Immunity 32:568–580

Yeung WC, Rawlinson WD, Craig ME (2011) Enterovirus infection and type 1 diabetes mellitus: systematic review and meta-analysis of observational molecular studies. BMJ 3:342–350

Ziegler AG, Nepom GT (2010) Prediction and pathogenesis in type 1 diabetes. Immunity 32:468–478

Ziegler AG, Bonifacio E; the BABYDIAB-BABYDIET Study Group (2012) Age-related islet autoantibody incidence in offspring of patients with type 1 diabetes. Diabetologia. [Epub ahead of print]

[illegible]

Chapter 32
Viruses as Major Environmental Factors in the Induction of Diabetes

Heikki Hyöty and Keith W. Taylor

Abstract Environmental factors play an important role in the pathogenesis of type 1 diabetes and their identification is essential for the prevention and cure of the disease. Viruses are among the prime environmental candidates showing association with diabetes both in animals and in man. Several viruses can modulate the risk of diabetes in animals having either a protective or a risk effect. In human studies viruses have mainly appeared as risk factors (especially enteroviruses). We believe that in some cases of type 1 diabetes there appears to be a simple and unique enterovirus attack on the islets (e.g. fulminant diabetes in Japan). However, in general the development of diabetes following the virus attack is considerably more complicated. The initiation of virus activity appears to take place long before clinical diabetes finally develops. We suggest that a number of factors may begin the process including viruses such as enteroviruses. These damage islet cells and a series of complicated processes make them more susceptible to a later virus insult such as appears to happen with enteroviruses. Viral persistence and interactions with the innate immune system are among the key factors driving this process. Most enteroviruses detected in diabetic patients appear to be coxsackie or echoviruses. Establishing the nucleotide sequence and the serotype of diabetogenic enteroviruses is highly important if attempts are made to produce antiviral vaccines. Although many observers have now reported evidence of enterovirus infection preceding type 1 diabetes, we would not wish to exclude the possibility that other viruses might be involved in the diabetic process.

H. Hyöty, M.D., Ph.D. (✉)
Department of Virology, University of Tampere and Tampere University Hospital,
Biokatu 10, 33520 Tampere, Finland
e-mail: heikki.hyoty@uta.fi

K.W. Taylor
† (Deceased)

K.W. Taylor, H. Hyoty, A. Toniolo, A.J. Zuckerman (eds.), *Diabetes and Viruses*,
DOI 10.1007/978-1-4614-4051-2_32, © Springer Science+Business Media New York 2013

It may surprise many readers of this book that so large a number of viruses are thought to be associated with the onset of diabetes both in man and in animals. The list covers a very wide range of viruses of all types and is illustrated in Table 32.1. In this table we have commented on what we believe to be the status and importance of each virus in relation to diabetes. Relatively few appear to be important in man. The situation is not like that associated with clinical syndromes such as pneumonia, where a large number of infectious agents may be involved as causal factors, although certain agents such as the pneumococcus may predominate.

The fact that a particular virus is associated with the onset of diabetes does not imply that it is a causal agent for the disease. In the case of diabetes of any kind, the situation is complex, since infection may non-specifically impair glucose tolerance, and will bring to light a latent diabetic condition. Thus, a patient with borderline diabetes may become temporarily diabetic during an attack of influenza. In such case, the influenza virus is not diabetogenic in the accepted sense. A number of viruses listed in Table 32.1 are in this category. In contrast to the viruses with this kind of non-specific effect, we define true *diabetogenic viruses* those which induce diabetes by observed specific effects on insulin-producing cells in the islets of Langerhans, either directly (say by cytolysis) or indirectly through, for example, autoimmune mechanisms.

The Clinical Course of Type 1 Diabetes

The role of viruses in the induction of diabetes in man cannot be interpreted without a detailed knowledge of how the disease develops and particularly possible changes in the course of the disease over an extended time period. Most of this work on the relationship between viruses and diabetes has been carried out on type 1 diabetes, although the possibility that virus infection might be associated with type 2 diabetes is becoming increasingly relevant.

Classically, type 1 diabetes is a disease of younger people including children. It may arise suddenly in otherwise healthy individuals but more usually there is a period of rapidly declining health over weeks or months, with rising blood glucose levels and weight loss. If untreated, it will culminate in a terminal and fatal ketoacidosis as insulin ceases entirely to be produced. This has been the generally accepted pattern for the severe form of the disease, ever since the end of the nineteenth century. However, some more recent work suggests a much slower progression towards complete insulin deficiency (Gorsuch et al. 1981; Tarn et al. 1987; Hekkala et al. 2007; Siljander et al. 2009). Another possibility is that the decline is episodic in nature with successive bouts of beta-cell damaging insults alternating with periods of recovery. This pattern might reflect a succession of attacks of an infective nature, where one or more viruses might be involved. In many cases the subclinical beta-cell damaging process starts several years prior to the diagnosis of clinical type 1 diabetes.

Table 32.1 Viruses associated with the onset of diabetes in man or animals

Virus	Host	Comments	References
Bovine viral diarrhoea virus	Cattle	Diabetes with ketosis	
Cytomegalovirus	Man	Clinical importance uncertain	
Encephalomyocarditis virus	Mice	Causes severe beta-cell damage and diabetes by direct viral effect and immune-mediated mechanisms	
Enteroviruses	Man, mice, primates	Coxsackieviruses, echoviruses, certain types and strains. Clinically significant	
Epstein–Barr	Man	Clinical relevance uncertain	
Foot-and-mouth disease	Cattle	Pancreas affected diabetes with ketosis	Barboni and Manochio (1962)
Infectious hepatitis	Man	Evidence incomplete	Oli and Nwokolo (1979)
Kilham rat virus	Rat	Causes diabetes by immune-mediated mechanisms	
Ljungan virus	Voles	Associated with diabetes, virus found in the pancreas	
Mengovirus	Mice	Direct beta-cell infection	Yoon et al. (1984)
Mumps	Man	Occasionally clinically important	
Poliomyelitis	Man	One case reported, example of enterovirus types not related to diabetes	
Reovirus	Mice	Induces endocrine autoimmunity	Onodera et al. (1978)
Retrovirus	Man	Evidence controversial	Marguarat et al. (2004)
Rotavirus	Man	Importance uncertain	
Rubella	Man, hamster, rabbit	Associated with congenital rubella syndrome. Diabetes onset delayed	
Varicella zoster	Man	Occasional case reports	

Note: References given refer to viruses not discussed elsewhere in this book

Investigations at the Time of Clinical Diagnosis of Type 1 Diabetes

Classical methods for detecting viruses and virus antibodies, neutralizing antibodies in particular, first suggested an association between type 1 diabetes and enterovirus infection. Very recently using these techniques, it has been shown that diabetes is associated with certain group B coxsackie viruses (Laitinen et al. 2012).

The use of molecular methods to identify viruses in blood specimens from newly diagnosed type 1 diabetics has indicated the presence of enteroviral RNA at the start of the disease in many cases (Clements et al. 1995; Andreoletti et al. 1997; Yin et al. 2002; Craig et al. 2003; Sarmiento et al. 2007; Toniolo et al. 2010). This is also discussed in Chap. 13, as well as the difficulties encountered in methodology. It is possible that enteroviruses detected in type 1 diabetic patients represent slowly replicating viruses which are able to persist in infected tissues. Their detection may require highly sensitive technologies and/or optimal processing of collected samples.

Enteroviruses have been especially detected in cases with diabetic ketoacidosis at onset. This was first seen in a study of newly diagnosed adult type 1 diabetics (Andreoletti et al. 1997) and later in child diabetics in Australia (Craig et al. 2003), as well as in Cuba (Sarmiento et al. 2007). This is similarly the case for the fulminant type 1 diabetes described in Japan which is thought to be associated with enteroviruses and in which ketosis is a cardinal feature (see Chap. 22). In addition there are growing numbers of individual case reports in which acute episodes of enterovirus infection are accompanied by an acute form of diabetes. Several of these cases are discussed in detail in Chap. 15. Diabetes in these individuals is therefore likely to be precipitated by the virus as a terminal event. As shown elsewhere, such enteroviruses readily replicate in human islets in tissue culture, or they may impair the function of islets or destroy them. Occasionally, these viruses produce diabetes when injected into animals.

A growing body of evidence suggests that enteroviral proteins and RNA are present in the islets of pancreases recovered post-mortem from newly diagnosed type 1 patients (see Chap. 17). The virus has been found also in the intestinal mucosa of these patients (Oikarinen et al. 2012). It seems unlikely that such viruses are mere innocent bystanders, and further studies on pancreas and other tissues of type 1 diabetic patients will be important for the elucidation of causal relationship. Currently, large international collaboration efforts have been started to carry out such studies (e.g. the nPOD and PEVNET studies, see Chap. 24).

Stages in the Development of Type 1 Diabetes

When an association between enteroviruses and type 1 diabetes was first suggested, it was thought that viral attack might result in direct and rapid destruction of the islets of Langerhans possibly by cytolytic mechanisms. If this has been then case, widespread outbreaks of diabetes might have been expected such as happens with many infectious agents such as influenza viruses. Such epidemics of diabetes have never been recorded, although there is occasional evidence of clustering. Instead, cases reported seem to occur sporadically, even at times of year when diabetes is more common such as the autumn in northern Europe. This could be explained if the virus was able only to attack the islets of individuals previously sensitized by an earlier insult.

We would further suggest that in some diabetes cases enteroviruses act rapidly leading to clinical diabetes in some weeks (as is suggested by the fulminant type of diabetes in Japan), while in other cases the process takes longer, even several years. The rapid process may be caused by a direct cytolytic effect and the slower process by a slowly replicating virus strains which can persist in the pancreas for prolonged periods causing inflammation and leading to an immune-mediated damage. This kind of dual mechanism has been described in virus-induced diabetes in mice (Jun and Yoon 2001). However, to understand the precise sequence of events one has to carry out prospective human studies where the exact timing of infections can be correlated with the appearance of diabetes-predictive autoantibodies and clinical type 1 diabetes.

The virus induces a cascade of immune responses which can include both antiviral and autoimmune components. The beta-cell damaging process can first be driven by rapid antiviral responses such as the secretion of pro-inflammatory mediators (e.g. IL-1 and TNF-alpha) and induction of cytotoxic T-cells and NK-cells, which both can directly damage beta-cells. In case that the virus persists, the autoimmune response may become increasingly important—the prolonged inflammation promotes the presentation of beta-cell autoantigens and the breakdown of tolerance. However, aggressive autoimmune process may only start if the host has the right genetic background (diabetes-related HLA and IFIH1 genes) and if boosted by additional factors, such as immunological cross-reactivity between viral and beta-cell proteins and/or serial infections by different enterovirus types. Such an aggressive autoimmune process may become self-perpetuating and progress even in the absence of the viral stimulus.

Strategies for Investigating Events Leading to the Induction of Diabetes

Prospective studies have proved to be crucially important in the evaluation of the viral aetiology of type 1 diabetes. Such studies follow initially non-diabetic individuals until they develop beta-cell damaging process or clinical type 1 diabetes. Therefore, they can identify viruses which occur long before clinical diabetes is diagnosed and which can be involved in the initiation of the beta-cell damaging process. Rubella is one clear example of such a virus since congenital rubella is associated with increased risk of diabetes in the child (see Chap. 1). Similarly,

Table 32.2 Key research strategies to confirm causal relationship between a virus and human type 1 diabetes

Prospective studies identifying virus infections among non-diabetic children who are followed longitudinally from birth until they develop type 1 diabetes
Studies evaluating the presence of viruses in the pancreas and other tissues of type 1 diabetic and pre-diabetic individuals
Studies exploring the mechanisms of virus-induced beta-cell damage (clinical and molecular studies, cell and animal models)
Genetic studies on diabetes-associated polymorphisms in host genes which are involved in virus replication or antiviral response
Intervention studies with vaccines or antiviral drugs

prospective studies have shown that enterovirus infections peak prior to the first detection of diabetes-predictive autoantibodies (Oikarinen et al. 2011). We would regard these viruses as *initiating viruses* since they seem to be able to start the process which leads to the clinical diabetes after a varying period of time ranging from some months to several years.

Both animal and human studies suggest that the activation of innate immune system plays a crucial role in virus-induced diabetes. One of the most important recent discoveries is the identification of diabetes-associated polymorphisms in genes which are involved in the innate immune system response to the virus (Chaps. 3 and 27). One of these genes, IFIH1, codes for a receptor recognizing dsRNA molecules which are produced in enterovirus-infected cells. The discovery of these kinds of gene polymorphisms provides additional evidence for the role of viruses in the pathogenesis of type 1 diabetes and makes it possible to explore the underlying mechanisms in man.

Conclusions

History has shown that it is difficult to confirm a causal relationship between a microbe and a slowly progressing chronic disease. For instance, this has been true in the case of papillomavirus-induced cervical cancer and in *Helicobacteria*-induced gastric cancer, where it took several decades to prove causality. Perhaps type 1 diabetes needs to be added to this list. It is evident that many viruses can cause diabetes in animals, and it seems likely that some of them cause human diabetes. Enteroviruses remain the main candidates for diabetogenic viruses in man (Yeung et al. 2011). One can hope that the rapidly expanding work in this field will eventually lead to a breakthrough and the development of new treatments for the prevention and cure of type 1 diabetes (Table 32.2). One attractive possibility would be a vaccine against the diabetogenic virus(es). In fact, the first attempts to explore this option are already in progress. Testing the efficacy of such vaccines in the prevention of type 1 diabetes would be a crucial part of proving causal relationship between a virus and type 1 diabetes.

References

Andreoletti L, Hober D, Hober-Vandenberghe C, Belaich S, Vantyghem MC, Lefebvre J, Wattre P (1997) Detection of coxsackie B virus sequences in whole blood samples from adult patients at the onset of type 1 diabetes. J Med Virol 52:121–127

Barboni E, Manochio J (1962) Alterazioni pancrea-tiche in bovini con diabete mellito. Arch Vet Ital 13:477–489

Clements GB, Galbraith DN, Taylor K (1995) Coxsackie B virus infection and the onset of childhood diabetes. Lancet 346:221–223

Craig ME, Howard NJ, Silink M, Rawlinson WD (2003) Reduced frequency of HLA DR1*03DQB1*D2 in children with type 1 diabetes associated with enterovirus RNA. J Infect Dis 187:1562–1570

Gorsuch AN, Spencer KM, Lister J, McNally JM, Dean BM, Bottazzo GF, Cudworth AG (1981) Evidence for a long prediabetic period in type 1 diabetes mellitus. Lancet ii:163–1365

Hekkala A, Knip M, Veijola R (2007) Ketoacidosis at diagnosis of type 1 diabetes in northern Finland: temporal changes over 20 years. Diabetes Care 30:861–866

Jun HS, Yoon JW (2001) The role of viruses in type I diabetes: two distinct cellular and molecular pathogenic mechanisms of virus-induced diabetes in animals. Diabetologia 44:271–285

Laitinen OH, Honkanen H, Pakkanen O, Oikarinen S, Pulkki MM, Huhtala H, Ruokoranta T, Lecouturier V, André P, Harju R, Almond JW, Simell T, Simell O, Ilonen J, Veijola R, Knip M, Hyöty H (2012) Coxsackievirus B1 induces beta-cell autoimmunity that portends type 1 diabetes europic meeting, saint rafael, France (885)

Marguarat S, Weng WY, Todd JA, Conrat B (2004) Association of human endogenous retrovirus K-18 polymorphisms with type 1 diabetes. Diabetes 53:852–854

Oikarinen S, Martiskainen M, Tauriainen S, Huhtala H, Ilonen J, Veijola R, Simell O, Knip M, Hyöty H (2011) Enterovirus RNA in blood is linked to the development of type 1 diabetes. Diabetes 60:276–279

Oikarinen M, Tauriainen S, Oikarinen S, Honkanen T, Collin P, Rantala I, Mäki M, Kaukinen K, Hyöty H (2012) Type 1 diabetes is associated with enterovirus infection in gut mucosa. Diabetes 61(3):687–691

Oli JM, Nwokolo C (1979) Diabetes following infectious hepatitis: a follow up. BMJ 1:926

Onodera T, Jenson AB, Yoon JW, Notkins AL (1978) Virus induced diabetes mellitus: reovirus infection of pancreatic beta cells in mice. Science 20:529–531

Sarmiento L, Cabrera Rode E, Lekuleni L, Cuba I, Molina G, Fonseca M, Heng-Hung L, Boroto AO, Gonzales P, Mas Lago P, Diaz Horta O (2007) Occurrence of enterovirus RNA in serum of children with newly diagnosed type 1 diabetes and islet cell antibody in a population with a low incidence of type 1 diabetes. Autoimmunity 40:540–545

Siljander HT, Simell S, Hekkala A, Lahde J, Simell T, Vahasalo P, Veijola R, Ilonen J, Simell O, Knip M (2009) Predictive characteristics of diabetes-associated autoantibodies among children with HLA conferred disease susceptibility. Diabetes 58:2835–2842

Tarn AC, Smith CP, Spencer KMG, Bottazzo GF, Gale EAM (1987) Type 1 insulin dependent diabetes is a disease of slow clinical onset. BMJ 294:342–345

Toniolo A, Maccari G, Federico G, Salvatori A, Bianchi G, Baj A (2010) Are enterovirus infections linked to the early stages of type 1 diabetes? In: Am. Soc. Microbiol. Meeting, San Diego, California (Abs.)

Yeung WC, Rawlinson WD, Craig ME (2011) Enterovirus infection and type 1 diabetes mellitus; systematic review and meta-analysis of observational molecular studies. BMJ 342:1136–1138

Yin H, Berg AK, Tuvemo T, Frisk G (2002) Enterovirus RNA is found in peripheral blood mononuclear cells in a majority of type 1 diabetic children at onset. Diabetes 51(6):1964–1971

Yoon JW, Morishima T, Mclintock P, Austin M, Notkins AL (1984) Virus induced diabetes mellitus: mengovirus infects beta cells in strains of mice resistant to the diabetogenic effects of encphalomyocarditis virus. J Virol 50:684–690

Chapter 33
Reflections on Viruses and Diabetes Mellitus

Arie J. Zuckerman

While the much publicised, unconfirmed and unfounded association of viruses, for example with multiple sclerosis, with benign myalgic encephalomyelitis (chronic fatigue syndrome or the Royal Free disease as designated originally) among others, serves as an imperative for considerable caution particularly in the case of "guilt by association," it is important to remember the propensity of viruses to surprise. Notable examples are persistent infection with hepatitis B virus and hepatitis C virus and chronic liver disease, cirrhosis and hepatocellular carcinoma; Epstein–Barr virus and Burkitt's lymphoma and various other lymphoid and epithelial tumours; high-risk papillomaviruses, particularly types 16 and 18 and cervical carcinoma; the human T lymphotropic virus 1 and adult T-cell leukaemia, human herpesvirus 8 and Kaposi's sarcoma, and other viruses and other syndromes. It would be unwise, therefore, to overlook or, worse still, to dismiss the role of virus infections directly or indirectly in the aetiology of diabetes mellitus.

The order of this final chapter has been reversed in that the Acknowledgements precede the body of the text. A précis of the reflections and conclusions of this chapter are based on the summaries as written and provided by the authors of this volume, who reviewed meticulously the results of numerous studies undertaken by many investigators throughout the world. For the sake of fluency and readability, specific references to my co-authors have been omitted, and any errors of interpretation and omission are strictly my own.

A.J. Zuckerman, M.D., D.Sc., F.R.C.P., F.R.C.Path, F.Med.Sci. (✉)
Emeritus Professor of Medical Microbiology, University College London Medical School, Royal Free Campus, London NW3 2PF, UK
e-mail: arie.zuckerman@ucl.ac.uk

K.W. Taylor, H. Hyoty, A. Toniolo, A.J. Zuckerman (eds.), *Diabetes and Viruses*, DOI 10.1007/978-1-4614-4051-2_33,

Autoimmunity

Type 1 diabetes is a disease characterised by immune cell-mediated destruction of the pancreatic islet cells leading to insufficient production of insulin. As in the case of other autoimmune diseases, the aetiology of diabetes is a complex interaction between genetic factors, the immune system and environmental factors. Certain genes particularly in the histocompatibility (HLA) region of chromosome 6 are pivotal in the development of diabetes. Non-genetic environmental factors are also important including disproportionate maternal influences which may operate in utero or very early in life, temperate climate, increasing wealth and industrial development, increased hygiene and decreased rates of infectious diseases, immunisation and antibiotics, certain drugs (methyl donors such as hydalazine), wheat consumption, iodine levels, and for type 1 diabetes specifically: overcrowding in childhood and viral and bacterial infections, early exposure to cow's milk, reduced rates or duration of breast feeding, and vitamin D and nitrite consumption.

Genetic Factors

Certain HLA class II haplotypes are associated with particular pancreatic beta-cell-specific autoantibody patterns. Risk modifier loci exist in the HLA class I region. Non-HLA genes may also play a role, and more than 50 gene variants display an association with type 1 diabetes outside the HLA region. Rare mutations in some of these genes may have potent effects, for example, the IFIH1 gene is an intracellular receptor for enterovirus RNA, representing a possible link between enterovirus and type 1 diabetes.

Immune Factors

The innate immune system is the first line of defence against infectious pathogens by responding rapidly to conserved structures expressed by pathogens, the so-called pathogen-associated molecular patterns (PAMPS). The innate immune system thereby prevents replication and spread of pathogens and promotes activation of specific immune responses. Extensive studies of genes involved in innate immunity identified several type 1 diabetes susceptibility loci which may be involved in the inflammatory response or the innate immune response to viruses.

When isolated, human pancreatic cells are infected with human enterovirus, IP-10 and MCP-1 are secreted from the islets. Human enterovirus infection of human islets induces many genes involved in the innate immune response or sensing viral dsRNA so that genes encoding for proteins with powerful biological activities such as IL-6, IL-8, MIP-1, MIP-2, RANTES, MCP-1, IP-10, TNF-alpha and interferon-beta are induced. If these proteins were expressed in the pancreas, they would promote directly or indirectly beta-cell death by attracting immune cells. It is known, for example, that the induction and secretion of the chemokine IP-10 have a prominent role in the induction of insulitis.

Antibodies can prevent viruses from infecting cells, or they may enhance infection—the so-called antibody enhancement (ADE) of infection. Antibody enhancement of Coxsackie B virus (an enterovirus) infection has been observed in animal experiments, and in vitro infection with Coxsackie B virus 4 in combination with non-neutralising antibody of human monocytes stimulating the production of interferon-alpha by interactions with a specific viral receptor (CAR) and receptors of the FC portion of IgG. It is thus possible that antibody-dependent enhancement of infection with Coxsackie B virus in humans can increase the infection of peripheral blood mononuclear cells (PBMCs) and may cause viral escape from the immune response, thereby contributing to the spread of Coxsackie B viruses in the host. Although the predominant importance of T cells in the pathogenesis of type 1 diabetes is recognised, several questions arise in relation to infection with enteroviruses, including the stage at which enteroviruses influence the process: initiation or acceleration? Do enteroviruses influence immune regulation negatively by inducing the secretion of cytokines known to inhibit Tregs, or is the effect mainly on augmenting the proinflammatory response? Is a single or a repeated exposure required?

As stated above, autoimmunity is the major factor in the development of diabetes mellitus. Studies in virus-induced type 1 diabetes in mice provide an insight into the pathogenesis of the disease. The RIP-lymphocytic choriomeningitis virus (LCMV) mouse model uses viral infection to initiate an aggressive immune response to a transgenically expressed beta-cell target antigen derived from the initiating virus. The mice express the glycoprotein or the nucleoprotein of LCMV under control of the rat insulin promoter (RIP) in the beta-cells of the islets of Langerhans. Infection of such mice with LCMV expressing the identical antigens as environmental triggers initiates an immune response against LCMV including the glycoprotein or nucleoprotein expressed transgenically in the beta-cells resulting in type 1 diabetes. It appears, therefore, that viral infection induces inflammation which accelerates autoimmunity by recruitment of excessive aggressive lymphocytes and augmented presentation of islet autoantigens. It should be noted that other transgenic mice have been developed and used experimentally, including the non-obese diabetes (NOD) mouse and the biobreeding (BB) rat.

Autoimmunity can develop as a result of molecular mimicry between viral antigens and host cell antigens. Interestingly, there is evidence of molecular mimicry between enterovirus antigens and pancreatic islet cell antigens, and repeated enterovirus infections may release repeatedly autoreactive T-cells islet cells which lead to cumulative damage of pancreatic islet cells.

In the same context, juvenile-onset type 1 insulin-dependent diabetes occurs in patients with the congenital rubella syndrome, and may be delayed for 9–20 years or even longer. Autoimmune molecular mimicry of immunoreactive epitopes in the capsid proteins of rubella virus with islet beta-cell autoantigens has been implicated; HLA-restricted cytotoxic responses to GAD65, a beta-cell autoantigen, and

HLA types typical of autoimmune disorders. However, a reduction in cases of juvenile type 1 diabetes did not occur following the implementation of mass immunisation against rubella virus.

There is increasing evidence that enteroviruses, particularly Coxsackie viruses, cause beta-cell damage and islet cell inflammation, at least in some individuals. Coxsackie virus may infect directly beta-cells, and by interaction with pattern recognition receptors may activate transcription factors leading to increase of MHC class I expression and/or stress expression by the endoplasmic reticulum of pro-inflammatory cytokines, which may contribute to the recruitment of pre-existing autoreactive effector $CD8^+$ T-cells or impair regulatory cell function. In addition, molecular mimicry induced by the enhancement of MHC class II expression on antigen-presenting cells may contribute to recruitment of virus-specific B- and T-cells, thereby favouring beta-cell destruction.

There is also a strong and temporal association between infection with another virus, rotavirus, in infants and children, and the appearance or exacerbation of antibodies to the pancreatic islet cells antigens GAD65 and IA2. Epidemics of rotavirus gastroenteritis peak each winter as is the case with the diagnosis of type 1 diabetes. Secondly, the dominant HLA-DR4-binding $CD4^+$ T-cell autoepitopes in GAD65 and IA2 are very similar to sequences in the VP7 (viral protein 7) of rotavirus. These VP7 sequences are also $CD4^+$ T-cell epitopes for T-cells from children with islet autoimmunity. Sequences of GAD65 and IA2 epitopes also contain $CD8^+$ T-cell epitopes which are dominant at the onset of type 1 diabetes, suggesting molecular mimicry between rotavirus sequences and islet cell antigens which could lead to type 1 diabetes. Also rotavirus can infect beta-cells in vitro and can accelerate the onset of diabetes in NOD mice with islet inflammation.

Environmental Factors

Epidemiological studies indicate that environmental factors such as temperate climate, industrial development and increasing wealth, increased hygiene and decreased rates of infection and other factors referred to above, as well as geographical variation, play a role in the distribution of type 1 diabetes. The prevalence of type 1 diabetes particularly in northern countries, with peaks in winter and low incidence in summer, is notable. This geographical variation appears to be related to the seasonal circulation of enteroviruses particularly Coxsackie virus B1, B2, B4 and B5, and echovirus 9 and 11, which are referred to by some investigators as "diabetogenic viruses." There are also reports of spatial and familial clustering of type 1 diabetes in smaller communities. However, it is also noted that some enteroviruses may be associated negatively with diabetes, particularly in the tropics by a mechanism which appears to be mediated by activation of immunoregulation. Alterations in the intestinal microbial flora in infancy and early childhood influence immune regulation. Viruses contribute to the microbial flora which may be altered by infection with enteroviruses or other viruses. It has also been suggested that it is possible that parasitic infections, which are common in less developed tropical and subtropical regions, may protect against the development of diabetes possibly by cytokines mediating the destruction of auto-aggressive T-cells.

More About Viruses

The following is a brief summary of studies in animals and humans. Among inbred strains of rats with spontaneous onset of diabetes, the prevalence of diabetes generally increases with progressive removal of viruses from the environment (see also the section on environmental factors referred to above). In contrast, in two rat strains with genetic but little or no spontaneous diabetes in clean environments, infection with various viruses including Kilham rat virus, rat cytomegalovirus, enteroviruses, parvoviruses, herpes viruses and poxviruses can initiate diabetes in juvenile animals depending on the state of innate immunity before infection. Importantly, maternal immunisation can prevent the onset of diabetes in infected offspring (but note in contrast, that in humans immunisation against rubella did not lead to a reduction in cases of human juvenile type 1 diabetes, as discussed above). It is noted that the potential significance of virus infection during pregnancy in the aetiology of type 1 diabetes in human offspring is under investigation (see rubella virus above). There is some evidence that transplacental enterovirus infection, with Coxsackie viruses in particular, is associated with the development of autoimmunity leading to destruction of beta-cells.

Experimental inoculation of mice with a number of Coxsackie B viruses including Coxsackie virus B4 leads to persistent infection of the pancreas. It is well established that in the heart in infected mice, Coxsackie viruses can persist as defective slow replicating viruses without inducing rapid cytolysis. Dendritic cells and other cells infected with defective viruses may persist so that activation of T regulating cells specific for pancreatic antigens may have a long-lasting effect. Results of studies in a number of cattle suggest that persistent infection with bovine diarrhoea virus may induce autoimmune insulin-dependent diabetes. Ljungan virus, a picornavirus, isolated originally from the bank vole is associated with several syndromes including diabetes in wild rodents, which can also be induced experimentally in mice.

Reoviruses type 1, 2 and 3 isolated from humans infect primarily the gut and, when passaged in primary beta-cell cultures of suckling mice, cause hypothyroidism and diabetes in newborn mice. Crystalline arrays of reovirus 1 have been demonstrated in the cytoplasm of pancreatic alpha and beta-cells. Reovirus type 1 also infected the anterior pituitary, reducing production of growth hormone and causing stunted growth. A reovirus type 2 isolated from a cow caused a diabetic immune-mediated syndrome in newborn mice infected without prior adaptation to pancreatic beta-cells. These studies show that reoviruses have an unexpected tropism for endocrine organs.

Encephalomyocarditis virus. This virus provides an interesting experimental model for virus-induced type 1 diabetes in animals in which autoimmunity is not likely to be a factor. Susceptibility to the diabetogenic encephalomyocarditis D strain depends on the genetic background of mice, which has not been identified yet.

Involvement of enteroviruses in the human pancreas in type 1 diabetes is supported by the presence of virus: namely, viral capsid protein(s), the finding of virus-like particles by electron microscopy or expression of viral RNA, and by the host response to viral infection. Differing host responses and possibly differing viral properties may determine whether infection of beta-cells by enteroviruses results in rapid cell destruction by lysis, or persistent infection leading to autoimmune reaction to beta-cells (see above), or persistent infection with little host response and consequently little damage. It is also of interest that there are reports of low titre enterovirus viraemia in the majority of children with type 1 diabetes at the time of clinical diagnosis. Asymptomatic enterovirus infection is common among family members at the same time.

In vitro studies confirm the significance of enteroviruses in human diabetes by the demonstration of enterovirus tropism for cultured human pancreatic cells. Both lytic and persistent enterovirus infections have been demonstrated in both insulin and non-insulin producing cells. Initial pyknosis followed by delayed necrosis is the major mechanism of cell death during lytic enterovirus infection in cultured pancreatic islets.

Finally, fulminant diabetes, which is a distinct form of type 1 diabetes, found predominantly in Asian patients (but three Caucasian patients were also described in France) with rapid onset, ketosis-prone diabetes in the absence of islet-related autoantibodies, is associated with a number of viral infections particularly with enteroviruses. Other viruses have also been found, albeit less frequently, in association with this type of diabetes, namely herpes viruses (Epstein–Barr virus, cytomegalovirus) and reactivation of herpes virus 6.

In summary, the evidence for viral involvement particularly for enteroviruses acting as a trigger for diabetes is persuasive, although much further work is required.

Concluding Remarks

The suggestions that immunisation against hepatitis B virus infection will prevent primary hepatocellular carcinoma, and that the more recent introduction of human papillomavirus vaccine will lead to reduction of cervical cancer in women, were relegated initially to science fiction. By analogy, it would not be a fantasy to conclude that there is much fertile ground to be explored in the case of diabetes and viruses, particularly in relation to the concept of molecular mimicry and the induction of autoimmunity. The search for antiviral drugs and vaccines against viruses implicated in diabetes must be pursued vigorously. While the view has been expressed that research collaboration with the pharmaceutical industry as "supping with the devil," ethical and transparent research together with industry will lead to more rapid translational progress from the test tube to the bedside, more so in the case of effective antiviral drugs and vaccines. Notable examples are the development of vaccines against many bacterial and viral infections, and antiviral drugs for the treatment of the human immunodeficiency virus and the acquired immune deficiency syndrome (HIV/AIDS), and the treatment of persistent hepatitis B and C virus infection.

Postscript

The task of summarising the summary and conclusions written by each of the authors was facilitated by the excellent reviews, so that I resorted mostly to reproducing, in part, their work.

The Editors wish to record their sincere thanks to Tanya Shennan who collated, corrected and copy-edited each chapter with meticulous care and attention to detail. We are indeed fortunate to have the benefit of her skills.

In Memoriam

This last chapter is dedicated in everlasting loving memory to the beautiful and talented Alice Zuckerman (28.1.1932 to 16.1.2011).

Index

K.W. Taylor, H. Hyoty, A. Toniolo, A.J. Zuckerman (eds.), *Diabetes and Viruses*, DOI 10.1007/978-1-4614-4051-2, © Springer Science+Business Media New York 2013

F

MIX
Papier aus verantwortungsvollen Quellen
Paper from responsible sources
FSC® C105338

If you have any concerns about our products,
you can contact us on
ProductSafety@springernature.com

In case Publisher is established outside the EU,
the EU authorized representative is:
Springer Nature Customer Service Center GmbH
Europaplatz 3, 69115 Heidelberg, Germany

Printed by Libri Plureos GmbH
in Hamburg, Germany